AF750938

Spatio-Temporal Abnormality Diagnosis for Industrial Distributed Parameter Systems

Yun Feng · Han-Xiong Li · Yaonan Wang

Spatio-Temporal Abnormality Diagnosis for Industrial Distributed Parameter Systems

Model-Based and Data-Driven Perspectives

Springer

Yun Feng
School of Artificial Intelligence
and Robotics
Hunan University
Changsha, China

Han-Xiong Li
Department of Systems Engineering
City University of Hong Kong
Hong Kong, China

Yaonan Wang
School of Artificial Intelligence
and Robotics
Hunan University
Changsha, China

ISBN 978-981-95-3749-5 ISBN 978-981-95-3750-1 (eBook)
https://doi.org/10.1007/978-981-95-3750-1

© The Editor(s) (if applicable) and The Author(s), under exclusive license to Springer Nature Singapore Pte Ltd. 2026

This work is subject to copyright. All rights are solely and exclusively licensed by the Publisher, whether the whole or part of the material is concerned, specifically the rights of translation, reprinting, reuse of illustrations, recitation, broadcasting, reproduction on microfilms or in any other physical way, and transmission or information storage and retrieval, electronic adaptation, computer software, or by similar or dissimilar methodology now known or hereafter developed.
The use of general descriptive names, registered names, trademarks, service marks, etc. in this publication does not imply, even in the absence of a specific statement, that such names are exempt from the relevant protective laws and regulations and therefore free for general use.
The publisher, the authors and the editors are safe to assume that the advice and information in this book are believed to be true and accurate at the date of publication. Neither the publisher nor the authors or the editors give a warranty, expressed or implied, with respect to the material contained herein or for any errors or omissions that may have been made. The publisher remains neutral with regard to jurisdictional claims in published maps and institutional affiliations.

This Springer imprint is published by the registered company Springer Nature Singapore Pte Ltd.
The registered company address is: 152 Beach Road, #21-01/04 Gateway East, Singapore 189721, Singapore

If disposing of this product, please recycle the paper.

To my parents and Yazhi

Preface

As a typical representative of industrial processes, distributed parameter systems (DPSs) widely exist in both process and discrete manufacturing and operations, such as the snap curing oven in chip manufacturing, the tubular reactor in chemical manufacturing, the soft robots in special operations, etc. With the increasing development of industrial distributed parameter systems (especially the electrical vehicles and hybrid electric vehicles), spatio-temporal (S-T) abnormality diagnosis has become a pain in the neck and has attracted a great amount of attention in recent years. Moreover, the rapid development of machine learning and big data techniques has shed new insights into data-driven fault diagnosis and promoted the enthusiasm for studying the industrial distributed parameter systems. However, nearly no book has addressed this issue well and most existing research only consider traditional actuator/sensor fault while neglecting the spatio-temporal distributed characteristic of abnormality for DPSs.

To fill this gap, in this book we intend to introduce recent developments and trends of S-T abnormality diagnosis for industrial DPSs. The main contents of this book include: (1) Model-based abnormality diagnosis and identification for completely known industrial DPSs ("white box"); (2) Combined model-based and data-driven abnormality detection and localization for partially known industrial DPSs ("grey box"); (3) Purely data-driven modeling and diagnosis for completely unknown DPSs ("black box"). In conclusion, this book summarizes the authors' works on both model-based and data-driven perspectives for S-T abnormality diagnosis of industrial DPSs. To be more precise, this book mainly focuses on the following challenges: space-time couple characteristic, limited sensing in space, and the dynamically varying abnormality in space. This book aims at post-graduate students, researchers, and engineers with background knowledge of industrial systems modeling and monitoring. Interesting readers can obtain state-of-the-art methods systematically in the last 5 years and have a general overview of recent developments and future direction of this specific research field.

In general, the chapters follow a natural manner according to the systems considered from easy to hard. To be more precise, Chap. 1 introduces the background information and the research motivation, Chaps. 2–7 mainly contain model-based abnormality diagnosis and identification for completely known industrial DPSs ("white box"), Chaps. 8–10 mainly contain combined model-based and data-driven abnormality detection and localization for partially known industrial DPSs ("grey box"), Chaps. 11–12 mainly contain purely data-driven modeling and diagnosis for completely unknown DPSs ("black box").

Changsha, China Yun Feng
Hong Kong, China Han-Xiong Li
Changsha, China Yaonan Wang
August 2025

Acknowledgements The work of this book is the continuous work of the authors over the last 8 years. The intuitive idea was motivated by Prof. Han-Xiong Li during Dr. Yun Feng's Ph.D. study at the City University of Hong Kong. During his visit to Duisburg, Germany, Prof. Steven X. Ding, Prof. Linlin Li, Prof. Daoyi Dong, and Dr. Changsheng Hua, just to name a few, have provided insightful discussions on this topic.

We would like to thank the collaborators on this interesting topic, which include but not limited to, Dr. Jun-Wei Wang, Dr. Bing-Chuan Wang, Prof. Wanke Yu, Prof. Wei He, Prof. Biao Luo, Prof. Zhijie Liu, Prof. Hui Zhang, Prof. Zheng-Guang Wu, Prof. Huaicheng Yan, and Prof. Wen Kang.

We would also like to thank those who help us to prepare this book, which include but not limited to Mr. Xingyu Zhu, Mr. Xiaohua Liu, Mr. Yixuan Tong, Mr. Yucong Shi, Miss Ya Li, and Mr. Haicheng Li.

This book was supported in part by the National Natural Science Foundation (NSFC) under Grants 62522311, 62203161, and 62473143; in part by the National Key Research and Development Program of China under Grant 2023YFB4706400; in part by the Natural Science Foundation of Hunan Province under Grant 2024JJ5087; in part by the Guangdong Basic and Applied Basic Research Foundation under Grant 2025A1515011482; and in part by the Natural Science Foundation of Jiangxi Province under Grant 20232BAB212024.

Competing Interests The authors have no conflicts of interest to declare that are relevant to the content of this book.

Contents

Acronyms

1-D	One-dimensional
2-D	Two-dimensional
3-D	Three-dimensional
ADF	Abnormality detection filter
AdNN	Adaptive neural network
AI	Artificial intelligence
ALF	Abnormality localization filter
AR	Auto-regression
BMS	Battery management system
BTS	Battery thermal system
DICA	Dynamic independent component analysis
DOBC	Disturbance observer-based control
DPCA	Dynamic principal component analysis
DPS	Distributed parameter system
EV	Electric vehicle
EVD	Eigenvalue decomposition
FAR	False alarm rate
FDF	Fault detection filter
FDLF	Fault detection and localization filter
FDM	Finite difference method
FDR	Fault detection rate
FD	Fault diagnosis
FLF	Fault localization filter
fMRI	Functional magnetic resonance imaging
HEV	Hybrid electric vehicle
ICA	Independent component analysis
ISC	Internal short circuit
I/O	Input-output
KDE	Kernel density estimation
Lasso	Least Absolute Shrinkage and Selection Operator
Li-ion	Lithium-ion

LMI	Linear matrix inequality
LPS	Lumped parameter system
LSCV	Least-squares cross-validation
LTI	Linear time invariant
ODE	Ordinary differential equation
PCA	Principal component analysis
PDE	Partial differential equation
PDF	Probability density function
PLS	Partial least squares
RMSE	Root mean square error
RMS	Root-mean-square
SICA	Spatial independent component analysis
SM-SPE	Spatial mean squared prediction error
SNR	Signal-to-noise ratio
SPC	Statistical process control
SPE	Squared prediction error
S-T	Spatio-temporal
TICA	Temporal independent component analysis
UUB	Uniformly ultimately bounded

Chapter 1
Introduction

Abstract In this chapter, we give the description of industrial distributed parameter system in the first place with real examples. Then we move to the main problems consider in this book, that is, abnormality detection in the time domain, abnormality localization in the space domain, and abnormality estimation. Basic concepts of these problems are introduced, which can help readers to understand the following chapters.

1.1 Background

As a representative category of industrial processes, distributed parameter systems (DPSs) [1, 2] are extensively encountered in both continuous and discrete manufacturing scenarios. Typical examples include snap curing ovens in semiconductor fabrication [3], tubular reactors in chemical production [4], batteries in energy storage applications [5], and soft robots employed in specialized tasks [6] (see Fig. 1.1). DPSs are characterized by dynamical behaviors that evolve not only over time but also across spatial domains. From a mathematical perspective, they are commonly modeled by partial differential equations (PDEs), with state variables defined as functions of both temporal and spatial coordinates.

With the rapid advancement of industrial manufacturing, increasing emphasis has been placed on the safe and precise management of processes and operations. Conventional approaches to improving system reliability primarily focus on enhancing the quality, robustness, and stability of individual components such as actuators, sensors, and controllers. Nevertheless, achieving fault-free operation is practically infeasible [7]. For instance, the recent fire incident of a Tesla battery at the Bouldercombe energy storage facility in Queensland, Australia resulted in substantial economic losses, while recurring electric vehicle fire accidents have posed serious risks to both human safety and property. Such events have drawn considerable attention from both academia and industry to the problem of abnormality diagnosis in DPSs. In general, once a fault or abnormality arises, rapid and accurate diagnosis becomes crucial for preventing further system degradation, a fact that has been validated by numerous successful industrial applications.

© The Author(s), under exclusive license to Springer Nature Singapore Pte Ltd. 2026

Y. Feng et al., *Spatio-Temporal Abnormality Diagnosis for Industrial Distributed Parameter Systems*, https://doi.org/10.1007/978-981-95-3750-1_1

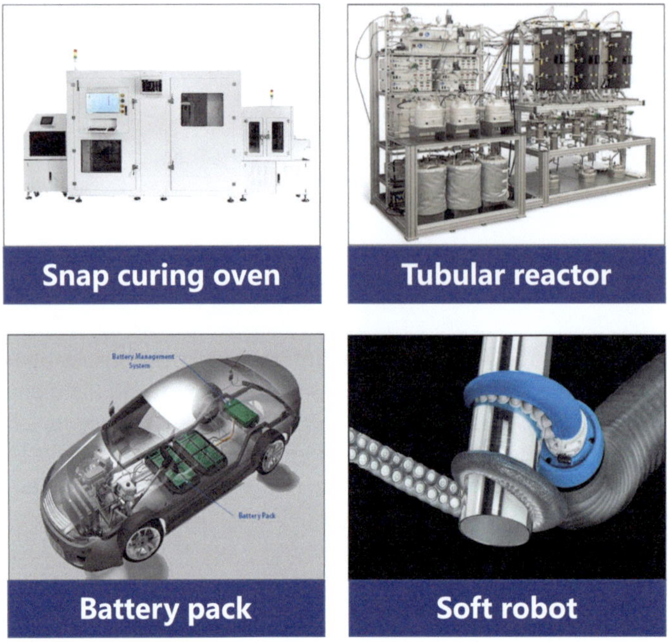

Fig. 1.1 Representative examples of industrial DPSs

Over the past decades, extensive research has been devoted to the diagnosis of faults and abnormalities in industrial systems [8–10]. Nonetheless, the majority of these studies have concentrated on dynamical systems modeled by ordinary differential equations (ODEs), also referred to as lumped parameter systems (LPSs), while overlooking spatially distributed features. This is mainly attributed to two factors. On the one hand, in practical scenarios, only a limited number of industrial processes can be naturally described by PDEs when compared with those modeled by ODEs. On the other hand, the intrinsic complexity of PDEs substantially increases the difficulty of analysis, making many well-established ODE-based tools—such as observer design and residual evaluation—unsuitable for direct application. It is worth noting that, to differentiate from fault diagnosis in ODE-based systems, the term "abnormality" is employed here to emphasize the inclusion of spatially distributed faults.

However, with the rapid advancement of industrial DPSs, particularly in applications such as hybrid electric vehicles (HEVs) and electric vehicles (EVs), abnormality diagnosis has become an increasingly critical challenge and has drawn substantial attention in recent years. In addition, the fast progress of big data and machine learning has provided new opportunities for data-driven fault diagnosis, further stimulating research interest in industrial DPSs.

1.1.1 Description of Industrial DPSs

To begin with, the description of DPSs is outlined from a fundamental physical perspective. For clearer demonstration, a thermal distribution model of Lithium-ion (Li-ion) batteries is considered. The Li-ion battery pack is critical in both EVs and HEVs, where effective thermal management is indispensable for safe operation. Consequently, the temperature field of the battery pack in three-dimensional (3-D) space can be represented as follows:

$$\rho C_p \frac{\partial T}{\partial t} = \frac{\partial}{\partial x}\left(\lambda_x \frac{\partial T}{\partial x}\right) + \frac{\partial}{\partial y}\left(\lambda_y \frac{\partial T}{\partial y}\right) + \frac{\partial}{\partial z}\left(\lambda_z \frac{\partial T}{\partial z}\right) + Q(x, y, z, t) + f(x, y, z, t). \tag{1.1}$$

In this context, $T(x, y, z, t)$ represents the temperature field of the battery pack at time t starting from 0, with x, y, and z denoting spatial coordinates. The parameters ρ and C_p correspond to the density and heat capacity of the pack, while λ_x, λ_y, and λ_z denote the thermal conductivities along the three spatial directions. Unlike conventional industrial processes modeled by ODEs, the thermal equation in (1.1) accounts for both temporal dynamics (partial derivative with respect to t) and spatial variation (partial derivatives with respect to x, y, z). The heat generation term $Q(x, y, z, t)$ is expressed as an *unknown nonlinear function* dependent on the terminal voltage $V(t)$, current $I(t)$, and the temperature distribution $T(x, y, z, t)$. Consequently, (1.1) describes a spatio-temporally nonlinear system. In addition, the extra term $f(x, y, z, t)$ captures location-dependent abnormalities, such as internal short circuit (ISC) or external mechanical damage [11]. It should be noted that the abnormality $f(x, y, z, t)$ is distinct from conventional faults, as the spatial variables x, y, z are incorporated into $f(\cdot)$, making it inherently location-dependent rather than a typical actuator or sensor fault [7].

Without loss of generality, the DPSs considered in this review can be mathematically described by a set of PDEs defined over the spatial domain $\Omega \triangleq (0, \bar{x}) \times (0, \bar{y}) \times (0, \bar{z}) \subset \mathbb{R}^3$.

$$\frac{\partial u}{\partial t} = g(\boldsymbol{q}(t), \boldsymbol{\rho}(x, y, z)) + f_P(x, y, z, t) + \mathcal{L}\left(u, \frac{\partial u}{\partial x}, \frac{\partial u}{\partial y}, \frac{\partial u}{\partial z}, \frac{\partial^2 u}{\partial x \partial y}, \frac{\partial^2 u}{\partial x \partial z}, \frac{\partial^2 u}{\partial y \partial z}, \frac{\partial^2 u}{\partial x^2}, \frac{\partial^2 u}{\partial y^2}, \frac{\partial^2 u}{\partial z^2}\right), \tag{1.2a}$$

$$\boldsymbol{p}(t) = \int_{\Omega} \boldsymbol{c}(x, y, z) u(x, y, z, t) \mathrm{d}x \mathrm{d}y \mathrm{d}z + \boldsymbol{f}_s(t), \tag{1.2b}$$

subject to the following boundary conditions:

$$h_1\left(u, \frac{\partial u}{\partial x}, \frac{\partial u}{\partial y}, \frac{\partial u}{\partial z}\right)\bigg|_{x=0} = 0, \tag{1.3a}$$

$$h_2\left(u, \frac{\partial u}{\partial x}, \frac{\partial u}{\partial y}, \frac{\partial u}{\partial z}\right)\bigg|_{x=\bar{x}} = 0, \tag{1.3b}$$

$$h_3\left(u, \frac{\partial u}{\partial x}, \frac{\partial u}{\partial y}, \frac{\partial u}{\partial z}\right)\bigg|_{y=0} = 0, \tag{1.3c}$$

$$h_4\left(u, \frac{\partial u}{\partial x}, \frac{\partial u}{\partial y}, \frac{\partial u}{\partial z}\right)\bigg|_{y=\bar{y}} = 0, \tag{1.3d}$$

$$h_5\left(u, \frac{\partial u}{\partial x}, \frac{\partial u}{\partial y}, \frac{\partial u}{\partial z}\right)\bigg|_{z=0} = 0, \tag{1.3e}$$

$$h_6\left(u, \frac{\partial u}{\partial x}, \frac{\partial u}{\partial y}, \frac{\partial u}{\partial z}\right)\bigg|_{z=\bar{z}} = 0, \tag{1.3f}$$

and with the following initial condition:

$$u(x, y, z, 0) = u_0(x, y, z). \tag{1.4}$$

In this formulation, the state variable is denoted by $u(x, y, z, t)$. The spatial coordinates are bounded as $x \in [0, \bar{x}]$, $y \in [0, \bar{y}]$, and $z \in [0, \bar{z}]$, while the temporal variable satisfies $t \geq 0$. The operator $\mathcal{L}(\cdot)$, mapping into $\mathbb{R}$, represents a nonlinear combination of spatial operators. The vector $\boldsymbol{q}(t) \triangleq [q_1(t), q_2(t), \cdots, q_n(t)]^{\mathrm{T}} \in \mathbb{R}^n$ corresponds to external system inputs. The shape function vector is defined as $\boldsymbol{\rho}(x, y, z) \triangleq [\rho_1(x, y, z), \rho_2(x, y, z), \cdots, \rho_n(x, y, z)]^{\mathrm{T}} \in \mathbb{R}^n$, and the nonlinear mapping $g(\cdot)$ describes how these inputs influence the system dynamics. The term $f_p(x, y, z, t)$ represents an additive distributed abnormality, including actuator faults with known shape functions as well as other process-related faults with unspecified shape characteristics. The output characteristics of the measurements are captured by $\boldsymbol{c}(x, y, z) \triangleq [c_1(x, y, z), c_2(x, y, z), \cdots, c_M(x, y, z)]^{\mathrm{T}} \in \mathbb{R}^m$, and the corresponding output measurements are given by $\boldsymbol{p}(t) \triangleq [p_1(t), p_2(t), \cdots, p_M(t)]^{\mathrm{T}} \in \mathbb{R}^m$. Nonlinear functions $h_1(\cdot)$ through $h_6(\cdot)$ describe the system boundary conditions. Finally, the vector $\boldsymbol{f}_s(t) \triangleq [f_{s1}(t), f_{s2}(t), \cdots, f_{sm}(t)]^{\mathrm{T}} \in \mathbb{R}^m$ accounts for sensor faults occurring at sensors with known positions.

1.1.2 Problem Statement

Analogous to fault diagnosis in ODE-based systems [7], the abnormality detection for DPSs typically involves three fundamental components:

- **Abnormality detection in the time domain**: Similar to the method used in LPSs fault detection, identify abnormal events and trigger alarms to prevent further damage or loss.

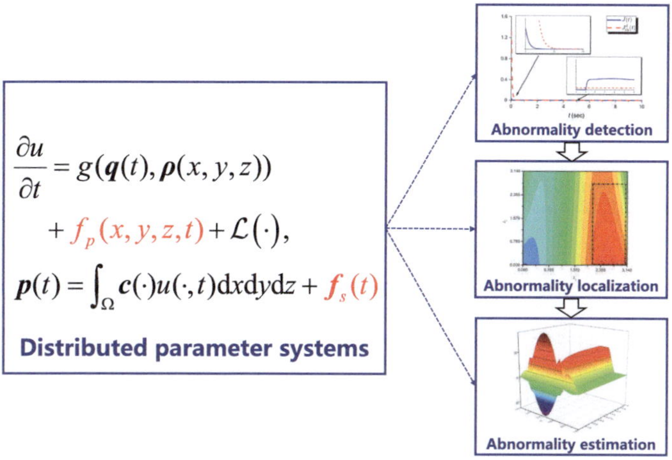

Fig. 1.2 Problem statement

- **Abnormality localization in the space domain**: Unlike fault isolation in LPSs, determine the *spatial region/location* where the abnormality occurs.
- **Abnormality estimation**: Similar to fault identification, quantify the magnitude of the abnormality.

For clarity, the system model in (1.2a)–(1.4) is used as an illustrative example, as shown in Fig. 1.2. Abnormality detection is responsible for issuing timely alarms for the distributed abnormality $f_p(x, y, z, t)$ and sensor faults $\boldsymbol{f}_s(t)$, while abnormality localization identifies the shape function of $f_p(x, y, z, t)$. Moreover, abnormality estimation determines the magnitude of $\hat{f}_p(x, y, z, t)$ and $\hat{\boldsymbol{f}}_s(t)$. Generally, these three tasks are performed sequentially, meaning that localization and estimation follow detection, and the results from localization can be used to facilitate estimation.

1.2 S-T Abnormality Diagnosis of Industrial DPSs

1.2.1 Abnormality Detection in the Time Domain

As illustrated in Fig. 1.2, the overall problem can be divided into three essential tasks: abnormality detection, fault localization, and quantitative estimation. This section begins with a discussion on abnormality detection. Within the field of fault diagnosis, the detection of faults in LPSs has long been recognized as a classical and fundamental topic. Over the past several decades, two principal research directions have emerged, namely model-based methods [7] and data-driven methods [12].

In more recent years, advances in PDE analysis techniques, together with the rapid development of artificial intelligence (AI), have drawn increasing attention to abnormality detection in DPSs. Depending on the methodological perspective, studies in

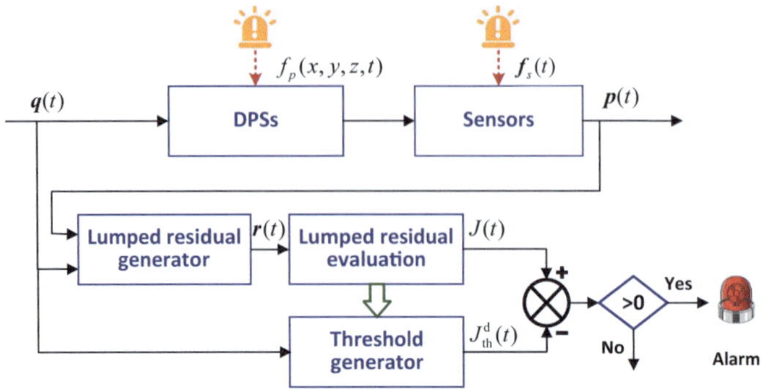

Fig. 1.3 Schematic plot of abnormality detection in the time domain

this area can broadly be classified into two categories: model-based approaches and data-driven approaches.

The overall framework for abnormality detection is illustrated in the flow chart shown in Fig. 1.3. In general, the process can be divided into three key stages: residual generation, residual evaluation, and threshold determination for detection. Unlike the conventional fault diagnosis of LPSs, the current framework employs both lumped residuals and distributed residuals, which serve the purposes of abnormality detection and localization, respectively. The central concept behind abnormality detection lies in constructing a residual signal, denoted as $\boldsymbol{r}(t)$, which represents the discrepancy between actual measurements and their corresponding estimates. From this residual, useful information about potential abnormalities can be extracted. The abnormality is further characterized by a feature index, $J(t)$, whereas $J^{\mathrm{d}}_{\mathrm{th}}(t)$ designates the detection threshold. The detection decision is therefore made by comparing the extracted feature with its corresponding threshold, which can be summarized in the following logic:

$$\begin{cases} J(t) > J^{\mathrm{d}}_{\mathrm{th}}(t) \Longrightarrow \text{ abnormal at time } t \\ \text{Otherwise} \Longrightarrow \text{ abnormality-free at time } t. \end{cases} \tag{1.5}$$

1.2.2 Abnormality Localization in the Space Domain

The problem of abnormality localization differs from the concept of fault isolation in LPSs [7], where the primary objective is to distinguish and classify different types of faults. In the case of spatio-temporal (S-T) systems, localization is instead concerned with determining the spatial position of the abnormality, which is a natural requirement for DPSs. In general, two possible strategies are available for this task. The most widely adopted approach relies on constructing a distributed residual $r(x, y, z, t)$ that is explicitly location-dependent. By analyzing this residual field, the abnormal regions can be identified, as illustrated in Fig. 1.4 and Eq. (1.6). An

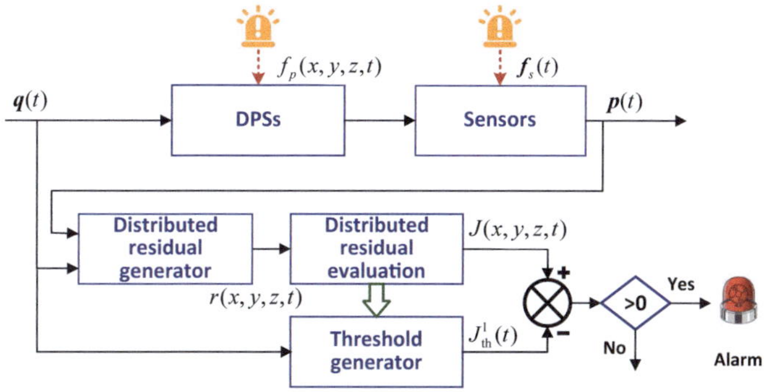

Fig. 1.4 Schematic plot of abnormality localization in the space domain

alternative approach seeks to directly estimate the spatial distribution function of the abnormality itself. This method is conceptually closer to abnormality estimation, which will be discussed in the following section. Nevertheless, it usually demands detailed system models and additional assumptions regarding the nature of the abnormality.

$$\begin{cases} J(x, y, z, t) > J^1_{\text{th}}(t) \Longrightarrow \text{ abnormal at location } (x, y, z) \\ \text{Otherwise} \Longrightarrow \text{ abnormality-free at location } (x, y, z). \end{cases} \tag{1.6}$$

1.2.3 Abnormality Estimation

In conventional LPSs, fault identification is typically concerned with estimating the magnitude of a fault. However, this task becomes considerably more challenging in S-T systems, since both the dynamics of the system and the abnormality itself exhibit spatial distribution. Consequently, the objective shifts toward estimating two types of unknowns: the abnormality term $f_p(x, y, z, t)$ and the sensor fault vector $\boldsymbol{f}_s(t)$. This is generally achieved by employing an abnormality identification filter, which parallels the concept of the fault identification filter described in [7], as illustrated in Fig. 1.5.

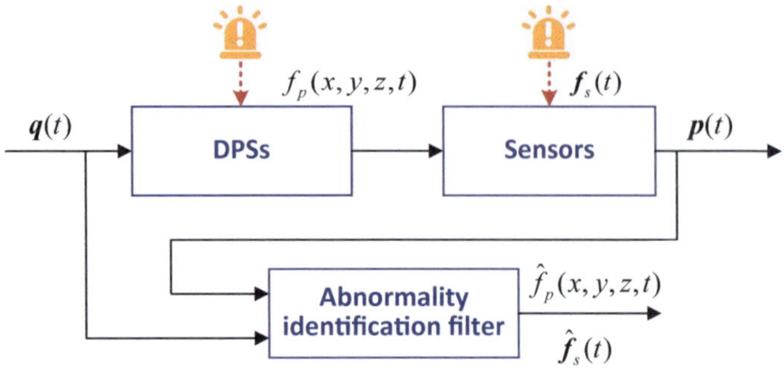

Fig. 1.5 Schematic plot of abnormality estimation

References

1. Li H-X, Qi C (2010) Modeling of distributed parameter systems for applications—a synthesized review from time-space separation. J Process Control 20(8):891–901
2. Li H-X, Qi C (2011) Spatio-temporal modeling of nonlinear distributed parameter systems: a time/space separation based approach. Springer, London
3. Deng H, Li H-X, Chen G (2005) Spectral-approximation-based intelligent modeling for distributed thermal processes. IEEE Trans Control Syst Technol 13(5):686–700
4. Christofides PD (2012) Nonlinear and robust control of PDE systems: methods and applications to transport-reaction processes. Springer, Berlin
5. Zhang D, Tang S-X, Couto LD, Wei Z, Viswanathan V (2025) State estimation of solid-state batteries via a distributed parameter electrochemical model. Automatica 173:112092
6. Han Z, Liu Z, Meurer T, He W (2024) PDE-based control synthesis for a planar cable-driven continuum arm. Automatica 163:111600
7. Ding S-X (2013) Model-based fault diagnosis techniques: design schemes, algorithms, and tools. Springer, Berlin
8. Qin SJ (2012) Survey on data-driven industrial process monitoring and diagnosis. Annu Rev Control 36(2):220–234
9. Ge Z, Song Z, Gao F (2013) Review of recent research on data-based process monitoring. Ind Eng Chem Res 52(10):3543–3562
10. Yin S, Ding SX, Xie X, Luo H (2014) A review on basic data-driven approaches for industrial process monitoring. IEEE Trans Ind Electron 61(11):6418–6428
11. Hu X, Zhang K, Liu K, Lin X, Dey S, Onori S (2020) Advanced fault diagnosis for Lithium-ion battery systems: a review of fault mechanisms, fault features, and diagnosis procedures. IEEE Ind Electron Mag 14(3):65–91
12. Ding SX (2014) Data-driven design of fault diagnosis and fault-tolerant control systems. Springer, Berlin

Part I
Model-Based Abnormality Diagnosis and Identification for Completely Known Industrial DPSs

Chapter 2
Spatial Basis Functions-Based Fault Localization for Linear Parabolic DPSs

Abstract Fault localization for distributed parameter systems is as important as fault detection but is seldom discussed in the literature. The main reason is that an infinite number of sensors in the space is needed to construct a *distributed residual signal*, which is nearly impossible in practice. In this chapter, a fault detection and localization filter which only uses a *finite number of sensors* is initiated based on an approximated ordinary differential equation model. Considering the limitations on computation resources for higher-order models in practice, a novel set of spatial basis functions is applied to the reduced-order fault detection and localization filter design. Under certain conditions, the novel spatial basis functions obtain smaller state truncation error while the order is lower compared to the mostly used eigenfunctions of the spatial operator.

2.1 Introduction

DPSs widely exist in industrial processes and consist of a large number of fundamental equations, such as the heat equation [1] in chemical engineering, string [2] and beam equation [3] in mechanical engineering. The nature of these systems is the infinite-dimensional system, which makes the analysis more involved compared to systems described by ODEs. Over the last few decades, researchers in the community of control theory and engineering have focused on the observer [4, 5] and controller [6, 7] design for DPSs. However, research works reported on fault diagnosis (FD) of DPSs are relatively fewer, although it has large potentials in industrial applications, for example, transport-reaction processes [8, 9], thermal processes [10], and fluid processes. To be more specific, the thermal faults constitutes a significant part of critical causes that lead to Li-ion batteries [11, 12] degradation and failure. It is of great significance to diagnose (detection, localization, and estimation) these thermal faults in real time to ensure battery safety, and the thermal model of batteries belong to the category of DPSs; Fault-tolerant control of vibrating string and beam [13] described by DPSs can be widely applied to mechanical systems.

Despite the fact that both model-based [14–16] and data-driven [17] FD of dynamic processes described by ODEs have been studied systematically in the last

© The Author(s), under exclusive license to Springer Nature Singapore Pte Ltd. 2026
Y. Feng et al., *Spatio-Temporal Abnormality Diagnosis for Industrial Distributed Parameter Systems*, https://doi.org/10.1007/978-981-95-3750-1_2

decades, research considering the fault diagnosis issues of DPSs were rather rare. The major challenge of applying the well-known design techniques in fault diagnosis of ODEs to that of DPSs lies in the infinite-dimensional nature and restrictions on sensors and computation resources. To be more specific, the number of sensors can be placed in the space is finite and sometimes they can only be placed at boundaries [12, 18]; Also, directly solving of PDEs is highly time-consuming and requires large amount of computation resources.

Existing works dealing with the FD issues of DPSs can be roughly classified into two categories based on the timing when the so-called *lumping* techniques were applied, i.e., *early-lumping* [19–22] and *late-lumping* [12, 18]. To be more specific, early-lumping [23] methods first transformed the original PDE into several ODEs and the following design is based on the approximated model; However, the late-lumping methods were directly based on the PDE model, which are more involved. For practical considerations, early-lumping-based methods are more favored since they are computationally more efficient, one accompanying disadvantage is the observation spillover [24] induced by the neglecting of the fast dynamics. Recently, another interesting direction on FD of DPSs was using the modulation function [25–27] and flatness property [28] to transform the PDE system into an algebraic equation and then analyze it. This modulation function-based design can not be classified into either early-lumping or late-lumping methods.

In summary, almost all existing results reported on the FD of DPSs only consider fault detection in the time domain, while fault localization issue is neglected. The main reason is that the distribution feature of the fault was not taken into account:

- On the one hand, for the actuator fault, it was predefined that the fault and the actuator share the same spatial distribution characteristic;
- On the other hand, for the sensor fault, since point-wise measurement sensors such as thermocouples are most frequently used in industrial applications and the locations of the sensors are known a priori, the spatial distribution of the sensor fault was not even well defined.

However, bearing the spatial distribution characteristic of DPSs in mind, it is very intuitive to study the fault localization problem. For example, for the thermal FD of batteries [11, 12], only detecting the time when a fault occurs is not sufficient for fault removal or safe operation, since the fault is *distributed* in a specific spatial area. Knowing the specific location of the fault will lead to a more efficient battery management system (BMS). This is the major difference between the fault diagnosis of systems described by PDEs and ODEs. Therefore, it is intuitive to consider the fault localization problem of DPSs. Fault localization for DPSs is inherently more complex than fault detection since the former requires a distributed residual generator in the whole space domain while the latter only needs a lumped one [12, 18]. Moreover, the limitations on the number of sensors in real industrial applications make the fault localization issue even harder. Continuous measurements in the space domain [29] is quite restrictive.

Motivated by the above considerations, a fault detection and localization filter (FDLF) is proposed to address detection and localization problem of S-T fault for industrial processes described by linear parabolic DPSs using limited sensor measurements. To further deal with the restriction on the number of sensors required, a novel set of spatial basis functions, which is deduced by the eigenfunctions is adopted. Based on the novel spatial basis functions, a reduced-order FDLF which requires less sensor measurements than that of the spectral approximation method [30] is designed.

The main contributions of this chapter can be summarized as follows:

- A reduced-order FDLF is constructed based on a set of novel basis functions which has a lower order without sacrificing the approximation performance compared to the traditional eigenfunctions;
- The proposed novel basis functions-based design superiors over eigenfunctions-based design both in computation efficiency and the required number of sensors.

The remainder of this chapter is organized as follows: system description and problem statement are introduced in Sect. 2.2. In Sect. 2.4, the proposed FDLF is investigated analytically. Moreover, an illustrative example is presented in Sect. 2.5 for illustration. Finally, the chapter is summarized in Sect. 2.6.

2.2 System Description and Problem Statement

Notations: $\mathbb{R}$ and $\mathbb{R}^n$ are the set of all real numbers and n-dimensional Euclidean space with the norm $||\cdot||$, respectively. $|\omega|$ denotes the absolute value of ω for any $\omega \in \mathbb{R}$. $\mathcal{L}^2([0,L]) \triangleq \mathcal{L}^2((0,L);\mathbb{R})$ is a real Hilbert space of square integrable functions $\omega(z):[0,L]\to\mathbb{R}$ with the spatial $\mathcal{L}^2$ norm $||\omega(\cdot)||_2 \triangleq \sqrt{\int_0^L \omega^2(z)\mathrm{d}z}$. Given a natural number $\bar{l}$, $\mathcal{H}^{\bar{l}}(0,L) \triangleq \mathcal{W}^{\bar{l},2}((0,L);\mathbb{R})$ is a real Sobolev space of absolutely continuous functions $\bar{\omega}(z):(0,L)\to\mathbb{R}$ with square integrable derivatives $\mathrm{d}^i\bar{\omega}(z)/\mathrm{d}z^i$ up to the order $\bar{l}\geq 1$ and with the norm $||\bar{\omega}(\cdot)||_{\mathcal{H}^{\bar{l}}} \triangleq \sqrt{\int_0^L \sum_{i=0}^{\bar{l}}\left(\frac{\mathrm{d}^i\bar{\omega}(z)}{\mathrm{d}z^i}\right)^2\mathrm{d}z}$. For any functions $\omega(\cdot)\in\mathcal{H}^1(0,L)$, the spatial $\mathcal{L}^\infty$ norm is defined as $||\omega(\cdot)||_\infty \triangleq \max_{z\in[0,L]}|\omega(z)|$. $\omega_t \triangleq \partial\omega/\partial t$, $\omega_{zz} \triangleq \partial^2\omega/\partial z^2$.

Consider a class of DPSs modeled by the following linear parabolic PDE:

$$\begin{aligned} x_t(z,t) &= x_{zz}(z,t) + \eta x(z,t) + u(z,t) + f(z,t) + d(z,t), \\ \boldsymbol{y}(t) &= \int_0^L \boldsymbol{c}(z)x(z,t)\mathrm{d}z, \end{aligned} \tag{2.1}$$

with

$$\begin{aligned} u(z,t) &= \boldsymbol{b}_u^{\mathrm{T}}(z)\boldsymbol{u}(t),\ f(z,t) = \boldsymbol{b}_f^{\mathrm{T}}(z)\boldsymbol{f}(t), \\ d(z,t) &= \boldsymbol{b}_d^{\mathrm{T}}(z)\boldsymbol{d}(t), \end{aligned} \tag{2.2}$$

subject to the following homogeneous Dirichlet boundary conditions:

$$x(0,t) = x(L,t) = 0, \tag{2.3}$$

and with the following initial condition:

$$x(z,0) = x_0(z), \tag{2.4}$$

where $x(z,t)$ denotes the state variable; $[0, L] \subset \mathbb{R}$ is the spatial domain of the system; $z \in [0, L]$ is the spatial coordinate; $t \in [0, \infty)$ denotes the time; $\boldsymbol{u}(t) \in \mathbb{R}^{n_u}$ denotes the vector of manipulated input; $\boldsymbol{f}(t) \in \mathbb{R}^{n_f}$ and $\boldsymbol{d}(t) \in \mathbb{R}^{n_d}$ denote the **unknown** fault and disturbance in the DPS, respectively; $\boldsymbol{y}(t) \in \mathbb{R}^{n_y}$ denotes the system output; η is a constant coefficient; the spatial distribution of the input and fault are characterized by $\boldsymbol{b}_u(z) \in \mathbb{R}^{n_u}$ and $\boldsymbol{b}_f(z) \in \mathbb{R}^{n_f}$, respectively. Note that $\boldsymbol{b}_u(z)$ and $\boldsymbol{b}_d(z)$ are known vector functions while $\boldsymbol{b}_f(z)$ is **unknown**. The i-th element of known integrable vector function $\boldsymbol{c}(z) \in \mathbb{R}^{n_y}$ is determined by the location and shape (point or distributed) of the i-th measurement sensor; and $x_0(z)$ is the initial condition.

Remark 2.1 It is worth noticing that both the fault $f(z,t)$ and disturbance $d(z,t)$ exist in the system. This is a common setting for both the ODE systems [14] and PDE systems [12, 22, 28]. In fact, the disturbance (unknown input) $d(z,t)$ consists of model uncertainty and process noises. As for the term $f(z,t)$, it represents all possible faults and will be zero in the fault-free case. This setting is essential for robust fault detection and localization, otherwise, false alarm may arise.

The problem investigated in this chapter can be summarized as follows:

Design a FDLF for fault detection and localization in the time and space domain under disturbances for systems which are subject to the mathematical model in (2.1) with the boundary conditions in (2.3) and the initial condition in (2.4) using as fewer sensors as possible.

2.3 Methodology Framework

To address the problem of fault localization, a general post-filter which can develop the residual signals at continuous locations with limited sensor measurements is designed. Moreover, distributed residual evaluation procedures should also be arranged for fault detection and localization.

As presented in Fig. 2.1, the proposed FDLF consists of an observer and a general post-filter. The observer generates the output estimation $\hat{\boldsymbol{y}}(t)$ while the post-filter generates a distributed S-T (spatial-temporal) residual signal $r(z,t)$ using the output estimation error $\boldsymbol{e}_y(t)$. After that, the residual signal is also evaluated in a distributed manner.

The major concerns of this chapter are twofold:

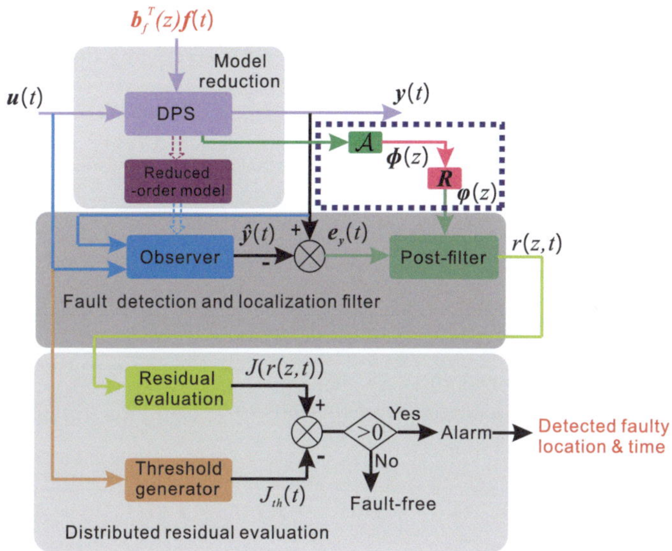

Fig. 2.1 Methodology framework

- Considering the fact that an *infinite number* of sensors is not possible in practice, how to design a general post-filter which maps *a finite number* of output estimation error $e_y(t)$ into the *infinite-dimensional* space, hence a distributed residual signal can be obtained for fault localization?
- A common restriction in applications is the numerical solving of PDEs, which requires a large amount of computation resources. An appropriate manner is to use several ODEs to approximate the original PDE. On the one hand, if the order of ODEs used for approximation is rather high, then the computation burden is still at large; On the other hand, if a lower-order ODE model is used, the approximation error may be large. How to find an approximated ODE model which has a *lower order* while maintaining *good approximation performance*?

2.4 Novel Basis Functions-Based Fault Localization

Modal approximation techniques are frequently used for model reduction of DPSs to transform the original PDE model into an approximated ODE model. In this manner, low-order observers can be designed based on the approximated ODE model. Moreover, since parabolic DPSs' spatial differential operator can be divided into a finite-dimensional slow subsystem and an infinite-dimensional fast subsystem [23], it is intuitive to apply modal approximation techniques to such systems.

To initiate the FDLF design, the spectral method was used for modal decomposition as the first step. For the convenience of notation, define the operator $\mathcal{A}$ in $\mathcal{L}^2([0, L])$ as:

$$\mathcal{A}x = \frac{\partial^2 x}{\partial z^2}, x \in \mathcal{S}(\mathcal{A}) \triangleq \{x \in \mathcal{L}^2([0, L]), \text{and the boundary conditions in (2.3) hold}\}.$$

Denote the eigenfunctions of the linear operator $\mathcal{A}$ as:

$$\{\phi_i(z)\}_{i=1}^{\infty} \triangleq \{\phi_1(z), \phi_2(z), \cdots, \phi_\infty(z)\},$$

and $\{\lambda(i)\}_{i=1}^{\infty}$ are the corresponding eigenvalues.

Moreover, applying orthogonal projection $\mathcal{P}$ to (2.1) yields the following infinite-dimensional ODE system:

$$\begin{aligned} \dot{\boldsymbol{x}}(t) &= \boldsymbol{A}\boldsymbol{x}(t) + \boldsymbol{B}\boldsymbol{u}(t) + \boldsymbol{E}\boldsymbol{f}(t) + \boldsymbol{F}\boldsymbol{d}(t), \\ \boldsymbol{y}(t) &= \boldsymbol{C}\boldsymbol{x}(t), \end{aligned} \tag{2.5}$$

where

$$\begin{aligned} \boldsymbol{x}(t) &= \mathcal{P}x(z, t) \triangleq [< \phi_1, x >, \cdots, < \phi_\infty, x >]^{\mathrm{T}} \in \mathbb{R}^\infty, \\ \boldsymbol{A} &= \mathcal{P}\mathcal{A}\boldsymbol{\phi}^{\mathrm{T}}(z) + \mathcal{P}\eta\boldsymbol{\phi}^{\mathrm{T}}(z) = diag\{\lambda_1 + \eta, \cdots, , \lambda_\infty + \eta\} \in \mathbb{R}^{\infty\times\infty}, \\ \boldsymbol{B} &= \mathcal{P}\boldsymbol{b}_u^{\mathrm{T}}(z) \in \mathbb{R}^{\infty\times n_u}, \ \boldsymbol{E} = \mathcal{P}\boldsymbol{b}_f^{\mathrm{T}}(z) \in \mathbb{R}^{\infty\times n_f}, \\ \boldsymbol{F} &= \mathcal{P}\boldsymbol{b}_d^{\mathrm{T}}(z) \in \mathbb{R}^{\infty\times n_d}, \\ \boldsymbol{C} &= \int_0^L \boldsymbol{c}(z)\boldsymbol{\phi}^{\mathrm{T}}(z)\mathrm{d}z \in \mathbb{R}^{n_y\times\infty}. \end{aligned}$$

First, similar to traditional fault detection filter (FDF) design, we can establish an infinite-dimensional FDF for system (2.5) as follows:

$$\begin{aligned} \dot{\hat{\boldsymbol{x}}}(t) &= \boldsymbol{A}\hat{\boldsymbol{x}}(t) + \boldsymbol{B}\boldsymbol{u}(t) + \bar{\boldsymbol{L}}(\boldsymbol{y}(t) - \hat{\boldsymbol{y}}(t)), \\ \hat{\boldsymbol{y}}(t) &= \boldsymbol{C}\hat{\boldsymbol{x}}(t), \end{aligned} \tag{2.6}$$

where $\bar{\boldsymbol{L}}$ denotes the observer gain to be determined.

Define the state estimation error $e_x(z, t)$ and the output estimation error $\boldsymbol{e}_y(t)$ as:

$$e_x(z, t) \triangleq x(z, t) - \hat{x}(z, t), \ \boldsymbol{e}_y(t) \triangleq \boldsymbol{y}(t) - \hat{\boldsymbol{y}}(t), \tag{2.7}$$

respectively, where $\hat{x}(z, t) \triangleq \sum_{i=1}^{\infty} \phi_i(z)\hat{x}_i(t) = \boldsymbol{\phi}^{\mathrm{T}}(z)\hat{\boldsymbol{x}}(t)$.

Combining (2.5) with (2.6) yields:

$$\boldsymbol{e}_y(t) = \boldsymbol{C}\boldsymbol{e}_x(t), \tag{2.8}$$

where $\boldsymbol{e}_x(t) \triangleq \boldsymbol{x}(t) - \hat{\boldsymbol{x}}(t)$. Moreover, it is easy to derive that:

$$e_x(z,t) = \boldsymbol{\phi}^{\mathrm{T}}(z)\boldsymbol{e}_x(t). \tag{2.9}$$

Remark 2.2 If $\boldsymbol{C} \in \mathbb{R}^{n_y \times \infty}$ is of full-column rank, then $\boldsymbol{e}_x(t)$ can be obtained as:

$$\boldsymbol{e}_x(t) = \boldsymbol{C}^{+}\boldsymbol{e}_y(t) = \left(\boldsymbol{C}^{\mathrm{T}}\boldsymbol{C}\right)^{-1}\boldsymbol{C}^{\mathrm{T}}\boldsymbol{e}_y(t),$$

where $\boldsymbol{C}^{+} = \left(\boldsymbol{C}^{\mathrm{T}}\boldsymbol{C}\right)^{-1}\boldsymbol{C}^{\mathrm{T}}$ denotes the pseudo-inverse of $\boldsymbol{C}$.

And the state estimation error $e_x(z,t)$ can be obtained as:

$$e_x(z,t) = \boldsymbol{\phi}^{\mathrm{T}}(z)\left(\boldsymbol{C}^{\mathrm{T}}\boldsymbol{C}\right)^{-1}\boldsymbol{C}^{\mathrm{T}}\boldsymbol{e}_y(t),$$

which can be used to construct the distributed S-T residual signal as:

$$\bar{r}(z,t) \triangleq e_x(z,t) = \boldsymbol{V}(z)\boldsymbol{e}_y(t),$$

where $\boldsymbol{V}(z) \triangleq \boldsymbol{\phi}^{\mathrm{T}}(z)\left(\boldsymbol{C}^{\mathrm{T}}\boldsymbol{C}\right)^{-1}\boldsymbol{C}^{\mathrm{T}}$ denotes the designed post-filter. In this manner, fault localization can be achieved as the spatial distribution characteristic is well preserved. However, it can only be achieved when $n_y \to \infty$, which means that an infinite-number of sensor measurements is needed, and is not possible in practice. This motivates us to design a finite-dimensional FDLF instead.

To this end, a finite-dimensional approximated ODE model should be obtained first. Separating (2.5) into fast and slow mode according to the value of the eigenvalue $\{\lambda(i)\}_{i=1}^{\infty}$ as:

$$\begin{aligned}
\dot{\boldsymbol{x}}_s(t) &= \boldsymbol{A}_s\boldsymbol{x}_s(t) + \boldsymbol{B}_s\boldsymbol{u}(t) + \boldsymbol{E}_s\boldsymbol{f}(t) + \boldsymbol{F}_s\boldsymbol{d}(t),\\
\dot{\boldsymbol{x}}_f(t) &= \boldsymbol{A}_f\boldsymbol{x}_f(t) + \boldsymbol{B}_f\boldsymbol{u}(t) + \boldsymbol{E}_f\boldsymbol{f}(t) + \boldsymbol{F}_f\boldsymbol{d}(t),\\
\boldsymbol{y}(t) &= \boldsymbol{y}_s(t) + \boldsymbol{y}_f(t) = \boldsymbol{C}_s\boldsymbol{x}_s(t) + \boldsymbol{C}_f\boldsymbol{x}_f(t),
\end{aligned} \tag{2.10}$$

where

$$\begin{aligned}
\boldsymbol{A}_s &\triangleq \operatorname{diag}\{\lambda_1 + \eta, \cdots, \lambda_m + \eta\},\\
\boldsymbol{A}_f &\triangleq \operatorname{diag}\{\lambda_{m+1} + \eta, \cdots, \lambda_\infty + \eta\},
\end{aligned}$$

and $\mathrm{Re}\{\lambda_1\} \geqslant \mathrm{Re}\{\lambda_2\} \geqslant \cdots \geqslant \mathrm{Re}\{\lambda_m\} \geqslant \cdots$, $\mathrm{Re}\{\lambda_{m+1}\} < 0$ and $|\mathrm{Re}\{\lambda_m\}| / |\mathrm{Re}\{\lambda_{m+1}\}| = O(\varepsilon)$, $\varepsilon < 1$ is a small positive number.

Remark 2.3 More details regarding the choice of the slow/dominant mode number can be referred to [23]. It is omitted due to triviality.

2.4.1 Novel Basis Functions for Model Reduction

Typically, the eigenfunctions of the linear operator $\mathcal{A}$ are chosen as the basis functions for model reduction, as discussed in [23]. However, these eigenfunctions are not optimal in the sense that the dimension of the reduced model is not lowest for a given accuracy. In this chapter, the novel basis functions [31] are adopted rather than the eigenfunctions for model reduction. The number of novel basis functions is smaller than that of eigenfunctions. To be more specific, the approximated reduced model has fewer modes while maintaining a higher performance compared with the eigenfunctions-based reduced model. The novel basis functions are induced by the eigenfunctions as follows:

$$\boldsymbol{\varphi}^{\mathrm{T}}(z) \triangleq \boldsymbol{\phi}_s^{\mathrm{T}}(z)\boldsymbol{R}, \tag{2.11}$$

where $\boldsymbol{R} \in \mathbb{R}^{m\times n}(n < m)$ denotes the constant transformation matrix. The following lemmas summarize two key characteristics of the novel basis functions.

Lemma 2.1 *[31] If the columns of the basis function transformation matrix R are orthogonal, n novel spatial basis functions $\{\varphi_i(z)\}_{i=1}^n$ are orthogonal toward one another.*

Lemma 2.2 *[31] Given an orthogonal matrix $\boldsymbol{R}$ obtained from the balanced truncation model reduction, then the state truncation error for the novel basis functions is smaller than that with the eigenfunctions $\boldsymbol{\phi}_s(z)$ for all t if $\boldsymbol{R}^{\mathrm{T}}\boldsymbol{E}_1\boldsymbol{E}_2\boldsymbol{R}$ is negative semi-definite, where:*

$$\begin{aligned} \boldsymbol{E}_1 &\triangleq \left(\begin{bmatrix} \boldsymbol{I}_n & \boldsymbol{0} \\ \boldsymbol{0} & \boldsymbol{0} \end{bmatrix} - \boldsymbol{R}\boldsymbol{R}^{\mathrm{T}}\right), \\ \boldsymbol{E}_2 &\triangleq \left(\begin{bmatrix} \boldsymbol{I}_n & \boldsymbol{0} \\ \boldsymbol{0} & 2\boldsymbol{I}_{m-n} \end{bmatrix} - \boldsymbol{R}\boldsymbol{R}^{\mathrm{T}}\right). \end{aligned} \tag{2.12}$$

These two lemmas claim the orthogonality of the novel basis functions and the conditions to satisfy for the superiority over the eigenfunctions. Details regarding the computation of $\boldsymbol{R}$ can be referred to [32].

2.4.2 Reduced-Order Observer Design

With the novel basis functions $\{\varphi_i(z)\}_{i=1}^n$, the slow subsystem in (2.10) is transformed into the following reduced-order ODE system:

$$\begin{aligned} \dot{\boldsymbol{x}}_r(t) &= \boldsymbol{A}_r\boldsymbol{x}_r(t) + \boldsymbol{B}_r\boldsymbol{u}(t) + \boldsymbol{E}_r\boldsymbol{f}(t) + \boldsymbol{F}_r\boldsymbol{d}(t), \\ \boldsymbol{y}_r(t) &= \boldsymbol{C}_r\boldsymbol{x}_r(t) = \boldsymbol{y}_s(t) = \boldsymbol{C}_s\boldsymbol{x}_s(t), \end{aligned} \tag{2.13}$$

where $\boldsymbol{A}_r = \boldsymbol{R}^{\mathrm{T}}\boldsymbol{A}_s\boldsymbol{R} \in \mathbb{R}^{n\times n}$, $\boldsymbol{B}_r = \boldsymbol{R}^{\mathrm{T}}\boldsymbol{B}_s \in \mathbb{R}^{\hat{n}\times n_u}$, $\boldsymbol{E}_r = \boldsymbol{R}^{\mathrm{T}}\boldsymbol{E}_s \in \mathbb{R}^{n\times n_f}$, $\boldsymbol{F}_r = \boldsymbol{R}^{\mathrm{T}}\boldsymbol{F}_s \in \mathbb{R}^{n\times n_d}$, and $\boldsymbol{C}_r = \boldsymbol{C}_s\boldsymbol{R} \in \mathbb{R}^{n_y\times n}$.

Due to the fact that $\boldsymbol{x}_r(t) \in \mathbb{R}^n$ and $n < m$, a reduced-order observer can be designed for distributed S-T residual generation as follows:

$$\begin{aligned} \dot{\hat{\boldsymbol{x}}}_r(t) &= \boldsymbol{A}_r\hat{\boldsymbol{x}}_r(t) + \boldsymbol{B}_r\boldsymbol{u}(t) + \boldsymbol{L}(\boldsymbol{y}(t) - \hat{\boldsymbol{y}}_r(t)), \\ \hat{\boldsymbol{y}}_r(t) &= \boldsymbol{C}_r\hat{\boldsymbol{x}}_r(t), \end{aligned} \tag{2.14}$$

where $\boldsymbol{L}$ denotes the designed observer gain matrix. Then the error dynamics can be obtained as:

$$\begin{aligned} \dot{\boldsymbol{e}}_r(t) &= (\boldsymbol{A}_r - \boldsymbol{L}\boldsymbol{C}_r)\boldsymbol{e}_r(t) + \boldsymbol{E}_r\boldsymbol{f}(t) + \boldsymbol{F}_r\boldsymbol{d}(t) - \boldsymbol{L}\boldsymbol{C}_f\boldsymbol{x}_f(t), \\ \boldsymbol{e}_{y_r}(t) &= \boldsymbol{C}_r\boldsymbol{e}_r(t) + \boldsymbol{y}_f(t), \end{aligned} \tag{2.15}$$

where $\boldsymbol{e}_r(t) \triangleq \boldsymbol{x}_r(t) - \hat{\boldsymbol{x}}_r(t)$, $\boldsymbol{e}_{y_r}(t) \triangleq \boldsymbol{y}(t) - \hat{\boldsymbol{y}}_r(t)$.

The following assumptions and lemma are needed for the reduced-order observer analysis.

Assumption 2.1 When no fault exists, the output of the $\boldsymbol{x}_f$-subsystem $\boldsymbol{y}_f(t)$ satisfies that:

$$\left\|\boldsymbol{y}_f(t)\right\|_{\text{peak}} \triangleq \sup_t ||\boldsymbol{y}_f(t)|| < \infty,$$

where $\|\cdot\|_{\text{peak}}$ and $\|\cdot\|$ denote the so-called peak-norm [14] and Euclidean norm, respectively.

Assumption 2.2 $(\boldsymbol{A}_r, \boldsymbol{C}_r)$ is observable and $\boldsymbol{C}_r$ is of full column rank.

Remark 2.4 The observability condition is a general assumption for the observer design of dynamic systems governed by ODEs. As for the full-column rank condition of $\boldsymbol{C}_r$, it can be satisfied by an appropriate choice of the order of the eigenfunctions m, the transformation matrix $\boldsymbol{R}$, and the number of the measurement sensors n_y.

Assumption 2.3 The disturbance satisfies that:

$$||d(\cdot, t)||_\infty \leq \bar{d},$$

where $\bar{d}$ is a known constant. This is a standard setting for FD of dynamical systems described by ODEs.

Note that the requirement of $\boldsymbol{C}_r$ being of full column rank is required by the post-filter design, which will be illustrated in the next subsection.

Lemma 2.3 *[33] Let $V(t)$ and $h(t)$ be regular real functions. Then*

$$\dot{V}(t) \leq -\alpha V(t) + h(t), \forall t \geq 0$$

implies that

$$V(t) \le e^{-\alpha t} V(0) + \int_0^t e^{-\alpha(t-\tau)} h(\tau)\mathrm{d}\tau, \quad \forall t \ge 0$$

for any finite constant α.

Lemma 2.4 *[20] For a given positive scalar* $\mu > 0$ *and a symmetric positive definite matrix* $\boldsymbol{G}$*, the following inequality holds:*

$$2\boldsymbol{x}^{\mathrm{T}}\boldsymbol{y} \leqslant \frac{1}{\mu}\boldsymbol{x}^{\mathrm{T}}\boldsymbol{G}\boldsymbol{x} + \mu\boldsymbol{y}^{\mathrm{T}}\boldsymbol{G}^{-1}\boldsymbol{y}, \ \boldsymbol{x}, \boldsymbol{y} \in \mathbb{R}^{p}.$$

Theorem 2.1 (Stability of the reduced-order observer) *Under Assumptions 2.1 and 2.2, if there exists symmetric positive definite matrices* $\boldsymbol{P} \in \mathbb{R}^{n\times n}$, $\boldsymbol{G}_1 \in \mathbb{R}^{n\times n}$ *matrix* $\boldsymbol{X} \in \mathbb{R}^{n\times n_y}$*, positive constant* ε *such that the following linear matrix inequality (LMI) is satisfied:*

$$\boldsymbol{\Xi} \triangleq \begin{bmatrix} \boldsymbol{\Xi}_{11} & * \\ -\boldsymbol{X}^{\mathrm{T}} & -\varepsilon\boldsymbol{I} \end{bmatrix} < 0, \tag{2.16}$$

where

$$\boldsymbol{\Xi}_{11} \triangleq \boldsymbol{P}\boldsymbol{A}_r + \boldsymbol{A}_r^{\mathrm{T}}\boldsymbol{P} - \boldsymbol{X}\boldsymbol{C}_r - \boldsymbol{C}_r^{\mathrm{T}}\boldsymbol{X}^{\mathrm{T}} + \boldsymbol{G}_1, \tag{2.17}$$

then the observer gain matrix $\boldsymbol{L} = \boldsymbol{P}^{-1}\boldsymbol{X}$*, and the reduced-order observer in (2.14) can realize* $\boldsymbol{e}_r(t)$ *uniformly ultimately bounded (UUB) when no fault exists.*

Proof Choose the Lyapunov candidate as follows:

$$V(t) = \boldsymbol{e}_r^{\mathrm{T}}(t)\boldsymbol{P}\boldsymbol{e}_r(t). \tag{2.18}$$

Recalling (2.15), its derivative with respect to time is:

$$\dot{V}(t) = \boldsymbol{e}_r^{\mathrm{T}}(t)(\boldsymbol{P}\boldsymbol{A}_r + \boldsymbol{A}_r^{\mathrm{T}}\boldsymbol{P} - \boldsymbol{X}\boldsymbol{C}_r - \boldsymbol{C}_r^{\mathrm{T}}\boldsymbol{X}^{\mathrm{T}})\boldsymbol{e}_r(t) - 2\boldsymbol{y}_f^{\mathrm{T}}(t)\boldsymbol{X}^{\mathrm{T}}\boldsymbol{e}_r(t) + 2\boldsymbol{e}_r^{\mathrm{T}}(t)\boldsymbol{P}\boldsymbol{F}_r\boldsymbol{d}(t). \tag{2.19}$$

Recalling Lemma 2.4 and selecting $\mu = 1$, it can be further derived that:

$$\begin{aligned}\dot{V}(t) &\le \boldsymbol{e}_r^{\mathrm{T}}(t)(\boldsymbol{P}\boldsymbol{A}_r + \boldsymbol{A}_r^{\mathrm{T}}\boldsymbol{P} - \boldsymbol{X}\boldsymbol{C}_r - \boldsymbol{C}_r^{\mathrm{T}}\boldsymbol{X}^{\mathrm{T}} + \boldsymbol{G}_1)\boldsymbol{e}_r(t) - 2\boldsymbol{y}_f^{\mathrm{T}}(t)\boldsymbol{X}^{\mathrm{T}}\boldsymbol{e}_r(t) + \boldsymbol{d}^{\mathrm{T}}(t)\boldsymbol{Y}\boldsymbol{d}(t) \\ &\le \boldsymbol{\xi}^{\mathrm{T}}(t)\boldsymbol{\Xi}\boldsymbol{\xi}(t) + \varepsilon\left\|\boldsymbol{y}_f(t)\right\|^2 + \lambda_{\max}(\boldsymbol{Y})\left\|\boldsymbol{d}(t)\right\|^2,\end{aligned} \tag{2.20}$$

where

$$\boldsymbol{\xi}(t) \triangleq \begin{bmatrix} \boldsymbol{e}_r(t) \\ \boldsymbol{y}_f(t) \end{bmatrix}, \boldsymbol{Y} \triangleq \boldsymbol{F}_r^{\mathrm{T}}\boldsymbol{P}\boldsymbol{G}_1^{-1}\boldsymbol{P}\boldsymbol{F}_r, \tag{2.21}$$

and $\lambda_{\max}(\boldsymbol{Y})$ denotes the maximum eigenvalue of matrix $\boldsymbol{Y}$.

Recalling Assumption 2.1 and 2.3, when $\boldsymbol{\Xi} < 0$, one can obtain that:

$$\begin{aligned}\dot{V}(t) &\leqslant -\lambda_{\min}(-\Xi)\|\boldsymbol{\xi}(t)\|^2 + \varepsilon\left\|\boldsymbol{y}_f(t)\right\|^2 + \lambda_{\max}(\boldsymbol{Y})\|\boldsymbol{d}(t)\|^2 \\ &\leqslant -\lambda_{\min}(-\Xi)\|\boldsymbol{e}_r(t)\|^2 + \varepsilon\left\|\boldsymbol{y}_f(t)\right\|_{\text{peak}}^2 + \lambda_{\max}(\boldsymbol{Y})\bar{d}^2,\end{aligned} \tag{2.22}$$

where $\lambda_{\min}(-\Xi)$ denotes the minimum eigenvalue of matrix $-\Xi$.

According to the definition of $V(t)$ in (2.18), it can be derived that:

$$V(t) \leqslant \lambda_{\max}(\boldsymbol{P})\|\boldsymbol{e}_r(t)\|^2. \tag{2.23}$$

Combining (2.22) and (2.23), it can be obtained that:

$$\dot{V}(t) \leqslant -\alpha V(t) + \varepsilon\left\|\boldsymbol{y}_f(t)\right\|_{\text{peak}}^2 + \lambda_{\max}(\boldsymbol{Y})\bar{d}^2, \tag{2.24}$$

where

$$\alpha \triangleq \frac{\lambda_{\min}(-\Xi)}{\lambda_{\max}(\boldsymbol{P})}.$$

By Lemma 2.3, it can be obtained that:

$$\begin{aligned}V(t) &\leqslant e^{-\alpha t}V(0) + (\varepsilon\left\|\boldsymbol{y}_f(t)\right\|_{\text{peak}}^2 + \lambda_{\max}(\boldsymbol{Y})\bar{d}^2)\int_0^t e^{-\alpha(t-\tau)}\mathrm{d}\tau \\ &\leqslant e^{-\alpha t}V(0) + \frac{\varepsilon}{\alpha}\left\|\boldsymbol{y}_f(t)\right\|_{\text{peak}}^2 + \frac{\lambda_{\max}(\boldsymbol{Y})}{\alpha}\bar{d}^2.\end{aligned} \tag{2.25}$$

Meanwhile, considering (2.18), it is intuitive that:

$$V(t) \geq \lambda_{\min}(\boldsymbol{P})\|\boldsymbol{e}_r(t)\|^2. \tag{2.26}$$

Combining (2.18), (2.25) and (2.26), it can be derived that:

$$\|\boldsymbol{e}_r(t)\|^2 \leqslant \frac{e^{-\alpha t}V(0)}{\lambda_{\min}(\boldsymbol{P})} + \rho^2 \leqslant (\sqrt{\frac{e^{-\alpha t}V(0)}{\lambda_{\min}(\boldsymbol{P})}} + \rho)^2, \tag{2.27}$$

where

$$\rho \triangleq \sqrt{\frac{\varepsilon\left\|\boldsymbol{y}_f(t)\right\|_{peak}^2 + \lambda_{\max}(\boldsymbol{Y})\bar{d}^2}{\alpha\lambda_{\min}(\boldsymbol{P})}}.$$

Hence it can be further derived that:

$$\|\boldsymbol{e}_r(t)\| \leqslant \sqrt{\frac{e^{-\alpha t}V(0)}{\lambda_{\min}(\boldsymbol{P})}} + \rho. \tag{2.28}$$

Recalling Assumption 2.1, it can be obtained that $0 \leqslant \rho < \infty$. Hence it implies that there exists a $T \to \infty$ such that $\|\boldsymbol{e}_r(t)\| \leqslant \rho$, for all $t > T$. That is to say, $\boldsymbol{e}_r(t)$ is UUB with ultimate bound ρ when no fault exists.

This completes the proof. □

2.4.3 General Post-filter Design

As discussed in Remark 2.1, the most restrictive constraint for the post-filter is the existence of the left-inverse of $\boldsymbol{C} \in \mathbb{R}^{n_y \times \infty}$, which requires continuous measurements [29] in the space domain and can not be satisfied in real situations.

However, with the help of spectral approximation, this restriction can be released. The intuitive idea is to construct a post-filter based on the slow sub-system (2.10) and the corresponding finite-dimensional observer as follows:

$$\boldsymbol{V}_s(z) \triangleq \boldsymbol{\phi}_s^{\mathrm{T}}(z) \left(\boldsymbol{C}_s^{\mathrm{T}} \boldsymbol{C}_s\right)^{-1} \boldsymbol{C}_s^{\mathrm{T}}, \tag{2.29}$$

where $\boldsymbol{C}_s \in \mathbb{R}^{n_y \times m}$. The necessary condition for the existence of the left inverse of $\boldsymbol{C}_s$ is that $m \leq n_y$, which can be satisfied by proper selection of the sensors and their corresponding locations. When applying the spectral approximation techniques, the order of the slow sub-system m can be rather high, hence it requires a large number of sensors n_y for the FDLF design.

To further release the requirement on the number of sensors, another post-filter based on the reduced-order model (2.13) is designed as follows:

$$\boldsymbol{V}_r(z) \triangleq \boldsymbol{\varphi}^{\mathrm{T}}(z) \left(\boldsymbol{C}_r^{\mathrm{T}} \boldsymbol{C}_r\right)^{-1} \boldsymbol{C}_r^{\mathrm{T}}. \tag{2.30}$$

Since $\boldsymbol{C}_r \in \mathbb{R}^{n_y \times n}$ and $m > n$ holds, the existence condition for the left-inverse of $\boldsymbol{C}_r$ is less restrictive than that of $\boldsymbol{C}_s$. Moreover, as it is discussed in Lemma 2.2, the reduced-order model (2.13) obtains less state truncation error compared to the slow sub-system (2.10) once the condition for the transformation matrix $\boldsymbol{R}$ is satisfied.

Motivated by the above considerations, the following distributed S-T residual signal is obtained for fault detection and localization:

$$r(z,t) \triangleq \boldsymbol{\varphi}^{\mathrm{T}}(z) \left(\boldsymbol{C}_r^{\mathrm{T}} \boldsymbol{C}_r\right)^{-1} \boldsymbol{C}_r^{\mathrm{T}} \boldsymbol{e}_{y_r}(t). \tag{2.31}$$

2.4.4 Distributed Residual Evaluation and Threshold Computation

The following distributed residual evaluation function and the corresponding threshold are designed:

$$
\begin{aligned}
J(r(z,t)) &\triangleq \int_t^{t+T} r^2(z,\tau)\mathrm{d}\tau, \\
J_{\text{th}}(t) &\triangleq \sup_{t\geq 0,\, f=0} J(r(z,t)),
\end{aligned}
\tag{2.32}
$$

where $T > 0$ denotes the length of the sliding time-evaluation window.

Hence the following decision logic will guarantee reliable fault detection and localization performance:

$$
\begin{cases}
J(r(z,t)) > J_{\text{th}}(t) \Longrightarrow \text{faulty for location } z \text{ at time } t \\
J(r(z,t)) \leq J_{\text{th}}(t) \Longrightarrow \text{fault-free for location } z \text{ at time } t.
\end{cases}
\tag{2.33}
$$

Denote z_f and t_f as the set of fault location and time obtained by (2.33). The following lemma is useful for computing the threshold $J_{\text{th}}(t)$.

Lemma 2.5 [34, 35] (A Variant of Agmon's Inequality in one-dimensional (1-D) Spatial Domain). *Let $\omega(z,t) \in \mathcal{H}^1(0,L)$ be a scalar function with $\omega(0,t) = 0$ or $\omega(L,t) = 0$, $t \geq 0$. The following inequality is fulfilled:*

$$
\|\omega(\cdot,t)\|_\infty^2 \leq 2\,\|\omega(\cdot,t)\|_2\,\|\omega_z(\cdot,t)\|_2 \leq \|\omega(\cdot,t)\|_{\mathcal{H}^1}^2, t \geq 0. \tag{2.34}
$$

Theorem 2.2 (Performance of the FDLF) *If the conditions in Theorem 2.1 are satisfied, the following* ***time-varying*** *threshold $J_{\text{th}}(t)$ can be defined and calculated off-line to guarantee the performance of the proposed FDLF under disturbance:*

$$
J_{\text{th}}(t) = \int_t^{t+T} \bar{\rho}(\tau)\mathrm{d}\tau, \tag{2.35}
$$

where

$$
\bar{\rho}(\tau) = 2n\pi L^{-1}\Big(\sqrt{\frac{e^{-\alpha\tau}V(0)}{\lambda_{\min}(\boldsymbol{P})}} + \sqrt{\frac{\varepsilon\,\|\boldsymbol{y}_f(t)\|_{peak}^2 + \lambda_{\max}(\boldsymbol{Y})\bar{d}^2}{\alpha\lambda_{\min}(\boldsymbol{P})}} + \sigma_{\max}(\boldsymbol{C}_r^+)\,\|\boldsymbol{y}_f(t)\|_{\text{peak}}\Big)^2. \tag{2.36}
$$

where $\boldsymbol{C}_r^+ = (\boldsymbol{C}_r^{\mathrm{T}}\boldsymbol{C}_r)^{-1}\boldsymbol{C}_r^{\mathrm{T}}$ denotes the pseudo-inverse of $\boldsymbol{C}_r$ and $\boldsymbol{Y}$ is defined in (2.21).

Proof First, the properties of the basis functions $\{\phi_i(z)\}_{i=1}^m$ for spectral method are analyzed. For the system model (2.1) and the boundary conditions in (2.3), one can select the following basis functions [34]:

$$
\phi_i(z) = a\sin(i\pi\frac{z}{L}), i = 1, \cdots, m. \tag{2.37}
$$

where a is a constant.

Then, for the novel basis functions $\{\varphi_i(z)\}_{i=1}^n$, it can be derived that:

$$
\begin{aligned}
&< \varphi_i(\cdot), \varphi_j(\cdot) > = 0, \|\varphi_i(\cdot)\|_2^2 = 1, \\
&< \dot{\varphi}_i(\cdot), \dot{\varphi}_j(\cdot) > = 0, \|\dot{\varphi}_i(\cdot)\|_2^2 = i^2\pi^2 L^{-2}, i \neq j.
\end{aligned} \tag{2.38}
$$

according to Lemma 2.1.

Note that since $\phi_i(z)$ is selected in the form (2.37), it can be verified that the distributed S-T residual signal satisfies that $r(0, t) = r(L, t) = 0$ by (2.31). According to Lemma 2.5, when no fault exists, it can be derived that:

$$
\|r(\cdot, t)\|_\infty^2 \le 2\,\|r(\cdot, t)\|_2\,\|r_z(\cdot, t)\|_2\,. \tag{2.39}
$$

Moreover, it can be derived that:

$$
\begin{aligned}
\|r(\cdot, t)\|_2 &= \left\|\boldsymbol{\varphi}_r^{\mathrm{T}}(z)\boldsymbol{C}_r^{+}\boldsymbol{e}_{y_r}(t)\right\|_2 \\
&\le \|\boldsymbol{e}_r(t)\| + \left\|\boldsymbol{C}_r^{+}\boldsymbol{y}_f(t)\right\| \\
&\le \|\boldsymbol{e}_r(t)\| + \sigma_{\max}(\boldsymbol{C}_r^{+})\left\|\boldsymbol{y}_f(t)\right\|_{\text{peak}},
\end{aligned} \tag{2.40}
$$

where $\sigma_{\max}(\boldsymbol{C}_r^{+})$ denotes the maximum singular value of matrix $\boldsymbol{C}_r^{+}$.

Recalling Theorem 2.1, it can be obtained that:

$$
\|r(\cdot, t)\|_2 \le \sqrt{\frac{e^{-\alpha t}V(0)}{\lambda_{\min}(\boldsymbol{P})}} + \rho + \sigma_{\max}(\boldsymbol{C}_r^{+})\left\|\boldsymbol{y}_f(t)\right\|_{\text{peak}}. \tag{2.41}
$$

Moreover, we have that:

$$
\begin{aligned}
\|r_z(\cdot, t)\|_2 &= \left\|\dot{\boldsymbol{\varphi}}_r^{\mathrm{T}}(z)\boldsymbol{C}_r^{+}\boldsymbol{e}_{y_r}(t)\right\|_2 = \left\|\dot{\boldsymbol{\varphi}}_r^{\mathrm{T}}(z)\boldsymbol{\zeta}(t)\right\|_2 \\
&= \sqrt{\zeta_1^2(t)\|\dot{\varphi}_1(\cdot)\|_2^2 + \cdots + \zeta_n^2(t)\|\dot{\varphi}_n(\cdot)\|_2^2},
\end{aligned} \tag{2.42}
$$

where

$$
\boldsymbol{\zeta}(t) \triangleq \boldsymbol{e}_r(t) + \boldsymbol{C}_r^{+}\boldsymbol{y}_f(t).
$$

Recalling (2.38), one can further derive that:

$$
\begin{aligned}
\|r_z(\cdot, t)\|_2 &\le \pi L^{-1}\sqrt{1^2\zeta_1^2(t) + \cdots + n^2\zeta_n^2(t)} \\
&\le n\pi L^{-1}\,\|\boldsymbol{\zeta}(t)\| \\
&\le n\pi L^{-1}(\|\boldsymbol{e}_r(t)\| + \sigma_{\max}(\boldsymbol{C}_r^{-})\left\|\boldsymbol{y}_f(t)\right\|_{\text{peak}}) \\
&\le n\pi L^{-1}(\sqrt{\frac{e^{-\alpha t}V(0)}{\lambda_{\min}(\boldsymbol{P})}} + \rho + \sigma_{\max}(\boldsymbol{C}_r^{+})\left\|\boldsymbol{y}_f(t)\right\|).
\end{aligned} \tag{2.43}
$$

Combining (2.39), (2.41) and (2.43), it can be obtained that:

$$\|r(\cdot,t)\|_{\infty}^{2} \le \bar{\rho}(t), \tag{2.44}$$

where

$$\bar{\rho}(t) \triangleq 2n\pi L^{-1}\left(\sqrt{\frac{e^{-\alpha t}V(0)}{\lambda_{\min}(\boldsymbol{P})}} + \rho + \sigma_{\max}(\boldsymbol{C}_r^{+})\left\|\boldsymbol{y}_f(t)\right\|_{\text{peak}}\right)^2.$$

Moreover, Combining (2.32) with (2.44), it can be derived that:

$$\begin{aligned} J(r(z,t)) &= \int_t^{t+T} r^2(z,\tau)\mathrm{d}\tau \le \int_t^{t+T} \|r(\cdot,\tau)\|_{\infty}^{2}\,\mathrm{d}\tau \\ &\le \int_t^{t+T} \bar{\rho}(\tau)\mathrm{d}\tau. \end{aligned} \tag{2.45}$$

Therefore, the time-varying threshold $J_{\text{th}}(t)$ for fault detection and localization can be selected as:

$$J_{\text{th}}(t) \triangleq \int_t^{t+T} \bar{\rho}(\tau)\mathrm{d}\tau.$$

This completes the proof. □

Remark 2.5 Note that the calculated threshold $J_{\text{th}}(t)$ is set to be time-varying rather than a deterministic which is mostly used in the literature. The reason is that the initial condition of the system $x_0(z)$ is **unknown** for the FDLF design and it is assigned an arbitrary real number. Therefore, the influence of this initial error should be taken into account when calculating the threshold. As stated in Theorem 2.1, the influence of this initial error is reflected in the term $V(0)$ and it is decreasing exponentially with a constant rate α. However, to avoid false alarm rate (FAR) in the initial stage, this term is also considered in the threshold setting (2.35) and (2.36). In this manner, false alarm can be well rejected for the proposed FDLF.

A summary of the proposed novel basis functions-based FDLF by pseudo code is shown in Algorithm 2.1.

2.5 Numerical Simulation

Consider a thin rod [20] whose temperature distribution can be modeled by the following parabolic PDE:

$$\frac{\partial T(z,t)}{\partial t} = \frac{\partial^2 T(z,t)}{\partial z^2} + \beta_U(b_u(z)u(t) - T(z,t)) + f(z,t) + d(z,t), \tag{2.46}$$

Algorithm 2.1: Novel basis functions-based FDLF

Input: $u(z,t), \bar{d}$
Output: Faulty location set z_f and time set t_f.
/****Initialization*****/
  Select the transformation matrix $\boldsymbol{R}$ satisfying Lemma 2.2 and T (evaluation time window length);
/****Off-line threshold generation*****/
  Observer gains: $\boldsymbol{L} = \boldsymbol{P}^{-1}\boldsymbol{X} \leftarrow$ LMI (2.16);
for $t = 1 : t_{\max}$ *(considered time length)* **do**
  Threshold calculation: $J_{\text{th}}(t) \leftarrow (2.35)$;
/****On-line distributed residual evaluation*****/
for $t = 1 : t_{\max}$ **do**
  for $z \in [0, L]$ **do**
    Residual evaluation: $J(r(z,t)) \leftarrow (2.32)$;
    if $J(r(z,t)) > J_{th}(t)$ **then**
      $z \in z_f, t \in t_f$
    else
      No alarm raised

$$
\boldsymbol{y}(t) = \begin{bmatrix} \int_0^{\pi} \delta(z - \frac{\pi}{3}) T(z,t) \mathrm{d}z \\ \int_0^{\pi} \delta(z - \frac{2\pi}{3}) T(z,t) \mathrm{d}z \end{bmatrix} = \begin{bmatrix} T(\frac{\pi}{3}, t) \\ T(\frac{2\pi}{3}, t) \end{bmatrix}, \tag{2.47}
$$

subject to the Dirichlet boundary conditions:

$$
T(0,t) = 0, \;\; T(\pi,t) = 0, \tag{2.48}
$$

where $T(z,t)$ denotes the dimensionless temperature of the rod; β_U denotes a dimensionless heat transfer coefficient; $u(t)$ is the manipulated input (temperature of the cooling medium); $\boldsymbol{y}(t)$ are the thermocouple measurements at point $z = \pi/3$ and $z = 2\pi/3$ with $\delta(\cdot)$ denoting the Dirac delta function; The typical value of β_U is given as 2 and the control input is selected as:

$$
u(t) = 0.
$$

Remark 2.6 Note that the controller design problem is not investigated in this chapter due to the fact that it is irrelevant to the FDLF design. Hence the above open-loop system is set for the purpose of simple illustration. However, as a future direction, it would be interesting to consider the fault-tolerant controller design for DPSs.

Moreover, the spatial distribution function of the S-T fault $f(z,t)$ is selected as:

$$
b_f(z) = H(z - \pi/4) - H(z - \pi/2), \tag{2.49}
$$

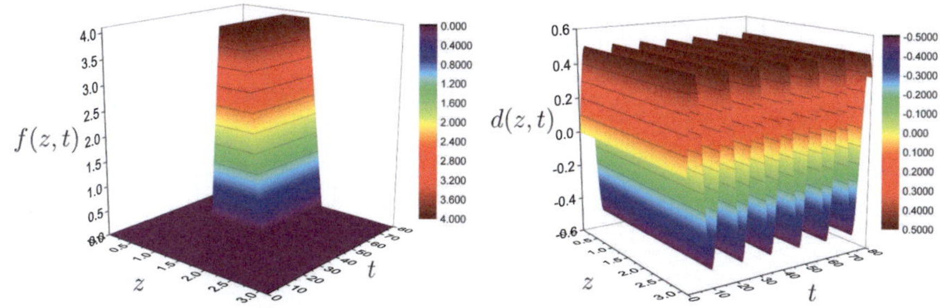

(a) Distributed fault $f(z, t)$.

(b) Distributed disturbance $d(z, t)$.

Fig. 2.2 Distributed fault and disturbance setting

for simple illustration, where $H(\cdot)$ is the standard Heaviside function. Note that when the designed FDLF is applied, no information of $b_f(z)$ is known. To evaluate the performance of the proposed FDLF, the following abrupt S-T fault is considered:

$$f(t) = \begin{cases} 0, 0 \leqslant t < 40\,(\mathrm{s}) \\ 4, 40 \leqslant t \leqslant 80\,(\mathrm{s}). \end{cases}$$

Moreover, the following distributed disturbance $d(z, t)$ is set:

$$b_d(z) = 1, d(t) = 0.5\sin(0.5t).$$

For the above setting, the distributed fault and disturbance are presented in Fig. 2.2a, b, respectively.

The original PDE system is solved numerically by the finite difference method (FDM) with the sampling time $\Delta t = 0.01$ s and 17 grid points with respect to the space. Based on the discussions in Sect. 2.4.1, the basis function transformation matrix $\boldsymbol{R}$ which satisfies conditions in Lemmas 2.1 and 2.2 is calculated first. To better illustrate the modal approximation performance, the performance index "root mean square error (RMSE)" of the state truncation is defined as:

$$\mathrm{RMSE}_s = \left(\int \sum (x(z,t) - \boldsymbol{\phi}_s^T(z)\boldsymbol{x}_s(t))^2 \mathrm{d}z \Big/ \int \mathrm{d}z \sum \Delta t\right)^{1/2},$$
$$\mathrm{RMSE}_r = \left(\int \sum (x(z,t) - \boldsymbol{\varphi}_r^T(z)\boldsymbol{x}_r(t))^2 \mathrm{d}z \Big/ \int \mathrm{d}z \sum \Delta t\right)^{1/2}.$$

As shown in Table 2.1, the RMSE value of the state truncation for the novel basis functions is smaller than that of the eigenfunctions of the same order.

Furthermore, the constructed distributed S-T residual $r(z, t)$ in the presence of distributed fault $f(z, t)$ and disturbance $d(z, t)$ is presented in Fig. 2.3. It can be observed that in the fault-free situation, $||r(z, t)||_2$ decreases quickly and is bounded

Table 2.1 RMSE for the novel basis functions and eigenfunctions

m or n	RMSE_s	RMSE_r
$m = n = 1$	9.70e-03	**5.90e-03**
$m = n = 2$	9.70e-03	**1.50e-03**
$m = n = 3$	2.50e-03	**1.50e-03**
$m = n = 4$	2.50e-03	**6.02e-04**
$m = n = 5$	1.00e-03	**5.36e-04**

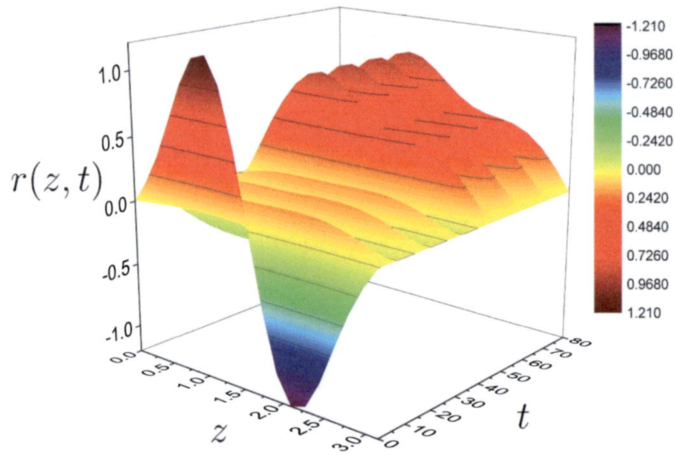

Fig. 2.3 Distributed S-T residual $r(z, t)$

before a fault occurs. This phenomena coincides with the results in Theorem 2.2. And the initial residual is large due to the initial error of the proposed FDLF, as discussed in Remark 2.5.

Moreover, the threshold $J_{\text{th}}(t)$ is computed numerically with the evaluation window $T = 0.16$ s based on the discussions in Sects. 2.4.2, 2.4.3 and 2.4.4. The number of novel basis functions is selected as $n = 2$ and the eigenfunctions and novel basis functions are presented in Fig. 2.4a, b, respectively.

The fault detection and localization performance base on both the novel basis functions and eigenfunctions are demonstrates in Fig. 2.5a, b, respectively. Moreover, to better illustrate the fault localization performance, after the S-T fault is detected in the time domain, the residual evaluation result at $t = 66$ s are presented in Fig. 2.6a, b, respectively.

In traditional statistical process control (SPC) methods, the "detection rate" is used to evaluate the performance of the diagnosis methods, which denotes the percentage of samples above the detection threshold for the samples obtained after the fault occurs. To show the superiority of the proposed novel basis functions over the eigenfunctions on fault localization, two performance indexes "detection rate" and

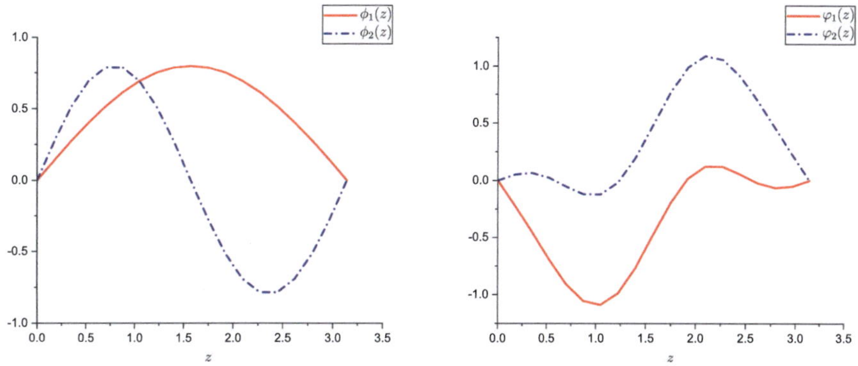

(a) Eigenfunctions (b) Novel basis functions

Fig. 2.4 The first two eigenfunctions and novel basis functions. The first two eigenfunctions and novel basis functions

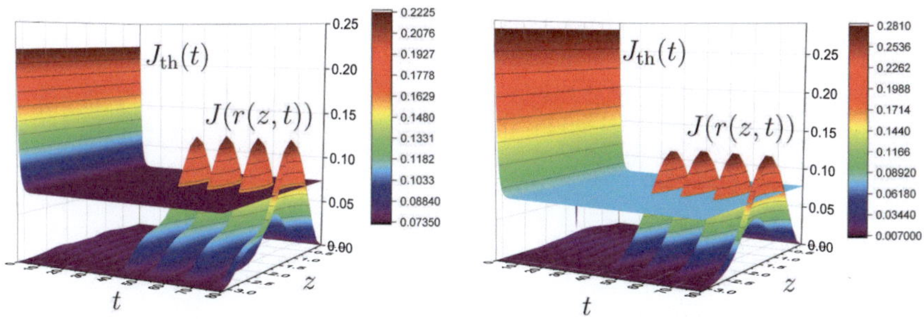

(a) Novel basis functions-based design (b) Eigenfunctions-based design

Fig. 2.5 Fault detection and localization performance

"misdetection rate" are defined as follows to evaluate the proposed fault localization method:

$$\text{Detection rate} = \frac{\text{Detected fault location area}}{\text{Actual fault location area}} - \frac{\text{Misdetected fault location area}}{\text{Actual fault location area}},$$
$$\text{Misdetection rate} = \frac{\text{Misdetected fault location area}}{\text{Detected fault location area}}.$$

As shown in Table 2.2, the novel basis functions-based design superiors over the eigenfunctions-based design with a more accurate localization performance.

Remark 2.7 For the eigenfunctions-based design, the procedures are very similar to the proposed novel basis functions-based method, with a slight difference in the approximated ODE model. Detailed design procedures are omitted due to triviality.

Remark 2.8 Comparing Fig. 2.3 with Fig. 2.5a, an interesting phenomena can be observed: the initial error of the proposed FDLF affects the distributed S-T residual

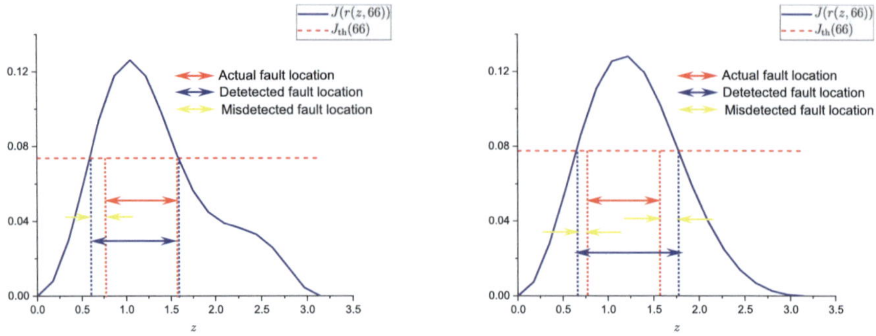

(a) Novel basis functions-based design

(b) Eigenfunctions-based design

Fig. 2.6 Fault localization result at $t = 66$ s

Table 2.2 Fault localization results using the novel basis functions and eigenfunctions

	Novel basis functions	Eigenfunctions
Number (m or n)	$n = 2$	$m = 2$
Detection rate (%)	**100**	**100**
Misdetection rate (%)	**21.18**	30.19

$r(z, t)$ more than that of the distributed evaluation function $J(r(z, t))$. The reason is that the initial error is mitigated by the sliding time-evaluation window of length T, which is the motivation of the proposed distributed evaluation function.

2.6 Conclusion

In this chapter, the design of a reduced-order FDLF for linear parabolic DPSs using limited sensor measurements is first discussed. To address the fault localization issue, a general post-filter which generates a distributed residual signal based on an approximated ODE model is proposed. One major challenge that comes with it is the requirement of a high order for the ODE model. To further address this issue, a set of novel basis functions which is a linear combination of the eigenfunctions is used for a further model reduction. Under certain circumstances, the novel basis functions possess smaller state truncation error than that using the eigenfunctions. Moreover, the localization performance of the proposed method superiors over that based on eigenfunctions.

However, the proposed method only considers parabolic DPSs due to the fact that the eigenspectrum for this kind of system is "separable", as well discussed in [23].

Therefore, how to expand this result to *hyperbolic* DPSs and investigate the fault localization problem is still an open problem.

References

1. Armaou A, Demetriou MA (2008) Robust detection and accommodation of incipient component and actuator faults in nonlinear distributed processes. AIChE J 54(10):2651–2662
2. He W, Ge SS (2011) Robust adaptive boundary control of a vibrating string under unknown time-varying disturbance. IEEE Trans Control Syst Technol 20(1):48–58
3. Wang J-W, Liu Y-Q, Sun C-Y (2018) Adaptive neural boundary control design for nonlinear flexible distributed parameter systems. IEEE Trans Control Syst Technol 27(5):2085–2099
4. Liu Y-Q, Wang J-W, Sun C-Y (2018) Observer-based output feedback compensator design for linear parabolic PDEs with local piecewise control and pointwise observation in space. IET Control Theory Appl 12(13):1812–1821
5. Zhang J, Cui B (2021) Mobile observation for distributed parameter system with moving boundary over mobile sensor networks. J Control Decis 8(2):124–134
6. Wang J-W, Liu Y-Q, Sun C-Y (2018) Observer-based dynamic local piecewise control of a linear parabolic PDE using non-collocated local piecewise observation. IET Control Theory Appl 12(3):346–358
7. Bao L, Huang P, Fei S (2021) Exponential stability and L2 gain analysis of switched distributed parameter systems with time delay. J Control Decis 8(4):385–393
8. Feng Y, Li H-X (2020) Dynamic spatial independent component analysis based abnormality localization for distributed parameter systems. IEEE Trans Ind Inf 16(5):2929–2936
9. El-Farra NH, Ghantasala S (2007) Actuator fault isolation and reconfiguration in transport-reaction processes. AIChE J 53(6):1518–1537
10. Feng Y, Li H-X (2018) Detection and spatial identification of fault for parabolic distributed parameter systems. IEEE Trans Ind Electron 66(9):7300–7309
11. Dey S, Perez HE, Moura SJ (2017) Model-based battery thermal fault diagnostics: algorithms, analysis, and experiments. IEEE Trans Control Syst Technol 27(2):576–587
12. Dey S, Perez HE, Moura SJ (2019) Robust fault detection of a class of uncertain linear parabolic PDEs. Automatica 107:502–510
13. Liu Z, Liu J, He W (2018) Robust adaptive fault tolerant control for a linear cascaded ODE-beam system. Automatica 98:42–50
14. Ding SX (2013) Model-based fault diagnosis techniques: design schemes, algorithms, and tools, 2nd edn. Springer, London
15. Zhu J-W, Yang G-H, Wang H et al (2015) Fault estimation for a class of nonlinear systems based on intermediate estimator. IEEE Trans Autom Control 61(9):2518–2524
16. Zhu J-W, Gu C-Y, Ding SX et al (2020) A new observer-based cooperative fault-tolerant tracking control method with application to networked multiaxis motion control system. IEEE Trans Ind Electron 68(8):7422–7432
17. Ding SX (2014) Data-driven design of fault diagnosis and fault-tolerant control systems. Springer, London
18. Cai J, Ferdowsi H, Sarangapani J (2016) Model-based fault detection, estimation, and prediction for a class of linear distributed parameter systems. Automatica 66:122–131
19. El-Farra NH (2006) Integrated fault detection and fault-tolerant control architectures for distributed processes. Ind Eng Chem Res 45(25):8338–8351
20. Feng Y, Li H-X, Yang H-D (2019) Abnormal source identification for parabolic distributed parameter systems. IEEE Trans Syst Man Cybern Syst 51(9):5698–5707
21. Ghantasala S, El-Farra NH (2009) Robust actuator fault isolation and management in constrained uncertain parabolic PDE systems. Automatica 45(10):2368–2373

22. Deutscher J (2016) Fault detection for linear distributed-parameter systems using finite-dimensional functional observers. Int J Control 89(3):550–563
23. Christofides PD (2012) Nonlinear and robust control of PDE systems: methods and applications to transport-reaction processes. Springer, Berlin
24. Deutscher J, Harkort C (2010) A parametric approach to finite-dimensional control of linear distributed-parameter systems. Int J Control 83(8):1674–1685
25. Fischer F, Deutscher J (2016) Algebraic fault detection and isolation for parabolic distributed-parameter systems using modulation functions. IFAC-PapersOnLine 49(8):162–167
26. Fischer F, Deutscher J (2017) Fault detection for parabolic systems with distributed inputs and outputs using the modulation function approach. IFAC-PapersOnLine 50(1):6774–6779
27. Fischer F, Deutscher J (2018) Modulating function based fault detection for parabolic systems with polynomial faults. IFAC-PapersOnLine 51(24):359–366
28. Fischer F, Deutscher J (2020) Flatness-based algebraic fault diagnosis for distributed-parameter systems. Automatica 117:108987
29. Li L, Ding SX, Peng K et al (2019) Fuzzy fault detection filter design for nonlinear distributed parameter systems. IEEE Access 7:11105–11113
30. Deng H, Li H-X, Chen G (2005) Spectral-approximation-based intelligent modeling for distributed thermal processes. IEEE Trans Control Syst Technol 13(5):686–700
31. Deng H, Jiang M, Huang C-Q (2012) New spatial basis functions for the model reduction of nonlinear distributed parameter systems. J Process Control 22(2):404–411
32. Laub A, Heath MT, Paige C et al (1987) Computation of system balancing transformations and other applications of simultaneous diagonalization algorithms. IEEE Trans Autom Control 32(2):115–122
33. Ioannou PA, Sun J (2012) Robust adaptive control. Courier Corporation, Chelmsford
34. Krstic M, Smyshlyaev A (2008) Boundary control of PDEs: a course on backstepping designs. SIAM, USA
35. Wang J-W, Wang J-M (2019) Mixed $\mathcal{H}_2/\mathcal{H}_\infty$ sampled-data output feedback control design for a semi-linear parabolic PDE in the sense of spatial $\mathcal{L}^\infty$ norm. Automatica 103:282–293

Chapter 3
Spectral Approximation-Based Fault Localization for Nonlinear DPSs: A Spatial Mapping Filter-Based Framework

Abstract Compared to the fault detection problem, fault localization for spatially distributed processes is physically intuitive but was seldom studied. The main challenge lies in the contradiction between the spatially distributed nature of the processes and a limited number of sensors that can be placed in the space for applications. In this chapter, we make the first attempt to systematically study the fault localization problem for a class of nonlinear spatially distributed processes without knowing any prior information of the spatial distribution function of the spatio-temporal fault. The main contribution is that a spatial mapping filter is first designed to generate a distributed residual signal which contains spatial information while using limited sensor measurements. The residual evaluation procedure is also conducted in a distributed manner. With these two parts, a systematic methodology framework is constructed and analyzed rigorously. Finally, the performance of the proposed method is evaluated on a faulty catalytic rod, which widely exists in chemical engineering.

3.1 Introduction

Many industrial processes (e.g. transport-reaction processes [1], thermal processes [2], and fluid processes) can be classified into spatially distributed processes. They can all be modeled by DPSs, whose system input, output, and parameters can change in both temporal and spatial domain, which can be concluded as the S-T dynamics [3]. Faults occurring in DPSs may cause the failure of controller or undesired system response, both are harmful to the safe and reliable operation of the system. On one hand, both model-based [4–6] and data-driven [7] FD for dynamic systems governed by ODEs have been investigated systematically during the last decades. On the other hand, researches focus on the FD for DPSs are rather limited.

During the last decades, some results have been reported on the FD issues of DPSs, see [8–14], most of these works belong to the category of model-based FD methods. Existing model-based approaches for FD of DPSs can be roughly classified into two categories: one was using a finite-dimensional ODE representation of DPSs; the other was based on the original PDE system. For example, motivated by the idea of using fictitious output observer [15, 16] to reduce the observation spillover, the

© The Author(s), under exclusive license to Springer Nature Singapore Pte Ltd. 2026
Y. Feng et al., *Spatio-Temporal Abnormality Diagnosis for Industrial Distributed Parameter Systems*, https://doi.org/10.1007/978-981-95-3750-1_3

authors proposed a finite-dimensional residual generator for the purpose of fault detection of linear DPSs in [17]. A novel model-based fault detection approach was developed in [18], where the state observer was designed based on the original PDE system.

Recently, the modulating functions-based approach that was introduced in [19] to the parameter identification for LPSs was applied to DPSs for FD [20–22]. Different from existing model-based approaches that were based on the observer design, the modulating functions-based approach transformed the PDEs into algebraic fault detection equations, that only depend on known signals and the fault. Moreover, since early-lumping technique was not involved, the observation spillover problem was well addressed.

Despite these innovative results, the studies on fault localization of DPSs are seldom, which are of great significance from the application viewpoint. Moreover, considering the spatial distribution characteristic of DPSs, it is natural to investigate the fault localization problem. Unfortunately, to the best of our knowledge, almost all existing results on FD of DPSs only focused on fault detection in the time domain due to the fact that the spatial distribution characteristic of the fault was not considered

- In the actuator fault case, the spatial distribution function of the fault was assumed to be the same as that of the actuator and was known a prior;
- In the sensor fault case, the spatial distribution characteristic of the fault was not considered for the most commonly used point-wise measurement sensors.

Fault localization of DPSs is inherently more challenging than fault detection in the time domain since the former aims at seeking a *distributed residual generator* (*continuous in space*) while the latter only needs a *lumped residual generator*. In this sense, fault localization requires an infinite number of sensors in the space literally, which is nearly impossible in applications. Therefore, the major challenge is to obtain a distributed residual signal while only a limited number of sensors (*discrete in space*) are applicable.

Motivated by the above-mentioned considerations, we make the first attempt to investigate the S-T fault detection and localization problem for spatially distributed processes modeled by nonlinear parabolic PDEs without knowing any prior information of the spatial distribution function of the fault. The S-T faults widely exist in industry applications. For example, the distributed thermal faults in Li-ion batteries [23, 24] that can lead to battery degradation and failure belong to the S-T faults. The in-domain actuator faults, for example, the actuator faults of diffusion-reaction processes [8, 9] that widely exist in chemical processes are special cases of the defined S-T faults. Therefore, the fault localization for such spatially distributed processes is of great significance in applications for fault alarm and removal. It is worth noticing that, considering limitations in real applications, we aim to solve this problem using *a limited number of sensors* rather than continuous measurements in the space domain as discussed in [25].

To explore the possibility of constructing a distributed residual generator for fault detection and localization, we apply the modal decomposition technique to the original system. One major discovered limitation is that the number of sensors should be infinity, which is nearly impossible for real applications. To release this constraint, the dominant dynamics of the original system is first extracted and a spatial mapping filter is initiated to obtain a distributed residual signal using a limited number of sensors. Moreover, the residual evaluation is also conducted in a distributed manner. Both theoretical analysis and numerical simulations are provided to validate the feasibility and effectiveness of the proposed method.

The main contributions of this chapter can be briefly summarized as

- For the first time, the fault localization problem for nonlinear spatially distributed processes is formulated and studied systematically in a model-based perspective.
- A spatial mapping filter which captures the spatial distribution characteristic of the unknown fault using *a finite number of sensors* is designed and analyzed rigorously.
- A systematic methodology framework consisting of the distributed residual signal construction and evaluation is established and discussed.
- The proposed methodology has potential to be applied to industrial spatially distributed processes due to several accompanying advantages of the early-lumping techniques used, including easy design and implementation, and a lower requirement on computation resources compared to late-lumping-based methods.

The rest of this chapter can be summarized as follows: preliminaries and problem descriptions are provided in Sect. 3.2. Detailed design and numerical simulations are provided in Sects. 3.4 and 3.5, respectively. Finally, this chapter is concluded in Sect. 3.6.

3.2 System Description

Notations: $\mathbb{R}$ and $\mathbb{R}^n$ are the set of all real numbers and n-dimensional Euclidean space with the norm $||\cdot||$, respectively. $|\omega|$ denotes the absolute value of ω for any $\omega \in \mathbb{R}$. $\mathcal{L}^2([0, L]) \triangleq \mathcal{L}^2((0, L); \mathbb{R})$ is a real Hilbert space of square integrable functions $\omega(z) : [0, L] \to \mathbb{R}$ with the spatial $\mathcal{L}^2$ norm $||\omega(\cdot)||_2 \triangleq \sqrt{\int_0^L \omega^2(z)\mathrm{d}z}$. Given a natural number $\bar{l}$, $\mathcal{H}^{\bar{l}}(0, L) \triangleq \mathcal{W}^{\bar{l},2}((0, L); \mathbb{R})$ is a real Sobolev space of absolutely continuous functions $\bar{\omega}(z) : (0, L) \to \mathbb{R}$ with square integrable derivatives $\mathrm{d}^i\bar{\omega}(z)/\mathrm{d}z^i$ up to the order $\bar{l} \geq 1$ and with the norm $||\bar{\omega}(\cdot)||_{\mathcal{H}^{\bar{l}}} \triangleq \sqrt{\int_0^L \sum_{i=0}^{\bar{l}} \left(\frac{\mathrm{d}^i\bar{\omega}(z)}{\mathrm{d}z^i}\right)^2 \mathrm{d}z}$. For any functions $\omega(\cdot) \in \mathcal{H}^1(0, L)$, the spatial $\mathcal{L}^\infty$ norm is defined as $||\omega(\cdot)||_\infty \triangleq \max_{z\in[0,L]} |\omega(z)|$.

Consider a class of nonlinear spatially distributed processes modeled by the following parabolic PDE:

$$\begin{aligned} \partial_t x(z, t) &= a_0\partial_z^2 x(z, t) + g(x) + \boldsymbol{b}_u^{\mathrm{T}}(z)\boldsymbol{u}(t) + \boldsymbol{b}_f^{\mathrm{T}}(z)\boldsymbol{f}(t), \\ \boldsymbol{y}(t) &= \int_0^L \boldsymbol{c}(z)x(z, t)\mathrm{d}z, \end{aligned} \tag{3.1}$$

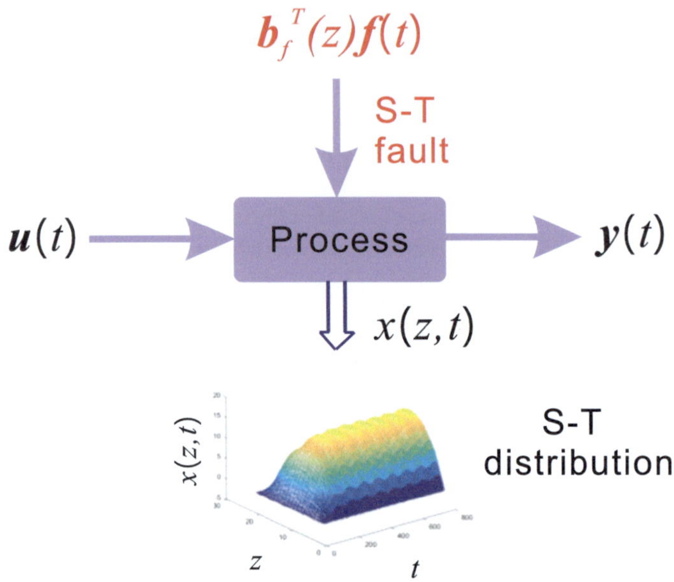

Fig. 3.1 System description

subject to the following homogeneous Dirichlet boundary conditions:

$$x(0, t) = x(L, t) = 0, \tag{3.2}$$

and with the following initial condition:

$$x(z, 0) = x_0(z), \tag{3.3}$$

where $x(z, t)$ denotes the state variable; $[0, L] \subset \mathbb{R}$ is the spatial domain of the system; $z \in [0, L]$ is the spatial coordinate; $t \in [0, \infty)$ denotes the time; $\boldsymbol{u}(t) \in \mathbb{R}^{n_u}$ denotes the vector of manipulated input; $\boldsymbol{f}(t) \in \mathbb{R}^{n_f}$ denotes the *unknown* fault in the DPS; $g(x)$ is a nonlinear function; $\boldsymbol{y}(t) \in \mathbb{R}^{n_y}$ denotes the system output; a_0 is a constant coefficient. The spatial distribution of the input and fault are characterized by $\boldsymbol{b}_u(z) \in \mathbb{R}^{n_u}$ and $\boldsymbol{b}_f(z) \in \mathbb{R}^{n_f}$, respectively. Note that $\boldsymbol{b}_u(z)$ is a known vector function while $\boldsymbol{b}_f(z)$ is *unknown*. The k-th element of known integrable vector function $\boldsymbol{c}(z) \in \mathbb{R}^{n_y}$ is determined by the location and shape (point or distributed) of the k-th measurement sensor; And $x_0(z)$ is the initial condition. A schematic graph of the system description is shown in Fig. 3.1.

The problem investigated in this chapter can be summarized as follows:

Use a limited number of sensors to design a FDLF and the corresponding residual evaluation and threshold computation procedures for fault detection and localization in the temporal and spatial domain for processes which subject to the mathematical model in (3.1) with the boundary conditions in (3.2) and the initial condition in (3.3).

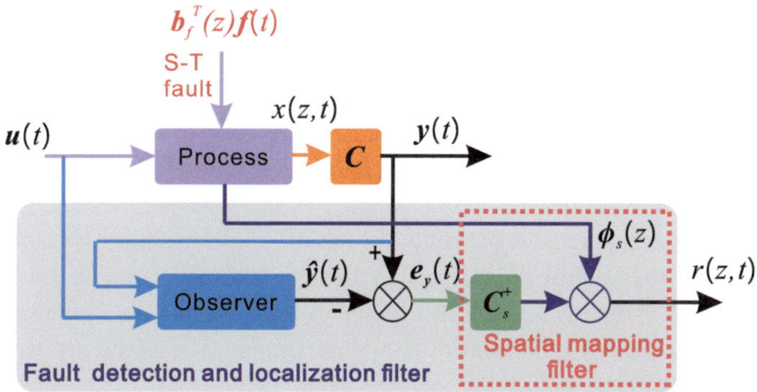

Fig. 3.2 FDLF framework

3.3 Methodology Framework

The core of model-based fault diagnosis lies in the construction of the analytical redundancy [4], which can be used for the residual generation. On one hand, to achieve the purpose of fault detection, the standard way is to construct the residual between the measurements and the analytical redundancy. On the other hand, to achieve the purpose of fault localization, one needs to use continuous measurements in the space, which is nearly impossible in practice.

To address the contradiction problem in fault localization, a spatial mapping filter which can develop the residual signals at continuous locations with a limited number of sensors is designed.

As presented in Fig. 3.2, the proposed FDLF consists of an observer and a spatial mapping filter. The observer generates the output estimation $\hat{\boldsymbol{y}}(t)$ while the spatial mapping filter generates the distributed S-T residual $r(z,t)$ using the output estimation error $\boldsymbol{e}_y(t)$. Detailed motivation and descriptions of the spatial mapping filter is illustrated in the following sections.

3.4 Spatial Mapping Filter-Based Fault Localization

In this section, detailed design procedures of the proposed FDLF and the corresponding residual evaluation as well as threshold computation are introduced as follows.

3.4.1 *Why Modal Approximation?*

Modal approximation techniques [26] are frequently used for model reduction of DPSs to transform the original PDE model into an approximated ODE model. In this manner, low-order observers can be designed based on the approximated ODE model, which is easy to implement and requires a lower number of computation resources compared to PDE observer. Moreover, since parabolic DPSs' spatial differential operator can be divided into a finite-dimensional slow subsystem and an infinite-dimensional fast subsystem [26], it is intuitive to apply modal approximation techniques to such systems.

The most important reason of using the modal approximation techniques is for the beneficial of the spatial mapping filter design, which generates the distributed S-T residual $r(z, t)$ using a limited number of sensors. To be more specific, the spectral method [27, 28] was used for modal decomposition as the first step. For the convenience of notation, define the operator $\mathcal{A}$ in $\mathcal{H}$ as

$$\mathcal{A}x \triangleq a_0 \frac{\partial^2 x}{\partial z^2},\ x \in \mathcal{S}(\mathcal{A}) \triangleq \{x \in \mathcal{H}, \text{and the boundary conditions in (3.2) hold}\}. \tag{3.4}$$

Denote the eigenfunctions of the linear operator $\mathcal{A}$ as

$$\{\phi_i(z)\}_{i=1}^{\infty} = \{\phi_1(z), \phi_2(z), \cdots, \phi_\infty(z)\}. \tag{3.5}$$

Moreover, applying orthogonal projection $\mathcal{P}$ to (3.1) yields the following infinite-dimensional ODE system:

$$\begin{aligned} \dot{\boldsymbol{x}}_M(t) &= \boldsymbol{A}\boldsymbol{x}_M(t) + g_M(\boldsymbol{x}_M(t)) + \boldsymbol{B}\boldsymbol{u}(t) + \boldsymbol{E}\boldsymbol{f}(t), \\ \boldsymbol{y}(t) &= \boldsymbol{C}\boldsymbol{x}_M(t), \end{aligned} \tag{3.6}$$

where

$$\boldsymbol{x}_M(t) \triangleq \mathcal{P}x(z,t) = \begin{bmatrix} <\phi_1, x> \\ \vdots \\ <\phi_\infty, x> \end{bmatrix} \in \mathbb{R}^\infty,$$

$$\boldsymbol{A} \triangleq \mathcal{P}\mathcal{A}\boldsymbol{\phi}^{\mathrm{T}}(z) = diag\{\lambda_1, \cdots, \lambda_\infty\} \in \mathbb{R}^{\infty\times\infty},$$

$$g_M(\boldsymbol{x}_M(t)) \triangleq \mathcal{P}g(x) = \begin{bmatrix} <\phi_1, g(x)> \\ \vdots \\ <\phi_\infty, g(x)> \end{bmatrix} \in \mathbb{R}^\infty,$$

$$\boldsymbol{B} \triangleq \mathcal{P}\boldsymbol{b}_u^{\mathrm{T}}(z) \in \mathbb{R}^{\infty\times n_u},\ \boldsymbol{E} \triangleq \mathcal{P}\boldsymbol{b}_f^{\mathrm{T}}(z) \in \mathbb{R}^{\infty\times n_f},$$

$$\boldsymbol{C} \triangleq \int_0^L \boldsymbol{c}(z)\boldsymbol{\phi}^{\mathrm{T}}(z)\mathrm{d}z \in \mathbb{R}^{n_y\times\infty}.$$

Similar to traditional FDF design, we can establish an infinite-dimensional FDF of (3.6) as follows:

$$\begin{aligned}\dot{\hat{\boldsymbol{x}}}_M(t) &= \boldsymbol{A}\hat{\boldsymbol{x}}_M(t) + g_M(\hat{\boldsymbol{x}}_M(t)) + \boldsymbol{B}\boldsymbol{u}(t) + \boldsymbol{L}_M(\boldsymbol{y}(t) - \hat{\boldsymbol{y}}(t)),\\ \hat{\boldsymbol{y}}(t) &= \boldsymbol{C}\hat{\boldsymbol{x}}_M(t).\end{aligned} \tag{3.7}$$

Define the state estimation error $e_{x_M}(z,t)$ and the output estimation error $\boldsymbol{e}_y(t)$ as

$$e_{x_M}(z,t) \triangleq x(z,t) - \hat{x}(z,t),\ \ \boldsymbol{e}_y(t) \triangleq \boldsymbol{y}(t) - \hat{\boldsymbol{y}}(t) \tag{3.8}$$

respectively, where $\hat{x}(z,t) \triangleq \sum_{i=1}^{\infty} \phi_i(z)\hat{x}_i(t) = \boldsymbol{\phi}^{\mathrm{T}}(z)\hat{\boldsymbol{x}}_M(t)$.

Combining (3.6) with (3.7) yields

$$\boldsymbol{e}_y(t) = \boldsymbol{C}\boldsymbol{e}_{\boldsymbol{x}_M}(t), \tag{3.9}$$

where $\boldsymbol{e}_{\boldsymbol{x}_M}(t) \triangleq \boldsymbol{x}_M(t) - \hat{\boldsymbol{x}}_M(t)$. Moreover, it is easy to derive that

$$e_{x_M}(z,t) = \boldsymbol{\phi}^{\mathrm{T}}(z)\boldsymbol{e}_{\boldsymbol{x}_M}(t). \tag{3.10}$$

Remark 3.1 If $\boldsymbol{C} \in \mathbb{R}^{n_y \times n}$ $(n \to \infty)$ is of full-column rank, then $\boldsymbol{e}_{\boldsymbol{x}_M}(t)$ can be obtained as

$$\boldsymbol{e}_{\boldsymbol{x}_M}(t) = \boldsymbol{C}^{+}\boldsymbol{e}_y(t) = \left(\boldsymbol{C}^{\mathrm{T}}\boldsymbol{C}\right)^{-1}\boldsymbol{C}^{\mathrm{T}}\boldsymbol{e}_y(t),$$

where $\boldsymbol{C}^{+} \triangleq (\boldsymbol{C}^{\mathrm{T}}\boldsymbol{C})^{-1}\boldsymbol{C}^{\mathrm{T}}$ denotes the pseudo-inverse of $\boldsymbol{C}$. And the state estimation error $e_{x_M}(z,t)$ can be obtained as

$$e_{x_M}(z,t) = \boldsymbol{\phi}^{\mathrm{T}}(z)\left(\boldsymbol{C}^{\mathrm{T}}\boldsymbol{C}\right)^{-1}\boldsymbol{C}^{\mathrm{T}}\boldsymbol{e}_y(t),$$

which can be used to construct the distributed S-T residual signal as

$$\bar{r}(z,t) \triangleq e_{x_M}(z,t) = \boldsymbol{V}(z)\boldsymbol{e}_y(t),$$

where $\boldsymbol{V}(z) \triangleq \boldsymbol{\phi}^{\mathrm{T}}(z)\left(\boldsymbol{C}^{\mathrm{T}}\boldsymbol{C}\right)^{-1}\boldsymbol{C}^{\mathrm{T}}$ denotes the designed spatial mapping filter, which maps the finite-dimensional output estimation error into the infinite-dimensional space. In this manner, fault localization can be achieved as the spatial distribution characteristic is well preserved. However, it can only be achieved when $n_y \to \infty$, which means an infinite number of sensors, and is not possible in practice. This motivates us to design a finite-dimensional FDLF instead.

To establish a finite-dimensional FDLF, a finite-dimensional approximated ODE model should first be obtained. Separate (3.6) into fast and slow mode according to the value of the eigenvalue $\{\lambda(i)\}_{i=1}^{\infty}$ as

$$\begin{aligned}\dot{\boldsymbol{x}}_s(t) &= \boldsymbol{A}_s\boldsymbol{x}_s(t) + g_s(\boldsymbol{x}_s(t), \boldsymbol{x}_f(t)) + \boldsymbol{B}_s\boldsymbol{u}(t) + \boldsymbol{E}_s\boldsymbol{f}(t),\\ \dot{\boldsymbol{x}}_f(t) &= \boldsymbol{A}_f\boldsymbol{x}_f(t) + g_f(\boldsymbol{x}_s(t), \boldsymbol{x}_f(t)) + \boldsymbol{B}_f\boldsymbol{u}(t) + \boldsymbol{E}_f\boldsymbol{f}(t),\\ \boldsymbol{y}(t) &= \boldsymbol{C}_s\boldsymbol{x}_s(t) + \boldsymbol{C}_f\boldsymbol{x}_f(t),\end{aligned} \tag{3.11}$$

where

$$
\begin{aligned}
\boldsymbol{A}_s &\triangleq \mathcal{P}_s \mathcal{A} \boldsymbol{\phi}^{\mathrm{T}}(z) = \operatorname{diag}\{\lambda_1, \cdots, \lambda_m\}, \\
\boldsymbol{A}_f &\triangleq \mathcal{P}_f \mathcal{A} \boldsymbol{\phi}^{\mathrm{T}}(z) = \operatorname{diag}\{\lambda_{m+1}, \cdots, \lambda_\infty\}, \\
g_s(\cdot) &\triangleq \mathcal{P}_s g(x), \, g_f(\cdot) \triangleq \mathcal{P}_f g(x), \\
\boldsymbol{B}_s &\triangleq \mathcal{P}_s \boldsymbol{b}_u^{\mathrm{T}}(z), \;\; \boldsymbol{B}_f \triangleq \mathcal{P}_f \boldsymbol{b}_u^{\mathrm{T}}(z), \\
\boldsymbol{E}_s &\triangleq \mathcal{P}_s \boldsymbol{b}_f^{\mathrm{T}}(z), \;\; \boldsymbol{E}_f(t) \triangleq \mathcal{P}_f \boldsymbol{b}_f^{\mathrm{T}}(z), \\
\boldsymbol{C}_s &\triangleq \int_0^L \boldsymbol{c}(z) \boldsymbol{\phi}_s^{\mathrm{T}}(z) \mathrm{d}z, \, \boldsymbol{C}_f \triangleq \int_0^L \boldsymbol{c}(z) \boldsymbol{\phi}_f^{\mathrm{T}}(z) \mathrm{d}z,
\end{aligned}
$$

$\operatorname{Re}\{\lambda_1\} \geqslant \operatorname{Re}\{\lambda_2\} \geqslant \cdots \geqslant \operatorname{Re}\{\lambda_m\} \geqslant \cdots, \operatorname{Re}\{\lambda_{m+1}\} < 0$ and $|\operatorname{Re}\{\lambda_m\}| / |\operatorname{Re}\{\lambda_{m+1}\}| = O(\epsilon)$, $\epsilon < 1$ is a small positive number. Based on the singular perturbation theory [29], the fast subsystem can be neglected. This modal approximation technique has been widely investigated and applied, detailed explanations can be referred to [26].

3.4.2 Finite-Dimensional Nonlinear Observer Design and Analysis

Based on the slow subsystem in (3.11), after neglecting the fast subsystem, a finite-dimensional observer is designed as follows for distributed S-T residual generation:

$$
\begin{aligned}
\dot{\hat{\boldsymbol{x}}}_s(t) &= \boldsymbol{A}_s \hat{\boldsymbol{x}}_s(t) + g_s(\hat{\boldsymbol{x}}_s(t), 0) + \boldsymbol{B}_s \boldsymbol{u}(t) + \boldsymbol{L}(\boldsymbol{y}(t) - \hat{\boldsymbol{y}}_s(t)), \\
\hat{\boldsymbol{y}}_s(t) &= \boldsymbol{C}_s \hat{\boldsymbol{x}}_s(t),
\end{aligned}
\tag{3.12}
$$

where $\boldsymbol{L}$ denotes the designed observer gain matrix. Then the error dynamics can be obtained as

$$
\begin{aligned}
\dot{\boldsymbol{e}}_s(t) &= (\boldsymbol{A}_s - \boldsymbol{L}\boldsymbol{C}_s)\boldsymbol{e}_s(t) + \tilde{g}_s(t) + \boldsymbol{E}_s \boldsymbol{f}(t) - \boldsymbol{L}\boldsymbol{C}_f \boldsymbol{x}_f(t), \\
\boldsymbol{e}_{y_s}(t) &= \boldsymbol{C}_s \boldsymbol{e}_s(t) + \boldsymbol{y}_f(t),
\end{aligned}
\tag{3.13}
$$

where $\boldsymbol{e}_s(t) \triangleq \boldsymbol{x}_s(t) - \hat{\boldsymbol{x}}_s(t)$, $\tilde{g}_s(t) \triangleq g_s(\boldsymbol{x}_s(t), \boldsymbol{x}_f(t)) - g_s(\hat{\boldsymbol{x}}_s(t), 0)$. The term $\boldsymbol{y}_f(t)$ denotes the so-called "observation spillover" [16], which was neglected in some of the early-lumping-based observer design for DPSs.

The following assumptions and lemma are needed for the finite-dimensional observer design.

Assumption 3.1 When no fault exists, the output of the $\boldsymbol{x}_f$-subsystem $\boldsymbol{y}_f(t)$ and $\boldsymbol{x}_f(t)$ satisfy that

$$\left\|\boldsymbol{y}_f(t)\right\|_{\text{peak}} \triangleq \sup_t ||\boldsymbol{y}_f(t)|| < \infty,$$
$$\left\|\boldsymbol{x}_f(t)\right\|_{\text{peak}} \triangleq \sup_t ||\boldsymbol{x}_f(t)|| < \infty, \ \forall t \geqslant 0$$

where $\|\cdot\|_{\text{peak}}$ denotes the so-called peak-norm [4].

Assumption 3.2 $(\boldsymbol{A}_s, \boldsymbol{C}_s)$ is observable and $\boldsymbol{C}_s$ is of full column rank.

Note that the requirement of $\boldsymbol{C}_s$ being of full column rank is required by the spatial mapping filter design, which will be illustrated in the next subsection.

Assumption 3.3 The nonlinear function $g(\cdot)$ is locally Lipschitz continuous with κ denoting the Lipschitz constant.

Lemma 3.1 *[30] Let $V(t)$ and $h(t)$ be regular real functions. Then*

$$\dot{V}(t) \leq -\alpha V(t) + h(t), \forall t \geq 0$$

implies that

$$V(t) \leq e^{-\alpha t} V(0) + \int_0^t e^{-\alpha(t-\tau)} h(\tau) \mathrm{d}\tau, \quad \forall t \geq 0$$

for any finite constant α.

Theorem 3.1 *Under Assumptions 3.1, 3.2, and 3.3, if there exists symmetric positive definite matrix $\boldsymbol{P} \in \mathbb{R}^{m\times m}$, matrix $\boldsymbol{X} \in \mathbb{R}^{m\times n_y}$, positive constant ε such that the following bilinear matrix inequality is satisfied:*

$$\boldsymbol{\Xi} \triangleq \begin{bmatrix} \boldsymbol{\Xi}_{11} & * \\ -\boldsymbol{X}^{\mathrm{T}} & -\varepsilon \boldsymbol{I} \end{bmatrix} < 0, \tag{3.14}$$

where

$$\boldsymbol{\Xi}_{11} \triangleq \boldsymbol{P}\boldsymbol{A}_s + \boldsymbol{A}_s^{\mathrm{T}}\boldsymbol{P} - \boldsymbol{X}\boldsymbol{C}_s - \boldsymbol{C}_s^{\mathrm{T}}\boldsymbol{X}^{\mathrm{T}} + \boldsymbol{P}^2 + \kappa^2 \boldsymbol{I}, \tag{3.15}$$

then the observer gain matrix $\boldsymbol{L} = \boldsymbol{P}^{-1}\boldsymbol{X}$, and the finite-dimensional observer (3.12) can realize $\boldsymbol{e}_s(t)$ UUB when no fault exists.

Proof Choose the Lyapunov candidate as follows:

$$V(t) = \boldsymbol{e}_s^{\mathrm{T}}(t)\boldsymbol{P}\boldsymbol{e}_s(t). \tag{3.16}$$

Recalling (3.13), its derivative with respect to time is

$$\dot{V}(t) = \boldsymbol{e}_s^{\mathrm{T}}(t)(\boldsymbol{P}\boldsymbol{A}_s + \boldsymbol{A}_s^{\mathrm{T}}\boldsymbol{P} - \boldsymbol{X}\boldsymbol{C}_s - \boldsymbol{C}_s^{\mathrm{T}}\boldsymbol{X}^{\mathrm{T}})\boldsymbol{e}_s(t) - 2\boldsymbol{y}_f^{\mathrm{T}}(t)\boldsymbol{X}^{\mathrm{T}}\boldsymbol{e}_s(t) + 2\tilde{g}_s^{\mathrm{T}}(t)\boldsymbol{P}\boldsymbol{e}_s(t). \tag{3.17}$$

Note that

$$2\tilde{g}_s^{\mathrm{T}}(t)\boldsymbol{P}\boldsymbol{e}_s(t) \leq \boldsymbol{e}_s{}^{\mathrm{T}}(t)\boldsymbol{P}^2\boldsymbol{e}_s(t) + \tilde{g}_s^{\mathrm{T}}(t)\tilde{g}_s(t).$$

Recalling the Lipschitz condition in Assumption 3.3, it can be obtained that

$$\begin{aligned}\tilde{g}_s^{\mathrm{T}}(t)\tilde{g}_s(t) &= \left\|\tilde{g}_s(t)\right\|^2 = \left\|g_s(\boldsymbol{x}_s(t), \boldsymbol{x}_f(t)) - g_s(\hat{\boldsymbol{x}}_s(t), 0)\right\|^2 = \| \, g_s(\boldsymbol{x}_s(t), \boldsymbol{x}_f(t)) - g_s(\boldsymbol{x}_s(t), 0) \\ &+ g_s(\boldsymbol{x}_s(t), 0) - g_s(\hat{\boldsymbol{x}}_s(t), 0)\right\|^2 \leq \kappa^2(\left\|\boldsymbol{x}_f(t)\right\|^2 + \|\boldsymbol{e}_s(t)\|^2),\end{aligned}$$

where κ is a positive constant. Hence it can be obtained that

$$\begin{aligned}\dot{V}(t) &\leq \boldsymbol{e}_s^{\mathrm{T}}(t)(\boldsymbol{P}\boldsymbol{A}_s + \boldsymbol{A}_s^{\mathrm{T}}\boldsymbol{P} - \boldsymbol{X}\boldsymbol{C}_s - \boldsymbol{C}_s^{\mathrm{T}}\boldsymbol{X}^{\mathrm{T}} + \boldsymbol{P}^2 + \kappa^2\boldsymbol{I})\boldsymbol{e}_s(t) - 2\boldsymbol{y}_f^{\mathrm{T}}(t)\boldsymbol{X}^{\mathrm{T}}\boldsymbol{e}_s(t) + \kappa^2\left\|\boldsymbol{x}_f(t)\right\|^2 \\ &= \boldsymbol{\xi}^{\mathrm{T}}(t)\Xi\boldsymbol{\xi}(t) + \kappa^2\left\|\boldsymbol{x}_f(t)\right\|^2 + \varepsilon\left\|\boldsymbol{y}_f(t)\right\|^2,\end{aligned} \tag{3.18}$$

where

$$\boldsymbol{\xi}(t) \triangleq \begin{bmatrix} \boldsymbol{e}_s(t) \\ \boldsymbol{y}_f(t) \end{bmatrix}.$$

Hence, when $\Xi < 0$, one can obtain that

$$\begin{aligned}\dot{V}(t) &\leqslant -\lambda_{\min}(-\Xi)\|\boldsymbol{\xi}(t)\|^2 + \kappa^2\left\|\boldsymbol{x}_f(t)\right\|^2 + \varepsilon\left\|\boldsymbol{y}_f(t)\right\|^2 \\ &= -\lambda_{\min}(-\Xi)\left(\|\boldsymbol{e}_s(t)\|^2 + \left\|\boldsymbol{y}_f(t)\right\|^2\right) + \kappa^2\left\|\boldsymbol{x}_f(t)\right\|^2 + \varepsilon\left\|\boldsymbol{y}_f(t)\right\|^2 \\ &\leqslant -\lambda_{\min}(-\Xi)\|\boldsymbol{e}_s(t)\|^2 + \kappa^2\left\|\boldsymbol{x}_f(t)\right\|^2 + \varepsilon\left\|\boldsymbol{y}_f(t)\right\|^2 \\ &\leqslant -\lambda_{\min}(-\Xi)\|\boldsymbol{e}_s(t)\|^2 + \kappa^2\left\|\boldsymbol{x}_f(t)\right\|_{\mathrm{peak}}^2 + \varepsilon\left\|\boldsymbol{y}_f(t)\right\|_{\mathrm{peak}}^2.\end{aligned} \tag{3.19}$$

According to the definition of $V(t)$ in (3.16), it can be derived that

$$V(t) \leqslant \lambda_{\max}(\boldsymbol{P})\|\boldsymbol{e}_s(t)\|^2. \tag{3.20}$$

Combining (3.19) and (3.20), it can be obtained that

$$\dot{V}(t) \leqslant -\alpha V(t) + \kappa^2\left\|\boldsymbol{x}_f(t)\right\|_{\mathrm{peak}}^2 + \varepsilon\left\|\boldsymbol{y}_f(t)\right\|_{\mathrm{peak}}^2, \tag{3.21}$$

where

$$\alpha \triangleq \frac{\lambda_{\min}(-\Xi)}{\lambda_{\max}(\boldsymbol{P})}.$$

By Lemma 3.1, it can be obtained that

$$\begin{aligned}
V(t) &\leqslant e^{-\alpha t} V(0) + (\kappa^2 \left\| \boldsymbol{x}_f(t) \right\|_{\text{peak}}^2 + \varepsilon \left\| \boldsymbol{y}_f(t) \right\|_{\text{peak}}^2) \int_0^t e^{-\alpha(t-\tau)} \mathrm{d}\tau \\
&= e^{-\alpha t} V(0) + (\kappa^2 \left\| \boldsymbol{x}_f(t) \right\|_{\text{peak}}^2 + \varepsilon \left\| \boldsymbol{y}_f(t) \right\|_{\text{peak}}^2) \frac{1}{\alpha} \left(1 - e^{-\alpha t}\right) \\
&\leqslant e^{-\alpha t} V(0) + (\frac{\kappa^2}{\alpha} \left\| \boldsymbol{x}_f(t) \right\|_{\text{peak}}^2 + \frac{\varepsilon}{\alpha} \left\| \boldsymbol{y}_f(t) \right\|_{\text{peak}}^2) \sup_{t \in [0,\infty)} \left\{1 - e^{-\alpha t}\right\} \\
&\leqslant e^{-\alpha t} V(0) + (\frac{\kappa^2}{\alpha} \left\| \boldsymbol{x}_f(t) \right\|_{\text{peak}}^2 + \frac{\varepsilon}{\alpha} \left\| \boldsymbol{y}_f(t) \right\|_{\text{peak}}^2).
\end{aligned} \tag{3.22}$$

Meanwhile, considering (3.16), it can be derived that

$$V(t) \geq \lambda_{\min}(\boldsymbol{P}) \left\| \boldsymbol{e}_s(t) \right\|^2. \tag{3.23}$$

Combining (3.22) with (3.16), it can be obtained that

$$\left\| \boldsymbol{e}_s(t) \right\|^2 \leqslant \frac{e^{-\alpha t} V(0)}{\lambda_{\min}(\boldsymbol{P})} + \rho^2 \leqslant (\sqrt{\frac{e^{-\alpha t} V(0)}{\lambda_{\min}(\boldsymbol{P})}} + \rho)^2, \tag{3.24}$$

where

$$\rho = \sqrt{\frac{(\frac{\kappa^2}{\alpha} \left\| \boldsymbol{x}_f(t) \right\|_{\text{peak}}^2 + \frac{\varepsilon}{\alpha} \left\| \boldsymbol{y}_f(t) \right\|_{\text{peak}}^2)}{\lambda_{\min}(\boldsymbol{P})}}.$$

Hence it can be further derived that

$$\left\| \boldsymbol{e}_s(t) \right\| \leqslant \sqrt{\frac{e^{-\alpha t} V(0)}{\lambda_{\min}(\boldsymbol{P})}} + \rho. \tag{3.25}$$

Recalling Assumption 3.1, it can be obtained that $0 \leqslant \rho < \infty$. Hence (3.25) implies that there exists a $T \to \infty$ such that $\left\| \boldsymbol{e}_s(t) \right\| \leqslant \rho$, for all $t > T$. That is to say, $\boldsymbol{e}_s(t)$ is UUB with ultimate bound ρ when no fault exists. This completes the proof. □

Remark 3.2 Note that the bilinear matrix inequality in (3.14) is hard to solve directly. As a matter of fact, it can be transformed to an equivalent LMI with the help of the Schur complement lemma [4] as follows:

$$\bar{\boldsymbol{\Xi}} \triangleq \begin{bmatrix} \bar{\boldsymbol{\Xi}}_{11} & * & * \\ \boldsymbol{P} & -\varepsilon \boldsymbol{I} & * \\ -\boldsymbol{X}^{\mathrm{T}} & 0 & -\varepsilon \boldsymbol{I} \end{bmatrix} < 0,$$
$$\bar{\boldsymbol{\Xi}}_{11} \triangleq \boldsymbol{P}\boldsymbol{A}_s + \boldsymbol{A}_s^{\mathrm{T}}\boldsymbol{P} - \boldsymbol{X}\boldsymbol{C}_s - \boldsymbol{C}_s^{\mathrm{T}}\boldsymbol{X}^{\mathrm{T}} + \kappa^2 \boldsymbol{I}.$$

3.4.3 Spatial Mapping Filter Design: From Finite-Dimensional Measurement Space to Infinite-Dimensional Space

As discussed in Remark 3.1, a spatial mapping filter is designed as follows:

$$\boldsymbol{V}(z) = \boldsymbol{\phi}^{\mathrm{T}}(z)\left(\boldsymbol{C}^{\mathrm{T}}\boldsymbol{C}\right)^{-1}\boldsymbol{C}^{\mathrm{T}}.$$

The most restrictive condition is the existence of the left-inverse of $\boldsymbol{C} \in \mathbb{R}^{n_y \times \infty}$, which requires an infinite number of measurements in the spatial domain. However, with the help of spectral approximation, this condition can be largely released. The intuitive idea is to construct a spatial mapping filter based on the approximated ODE model (3.11) and the finite-dimensional observer (3.12) as follows:

$$\boldsymbol{V}_s(z) \triangleq \boldsymbol{\phi}_s^{\mathrm{T}}(z)\left(\boldsymbol{C}_s^{\mathrm{T}}\boldsymbol{C}_s\right)^{-1}\boldsymbol{C}_s^{\mathrm{T}}. \tag{3.26}$$

Remark 3.3 Note that the existence of the spatial mapping filter in (3.26) can be fulfilled as long as $\boldsymbol{C}_s \in \mathbb{R}^{n_y \times m}$ is of full-column rank (Assumption 3.2). This can be done by selecting an appropriate number of sensors and dominant modes. For the parabolic PDE discussed in this chapter, the number of dominant modes for spectral approximation is small, hence this condition is not hard to be satisfied.

Based on the designed finite-dimensional observer in (3.12) and the spatial mapping filter in (3.26), the following distributed S-T residual signal is obtained for fault detection and localization:

$$r(z,t) \triangleq \boldsymbol{\phi}_s^{\mathrm{T}}(z)\left(\boldsymbol{C}_s^{\mathrm{T}}\boldsymbol{C}_s\right)^{-1}\boldsymbol{C}_s^{\mathrm{T}}\boldsymbol{e}_{y_s}(t). \tag{3.27}$$

Recalling the discussions on the finite-dimensional observer design in Sect. 3.4.2, the state estimation error $e_{x_s}(z,t)$ is obtained as

$$e_{x_s}(z,t) \triangleq x(z,t) - \hat{x}_s(z,t) = \boldsymbol{\phi}_s^{\mathrm{T}}(z)\boldsymbol{x}_s(t) + \boldsymbol{\phi}_f^{\mathrm{T}}(z)\boldsymbol{x}_f(t) - \boldsymbol{\phi}_s^{\mathrm{T}}(z)\hat{\boldsymbol{x}}_s(t) = \boldsymbol{\phi}_s^{\mathrm{T}}(z)\boldsymbol{e}_s(t) + \boldsymbol{\phi}_f^{\mathrm{T}}(z)\boldsymbol{x}_f(t). \tag{3.28}$$

Combine (3.13), (3.27) with (3.28), it can be derived that

$$r(z,t) = e_{x_s}(z,t) + \left(\boldsymbol{\phi}_s^{\mathrm{T}}(z)\boldsymbol{C}_s^{+}\boldsymbol{C}_f - \boldsymbol{\phi}_f^{\mathrm{T}}(z)\right)\boldsymbol{x}_f(t), \tag{3.29}$$

where $\boldsymbol{C}_s^{+} \triangleq (\boldsymbol{C}_s^{\mathrm{T}}\boldsymbol{C}_s)^{-1}\boldsymbol{C}_s^{\mathrm{T}}$ denotes the pseudo-inverse of $\boldsymbol{C}_s$. It indicates that the designed distributed S-T residual signal contains two terms, the first term $e_{x_s}(z,t)$ can reveal the difference between the state variable $x(z,t)$ and its estimation $\hat{x}(z,t)$. This term is beneficial to the fault localization since the spatial distribution characteristic of the S-T fault is well preserved. However, due to the spectral approximation techniques used in Sect. 3.4.1, extra errors are introduced by the neglecting of the fast subsystem, which is reflected in the second term of $r(z,t)$.

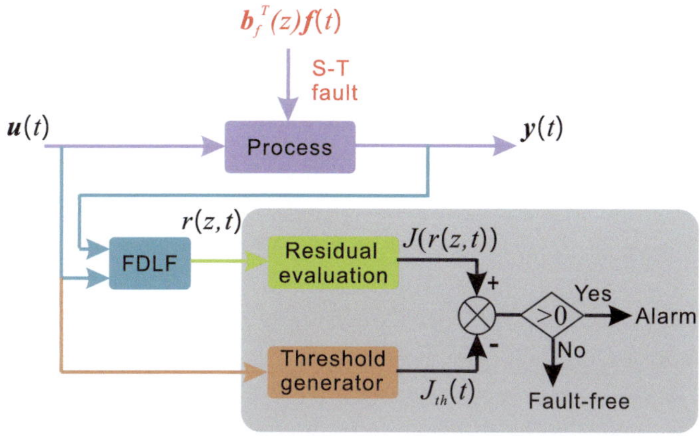

Fig. 3.3 Distributed residual evaluation and threshold computation framework

3.4.4 Distributed Residual Evaluation and Threshold Computation

Following the distributed residual generation, the residual evaluation aims at providing a feature of the distributed S-T residual signal $r(z, t)$. Along with the threshold setting, we are able to detect and localize the faults occurring to the system. Figure 3.3 summarizes the general framework of the distributed residual evaluation and threshold computation.

Remark 3.4 To achieve the purpose of fault localization, the distributed residual evaluation function $J(r(z, t))$ should be designed to be location-specific, rather than a lumped function of all locations as shown in [25].

To be specific, the distributed residual evaluation function and the threshold are designed as follows:

$$\begin{aligned} J(r(z,t)) &\triangleq \int_t^{t+T} r(z,\tau)^2 \mathrm{d}\tau, \\ J_{\mathrm{th}}(t) &\triangleq \sup_{t\geq 0,\, f(t)=0} J(r(z,t)), \end{aligned} \tag{3.30}$$

where $T > 0$ denotes the length of the evaluation time window.

Remark 3.5 Note that the threshold $J_{\mathrm{th}}(t)$ is irrelevant to the space variable z while $J(r(z, t))$ is location specific. The main reason is that false alarm can be largely avoided at the cost of detectability since the $\mathcal{L}_\infty$ norm was used in the derivation of the threshold. However, it is an interesting topic to consider a more "sensitive" fault localization scheme in the future.

Hence the following decision logic will guarantee reliable fault detection and localization performance:

$$\begin{cases} J(r(z,t)) > J_{\text{th}}(t) \Longrightarrow \text{faulty for location } z \text{ at time } t \\ J(r(z,t)) \leq J_{\text{th}}(t) \Longrightarrow \text{fault-free for location } z \text{ at time } t. \end{cases} \tag{3.31}$$

The following lemma is useful for computing the threshold $J_{\text{th}}(t)$.

Lemma 3.2 [31, 32] (A Variant of Agmon's Inequality in 1-D Spatial Domain) *Let $\omega(z,t) \in \mathcal{H}^1(0,L)$ be a scalar function with $\omega(0,t) = 0$ or $\omega(L,t) = 0$, $t \geq 0$. The following inequality is fulfilled:*

$$\|\omega(\cdot,t)\|_\infty^2 \leq 2\,\|\omega(\cdot,t)\|_2\,\|\omega_z(\cdot,t)\|_2 \leq \|\omega(\cdot,t)\|_{\mathcal{H}^1}^2, t \geq 0. \tag{3.32}$$

Theorem 3.2 (Performance of the FDLF) *If the conditions in Theorem 3.1 are satisfied, the following **time-varying** threshold $J_{\text{th}}(t)$ can be defined and calculated **off-line** to guarantee the performance of the proposed FDLF:*

$$J_{\text{th}}(t) = \int_t^{t+T} \bar{\rho}(\tau)\mathrm{d}\tau, \tag{3.33}$$

where

$$\bar{\rho}(\tau) \triangleq 2m\pi L^{-1}\Big(\sqrt{\frac{e^{-\alpha\tau}V(0)}{\lambda_{\min}(\boldsymbol{P})}} + \rho + \sigma_{\max}(\boldsymbol{C}_s^+)\,\big\|\boldsymbol{y}_f(t)\big\|_{\text{peak}}\Big)^2, \tag{3.34}$$

and $\sigma_{\max}(\boldsymbol{C}_s^+)$ denotes the maximum singular value of matrix $\boldsymbol{C}_s^+$.

Proof First, the properties of the basis functions $\{\phi_i(z)\}_{i=1}^m$ are analyzed for the threshold computation. For the system model (3.1) and the boundary conditions in (3.2), one can calculate the following m number of basis functions [31]:

$$\phi_i(z) = a\sin(i\pi\frac{z}{L}), i = 1, \cdots, m, \tag{3.35}$$

where a is a constant determined by

$$\|\phi_i(\cdot)\|_2^2 = 1. \tag{3.36}$$

Then, by the orthogonality we have that

$$< \phi_i(\cdot), \phi_j(\cdot) >= 0, i \neq j. \tag{3.37}$$

Combining (3.35) and (3.36), it can be obtained that

$$\big\|\dot{\phi}_i(\cdot)\big\|_2^2 = i^2\pi^2L^{-2}, < \dot{\phi}_i(\cdot), \dot{\phi}_j(\cdot) >= 0, i \neq j. \tag{3.38}$$

where $\dot{\phi}_i \triangleq \mathrm{d}\phi_i/\mathrm{d}z$.

Note that since $\phi_i(z)$ is selected in the form (3.35), it can be verified that the distributed S-T residual signal satisfies that $r(0,t) = r(L,t) = 0$ by (3.27). According to Lemma 3.2, when no fault exists, it can be derived that

$$\|r(\cdot,t)\|_\infty^2 \le 2\,\|r(\cdot,t)\|_2\,\|r_z(\cdot,t)\|_2\,. \tag{3.39}$$

Moreover, it can be obtained that

$$\begin{aligned}\|r(\cdot,t)\|_2 &= \left\|\boldsymbol{\phi}_s^{\mathrm{T}}(\cdot)\boldsymbol{C}_s^{+}\boldsymbol{e}_{y_s}(t)\right\|_2 = \left\|\boldsymbol{\phi}_s^{\mathrm{T}}(\cdot)(\boldsymbol{e}_s(t)+\boldsymbol{C}_s^{+}\boldsymbol{y}_f(t))\right\|_2 = \left\|\boldsymbol{e}_s(t)+\boldsymbol{C}_s^{+}\boldsymbol{y}_f(t)\right\| \\ &\le \|\boldsymbol{e}_s(t)\| + \left\|\boldsymbol{C}_s^{+}\boldsymbol{y}_f(t)\right\| \le \|\boldsymbol{e}_s(t)\| + \sigma_{\max}(\boldsymbol{C}_s^{+})\left\|\boldsymbol{y}_f(t)\right\|_{\text{peak}}\,.\end{aligned} \tag{3.40}$$

Recalling Theorem 3.1, it can be obtained that

$$\|r(\cdot,t)\|_2 \le \sqrt{\frac{e^{-\alpha t}V(0)}{\lambda_{\min}(\boldsymbol{P})}} + \rho + \sigma_{\max}(\boldsymbol{C}_s^{+})\left\|\boldsymbol{y}_f(t)\right\|_{\text{peak}}\,. \tag{3.41}$$

Moreover, we have that

$$\begin{aligned}\|r_z(\cdot,t)\|_2 &= \left\|\dot{\boldsymbol{\phi}}_s^{\mathrm{T}}(\cdot)\boldsymbol{C}_s^{+}\boldsymbol{e}_{y_s}(t)\right\|_2 = \left\|\dot{\boldsymbol{\phi}}_s^{\mathrm{T}}(\cdot)(\boldsymbol{e}_s(t)+\boldsymbol{C}_s^{+}\boldsymbol{y}_f(t))\right\|_2 = \left\|\dot{\boldsymbol{\phi}}_s^{\mathrm{T}}(\cdot)\boldsymbol{\zeta}(t)\right\|_2 \\ &= \sqrt{\zeta_1^2(t)\left\|\dot{\phi}_1(\cdot)\right\|_2^2 + \cdots + \zeta_m^2(t)\left\|\dot{\phi}_m(\cdot)\right\|_2^2},\end{aligned} \tag{3.42}$$

where

$$\boldsymbol{\zeta}(t) \triangleq \boldsymbol{e}_s(t) + \boldsymbol{C}_s^{+}\boldsymbol{y}_f(t).$$

Recalling (3.38), one can further derive that

$$\begin{aligned}\|r_z(\cdot,t)\|_2 &\le \pi L^{-1}\sqrt{1^2\zeta_1^2(t) + \cdots + m^2\zeta_m^2(t)} \le m\pi L^{-1}\,\|\boldsymbol{\zeta}(t)\| \\ &\le m\pi L^{-1}(\|\boldsymbol{e}_s(t)\| + \sigma_{\max}(\boldsymbol{C}_s^{+})\left\|\boldsymbol{y}_f(t)\right\|_{\text{peak}}) \\ &\le m\pi L^{-1}(\sqrt{\frac{e^{-\alpha t}V(0)}{\lambda_{\min}(\boldsymbol{P})}} + \rho + \sigma_{\max}(\boldsymbol{C}_s^{+})\left\|\boldsymbol{y}_f(t)\right\|).\end{aligned} \tag{3.43}$$

Combine (3.39), (3.41) and (3.43), it can be obtained that

$$\|r(\cdot,t)\|_\infty^2 \le \bar{\rho}(t), \tag{3.44}$$

where

$$\bar{\rho}(t) = 2m\pi L^{-1}(\sqrt{\frac{e^{-\alpha t}V(0)}{\lambda_{\min}(\boldsymbol{P})}} + \rho + \sigma_{\max}(\boldsymbol{C}_s^{+})\left\|\boldsymbol{y}_f(t)\right\|_{\text{peak}})^2.$$

Moreover, combining (3.30) with (3.44), it can be derived that

$$J(r(z,t)) = \int_t^{t+T} r(z,\tau)^2 \mathrm{d}\tau \le \int_t^{t+T} \|r(\cdot,\tau)\|_\infty^2 \,\mathrm{d}\tau \le \int_t^{t+T} \bar{\rho}(\tau)\mathrm{d}\tau. \tag{3.45}$$

Therefore, the time-varying threshold $J_{\text{th}}(t)$ for fault detection and localization can be selected as

$$J_{\text{th}}(t) = \int_t^{t+T} \bar{\rho}(\tau)\mathrm{d}\tau. \tag{3.46}$$

This completes the proof. □

3.5 Illustrative Example

As presented in Fig. 3.4, a long, thin rod in a reactor is considered [26]. This reactor is fed with pure species A. Along the rod, a zero-th order exothermic catalytic reaction with the form A → B occurs. In addition, by contacting with the rod, a cooling medium is playing the role of cooling. For the investigation, some general assumptions are needed, which are excess of species A in the furnace, constant conductivity of the rod, constant density and heat capacity of the rod, and constant temperature at both ends of the rod. Then, the temperature distribution equation along this catalytic rod can be represented by the following parabolic PDE:

$$\frac{\partial x(z,t)}{\partial t} = \frac{\partial^2 x(z,t)}{\partial z^2} + \beta_T \left(e^{-(\gamma/(1+x(z,t)))} - e^{-\gamma}\right) + \beta_U (b_u(z)u(t) - x(z,t)) + b_f(z)f(t), \tag{3.47}$$

$$\mathbf{y}(t) = \begin{bmatrix} \int_0^\pi \delta(z - \frac{\pi}{4})x(z,t)\mathrm{d}z \\ \int_0^\pi \delta(z - \frac{3\pi}{4})x(z,t)\mathrm{d}z \end{bmatrix} = \begin{bmatrix} x(\frac{\pi}{4}, t) \\ x(\frac{3\pi}{4}, t) \end{bmatrix}, \tag{3.48}$$

subject to the Dirichlet boundary conditions

$$x(0,t) = 0, \; x(\pi,t) = 0, \tag{3.49}$$

where $x(z,t)$ denotes the dimensionless temperature of the rod; β_T denotes a dimensionless heat of reaction; γ denotes a dimensionless activation energy; β_U denotes a dimensionless heat transfer coefficient; $u(t)$ is the manipulated input (temperature of the cooling medium); $f(t)$ denotes possible faults that may occur to the process with unknown shape function $b_f(z)$; $\delta(\cdot)$ denotes the Dirac delta function and $\mathbf{y}(t)$ are the thermocouple measurements at point $z = \pi/4$ and $z = 3\pi/4$. The catalytic rod is one of the most representative nonlinear transport-reaction process in chemical engineering [2, 26].

The typical values of the process parameters are

$$\beta_T = 50.0, \quad \beta_U = 2.0, \quad \gamma = 4.0.$$

The actuator distribution function is set as

$$b_u(z) = \sqrt{\frac{2}{\pi}} \sin(z).$$

And the manipulated input is selected as

$$u(t) = 1 + \sin(t).$$

The eigenvalue problem for the spatial differential operator

$$\mathcal{A}x = \frac{\partial^2 x}{\partial z^2}, x \in \mathcal{S}(\mathcal{A}) = \{x \in \mathcal{L}_2([0, \pi]; \mathbb{R}); x(0, t) = 0, x(\pi, t) = 0\}$$

can be directly solved as

$$\lambda_j = -j^2, \ \phi_j(z) = \sqrt{\frac{2}{\pi}} \sin(jz), \ j = 1, \cdots, \infty.$$

Details of procedures for solving this kind of eigenvalue problem can be referred to Chap. 4 of [33]. The first two eigenvalues are considered as the dominant ones, thus $\epsilon = |\lambda_1| / |\lambda_3| \approx 0.111 \ll 1$. Also, recalling Assumption 3.3, the value of κ is selected as 0.2512.

Recalling the modal approximation techniques mentioned in Sect. 3.4.1, the following two-dimensional (2-D) slow sub-system is derived:

$$\begin{aligned} \dot{\boldsymbol{x}}_s(t) &= \boldsymbol{A}_s \boldsymbol{x}_s(t) + g_s(\boldsymbol{x}_s(t), \boldsymbol{x}_f(t)) + \boldsymbol{B}_s \boldsymbol{u}(t) + \boldsymbol{E}_s \boldsymbol{f}(t), \\ \boldsymbol{y}_s(t) &= \boldsymbol{C}_s \boldsymbol{x}_s(t), \end{aligned} \tag{3.50}$$

where

$$\boldsymbol{x}_s(t) = \begin{bmatrix} x_1(t) \\ x_2(t) \end{bmatrix}, \boldsymbol{A}_s = \begin{bmatrix} -1 & 0 \\ 0 & -4 \end{bmatrix}, \boldsymbol{B}_s = \begin{bmatrix} 2 \\ 0 \end{bmatrix},$$

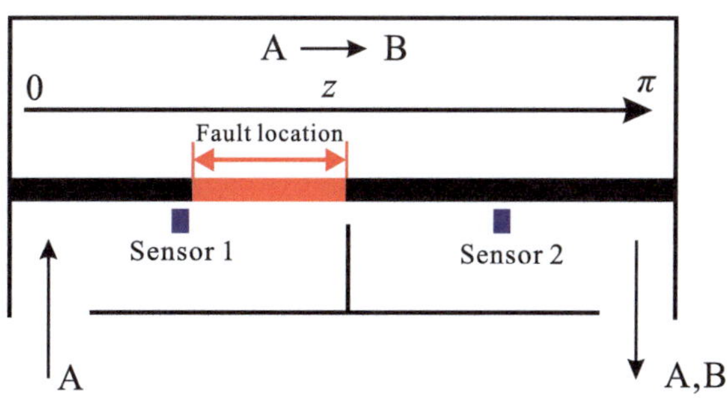

Fig. 3.4 A faulty catalytic rod with two thermocouples

$$\boldsymbol{C}_s = \int_0^{\pi} \begin{bmatrix} \delta\left(z - \frac{\pi}{4}\right) \\ \delta\left(z - \frac{3\pi}{4}\right) \end{bmatrix} [\phi_1(z)\phi_2(z)]\,\mathrm{d}z = \begin{bmatrix} \sqrt{\frac{1}{\pi}} & \sqrt{\frac{2}{\pi}} \\ \sqrt{\frac{1}{\pi}} & -\sqrt{\frac{2}{\pi}} \end{bmatrix},$$

and

$$g_s(\boldsymbol{x}_s(t), \boldsymbol{x}_f(t)) = \begin{bmatrix} -\beta_U x_1(t) + \beta_T \int_0^{\pi} \phi_1(z)(e^{-(\upsilon/(1+x_1(t)\phi_1(z)+x_2(t)\phi_2(z)+\boldsymbol{\phi}_f^{\mathrm{T}}(z)\boldsymbol{x}_f(t)))} - e^{-\upsilon})\mathrm{d}z \\ -\beta_U x_2(t) + \beta_T \int_0^{\pi} \phi_2(z)(e^{-(\upsilon/(1+x_1(t)\phi_1(z)+x_2(t)\phi_2(z)+\boldsymbol{\phi}_f^{\mathrm{T}}(z)\boldsymbol{x}_f(t)))} - e^{-\upsilon})\mathrm{d}z \end{bmatrix}. \tag{3.51}$$

In this numerical simulation, the spatial distribution function of the S-T fault is selected as

$$b_f(z) = H(z - \pi/4) - H(z - \pi/2) \tag{3.52}$$

for simple illustration, where $H(\cdot)$ is the standard Heaviside function. This indicates that the fault occurs in the area $[\pi/4, \pi/2]$. Note that when the designed FDLF is applied, **no** information of $b_f(z)$ is known a prior. To evaluate the performance of the proposed FDLF, the following abrupt S-T fault is considered:

$$f(t) = \begin{cases} 0, & 0 \leqslant t < 40 \text{ (s)} \\ 30, & 40 \leqslant t \leqslant 80 \text{ (s)}. \end{cases}$$

Recalling Theorem 3.1, choosing $\varepsilon = 1$, and solving (3.14) and (3.15) with the help of Schur complement lemma [4] discussed in Remark 3.2, one can obtain that

$$\boldsymbol{P} = \begin{bmatrix} 0.4481 & 0 \\ 0 & 0.1777 \end{bmatrix},$$

$$\boldsymbol{X} = \begin{bmatrix} 0.0675 & 0.0675 \\ -0.0570 & 0.0570 \end{bmatrix},$$

$$\boldsymbol{L} = \begin{bmatrix} 0.1506 & 0.1506 \\ -0.3210 & 0.3210 \end{bmatrix}$$

by the MATLAB LMI toolbox.

The original PDE system in (3.47) is solved numerically by the FDM [34] with the sampling time $\Delta t = 0.01$ s and 19 grid points with respect to the space. Recalling the discussions in Sects. 3.4.2, 3.4.3 and 3.4.4 the threshold $J_{\mathrm{th}}(t)$ can be computed numerically with the evaluation window $T = 0.16$ s. Table 3.1 summarizes the parameters setting of catalytic rod and the designed FDLF.

The trajectory of reduced-order state estimation error $\boldsymbol{e}_s(t)$ and nonlinear term estimation error $\tilde{g}_s(t)$ under faulty conditions are shown in Figs. 3.5 and 3.6, respectively. It can be observed that both $\boldsymbol{e}_s(t)$ and $\tilde{g}_s(t)$ are not equal to zero in the fault-free condition, this is mainly due to the modal approximation technique used. Analytic analysis of this issue can be refereed to Theorem 3.1.

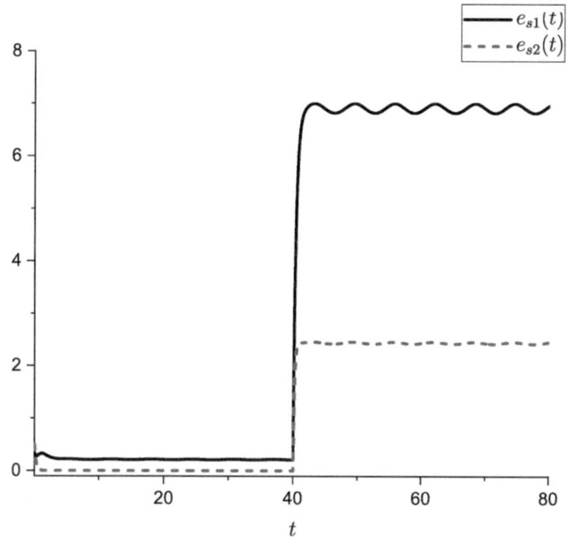

Fig. 3.5 Trajectory of reduced-order state estimation error $\boldsymbol{e}_s(t)$ under faulty conditions

The state estimation error $e_{x_s}(z, t)$ and the constructed residual signal for faulty case are shown in Fig. 3.7a, b, respectively. It can be found that the distributed residual signal can reveal the spatial distribution characteristic of the fault. The corresponding fault detection and localization performance is demonstrates in Fig. 3.8. The fault occurring time $t_f = 40$ s is well indicated. Moreover, to better illustrate the fault localization performance, after the S-T fault is detected in the time domain, the residual evaluation result at $t = 50$ s is presented in Fig. 3.9, the considered fault can be localized with a satisfied performance. The fault detection rate (FDR) and FAR [35] are two key performance indexes for evaluating the fault detection results.

Table 3.1 Parameters of catalytic rod and FDLF

Parameters	Value
Heat of reaction (β_T)	50.0
Activation energy (γ)	4.0
Heat transfer coefficient (β_U)	2.0
Number of sensors (n_y)	2
Sensor location	$z = \pi/4$ and $z = 3\pi/4$
Fault location	$[\pi/4, \pi/2]$
Number of dominant mode (m)	2
Observer gain	$\boldsymbol{L} = \begin{bmatrix} 0.1506 & 0.1506 \\ -0.3210 & 0.3210 \end{bmatrix}$
Evaluation window (T)	0.16

Fig. 3.6 Trajectory of nonlinear term estimation error $\tilde{g}_s(t)$ under faulty conditions

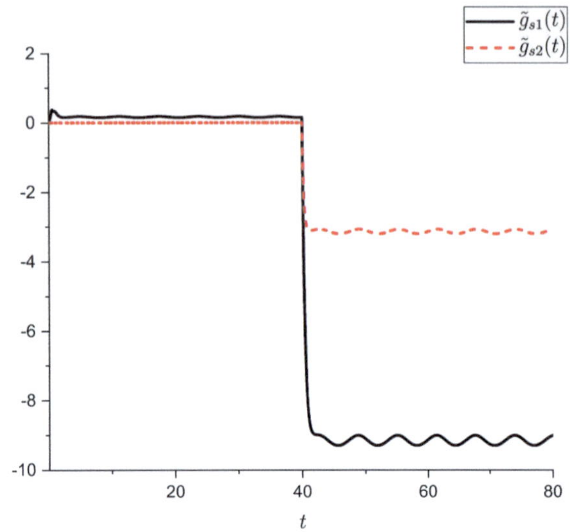

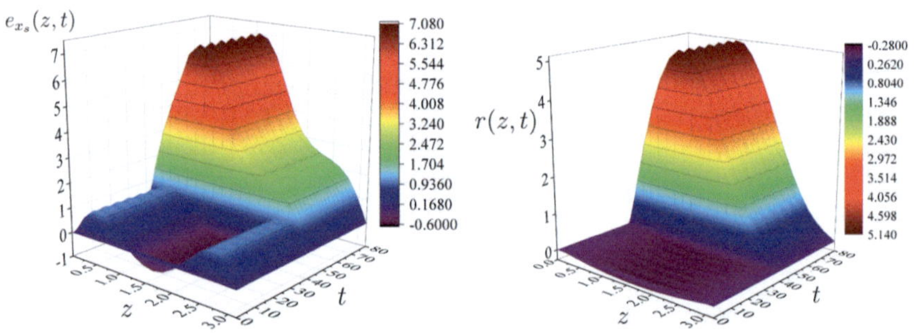

(a) Profile of evolution of state estimation error $e_{x_s}(z,t)$. (b) Profile of evolution of distributed residual signal $r(z,t)$.

Fig. 3.7 State estimation error $e_{x_s}(z,t)$ and distributed residual signal $r(z,t)$ under faulty conditions

Similarly, two performance indexes "detection rate" and "misdetection rate" are defined as follows to evaluate the proposed fault localization method:

$$\text{Detection rate} = \frac{\text{Detected fault location area}}{\text{Actual fault location area}} - \frac{\text{Misdetected fault location area}}{\text{Actual fault location area}},$$

$$\text{Misdetection rate} = \frac{\text{Misdetected fault location area}}{\text{Detected fault location area}}.$$

For the considered abrupt fault, the detection rate and misdetection rate are 91.59 and 8.05%, respectively.

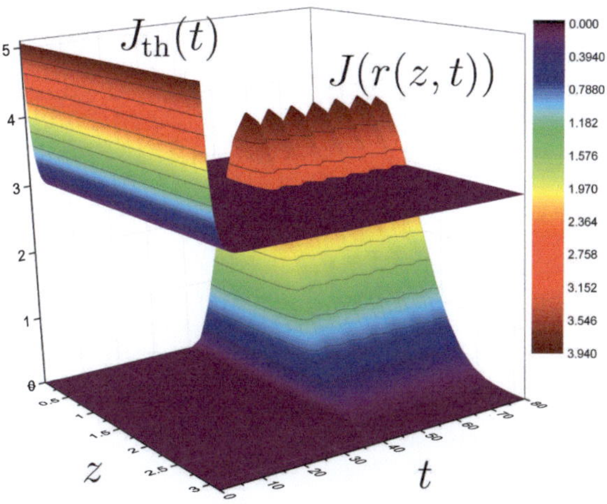

Fig. 3.8 Fault detection and localization performance

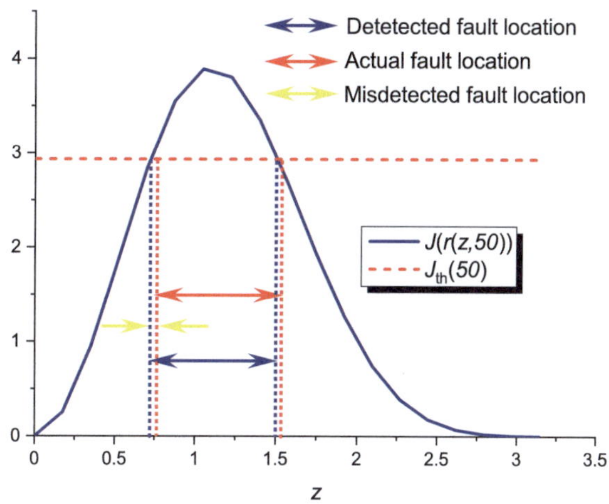

Fig. 3.9 Fault localization result at $t = 50\,\text{s}$

3.6 Conclusion

In this chapter, the S-T fault detection and localization problem for a class of nonlinear spatially distributed processes is discussed in a systematic way. By designing a spatial mapping filter which maps the finite-dimensional output estimation error into the infinite-dimensional space, a distributed residual signal is obtained using a limited number of sensors. Considering the spatial distribution characteristic of the process, the residual evaluation is conducted in a distributed manner. Since the early-lumping techniques are employed for modal approximation, the observation spillover is inevitable. Moreover, to obtain a satisfied approximation performance, the order of the finite-dimensional system may be rather high.

References

1. Wang J-W, Wu H-N, Li H-X (2013) Guaranteed cost distributed fuzzy observer-based control for a class of nonlinear spatially distributed processes. AIChE J 59(7):2366–2378
2. Qing X, Jin J, Niu Y et al (2020) Time-space coupled learning method for model reduction of distributed parameter systems with encoder-decoder and RNN. AIChE J 66(8):e16251
3. Li H-X, Qi C (2010) Modeling of distributed parameter systems for applications-A synthesized review from time-space separation. J Process Control 20(8):891–901
4. Ding SX (2013) Model-based fault diagnosis techniques: design schemes, algorithms, and tools, 2nd edn. Springer-Verlag, London
5. Liu X, Gao Z (2018) Integrated fault estimation and fault-tolerant control for stochastic systems with Brownian motions. Int J Robust Nonlinear Control 28(6):1915–1941
6. Ferranti L, Wan Y, Keviczky T (2019) Fault-tolerant reference generation for model predictive control with active diagnosis of elevator jamming faults. Int J Robust Nonlinear Control 29(16):5412–5428
7. Ding SX (2014) Data-driven design of fault diagnosis and fault-tolerant control systems. Springer, London, London
8. El-Farra NH (2006) Integrated fault detection and fault-tolerant control architectures for distributed processes. Ind Eng Chem Res 45(25):8338–8351
9. El-Farra NH, Ghantasala S (2007) Actuator fault isolation and reconfiguration in transport-reaction processes. AIChE J 53(6):1518–1537
10. Demetriou MA (2002) A model-based fault detection and diagnosis scheme for distributed parameter systems: a learning systems approach. ESAIM Control Optim Calc Var 7:43–67
11. Demetriou MA, Ito K, Smith RC (2007) Adaptive monitoring and accommodation of nonlinear actuator faults in positive real infinite dimensional systems. IEEE Trans Autom Control 52(12):2332–2338
12. Armaou A, Demetriou MA (2008) Robust detection and accommodation of incipient component and actuator faults in nonlinear distributed processes. AIChE J 54(10):2651–2662
13. Ghantasala S, El-Farra NH (2009) Robust actuator fault isolation and management in constrained uncertain parabolic PDE systems. Automatica 45(10):2368–2373
14. Xu X, Yuan Y, Wang M et al (2019) Fault detection and estimation for a class of PIDE systems based on boundary observers. Int J Robust Nonlinear Control 29(17):5867–5885
15. Deutscher J, Harkort C (2010) A parametric approach to finite-dimensional control of linear distributed-parameter systems. Int J Control 83(8):1674–1685
16. Harkort C (2014) Early-lumping based controller synthesis for linear infinite-dimensional systems. FAU Forschungen, Reihe B, Medizin, Naturwissenschaft, Technik 1. FAU University Press, Erlangen
17. Deutscher J (2016) Fault detection for linear distributed-parameter systems using finite-dimensional functional observers. Int J Control 89(3):550–563
18. Cai J, Ferdowsi H, Sarangapani J (2016) Model-based fault detection, estimation, and prediction for a class of linear distributed parameter systems. Automatica 66:122–131
19. Shinbrot M (1954) On the analysis of linear and nonlinear dynamical systems from transient-response data. NACA TN 3288
20. Fischer F, Deutscher J (2016) Algebraic fault detection and isolation for parabolic distributed-parameter systems using modulation functions. IFAC-PapersOnLine 49(8):162–167
21. Fischer F, Deutscher J (2017) Fault detection for parabolic systems with distributed inputs and outputs using the modulation function approach. IFAC-PapersOnLine 50(1):6774–6779
22. Fischer F, Deutscher J (2018) Modulating function based fault detection for parabolic systems with polynomial faults. IFAC-PapersOnLine 51(24):359–366
23. Dey S, Perez HE, Moura SJ (2017) Model-based battery thermal fault diagnostics: algorithms, analysis, and experiments. IEEE Trans Control Syst Technol 27(2):576–587
24. Dey S, Perez HE, Moura SJ (2019) Robust fault detection of a class of uncertain linear parabolic PDEs. Automatica 107:502–510

25. Li L, Ding SX, Peng K et al (2019) Fuzzy fault detection filter design for nonlinear distributed parameter systems. IEEE Access 7:11105–11113
26. Christofides PD (2012) Nonlinear and robust control of PDE systems: methods and applications to transport-reaction processes. Springer, Berlin
27. Deng H, Li H-X, Chen G (2005) Spectral-approximation-based intelligent modeling for distributed thermal processes. IEEE Trans Control Syst Technol 13(5):686–700
28. Wu H-N, Li H-X (2011) Robust adaptive neural observer design for a class of nonlinear parabolic PDE systems. J Process Control 21(8):1172–1182
29. Khalil HK, Grizzle JW (2002) Nonlinear systems. Prentice Hall, Upper Saddle River
30. Ioannou PA, Sun J (2012) Robust adaptive control. Courier Corporation, Chelmsford
31. Krstic M, Smyshlyaev A (2008) Boundary control of PDEs: a course on backstepping designs. SIAM
32. Wang J-W, Wang J-M (2019) Mixed H_2/H_∞ sampled-data output feedback control design for a semi-linear parabolic PDE in the sense of spatial L^∞ norm. Automatica 103:282–293
33. Ray WH (1981) Advanced process control. McGraw-Hill, New York
34. Strikwerda JC (2004) Finite difference schemes and partial differential equations. SIAM
35. Ding SX, Shen B, Wang Z et al (2014) A fault detection scheme for linear discrete-time systems with an integrated online performance evaluation. Int J Control 87(12):2511–2521

Chapter 4
PDE Backstepping-Based Abnormality Detection and Localization for Linear Parabolic DPSs

Abstract Abnormality localization for industrial processes described by distributed parameter systems is vital but seldom studied to the best of our knowledge. In this chapter, a systematic framework is proposed to address this problem for a class of linear parabolic distributed parameter systems using a limited number of in-domain measurements plus one boundary measurement rather than the full-state measurement. The proposed methodology consists of an abnormality detection filter and an abnormality localization filter design based on the backstepping techniques and eigenspectrum. For the detection purpose, the residual is evaluated in a lumped manner; For the localization purpose, a distributed residual is first constructed and then evaluated in a distributed manner. Numerical simulations on a heat transfer rod are conducted to demonstrate the effectiveness of the proposed method.

4.1 Introduction

DPSs widely exist in industrial processes (e.g. transport-reaction processes, thermal processes, and fluid processes), whose system dynamics are modeled by PDEs. The most unique feature of DPSs is that the state variable varies in both time and space domain, which makes ODEs unable to describe their dynamic behaviors accurately.

Abnormal behaviors or events in DPSs may cause the failure of controller or undesired system response, and are harmful to the safe and reliable operation of the system. Without loss of generality, these abnormal behaviors or events can be considered as the result of an unknown distributed abnormality term $s(x, t)$ in the system dynamics [1, 2]. Considering the spatial distribution characteristic, the localization of abnormality in the space domain is as important as the detection in the time domain. The distributed abnormality can also be treated equivalently as a general *distributed fault* in the process. Despite the fact that both model-based [3] and data-driven [4] methods for FD of dynamic systems described by ODEs have been studied systematically over the past few decades, works reported on the FD of DPSs are rather limited.

Recently, some results have been reported on the FD issues of DPSs, see [2, 5] for reference. Without exception, the aforementioned works all belong to the category

© The Author(s), under exclusive license to Springer Nature Singapore Pte Ltd. 2026

Y. Feng et al., *Spatio-Temporal Abnormality Diagnosis for Industrial Distributed Parameter Systems*, https://doi.org/10.1007/978-981-95-3750-1_4

of model-based FD. Existing model-based approaches for fault detection and diagnosis of DPSs can be roughly classified into two categories: one was using a finite-dimensional ODE representation of DPSs, i.e., early-lumping design; the other was based on the original PDE system, i.e., late-lumping design. For example, motivated by the idea of using fictitious output observer [6] to reduce the observation spillover, the authors proposed a finite-dimensional residual generator for fault detection of linear DPSs in [7]. A novel model-based fault detection approach was developed in [8], where the state observer was constructed based on the original PDE system. The modulating functions-based approach that was introduced in [9] to solve the parameter identification problem for LPSs was applied to DPSs for FD [10]. Since modal approximation technique was not involved, the observation spillover problem was avoided in the design stage.

To the best of our knowledge, almost all existing results on FD of DPSs only focused on fault detection in the time domain while the localization in the space domain was not considered for either actuator or sensor fault. However, considering the spatial distribution characteristics of DPSs, it is intuitive to investigate the fault localization problem, which is of great significance from the application viewpoint. More recently, a similar interesting problem to the fault localization-the parameters and source estimation of PDEs was studied via the modulating functions-based design [11, 12]. Despite the fact that the full-state measurement was employed, these inspiring works opened up a new way for the fault estimation of PDEs out of the well-studied observer-based framework. The distributed fault estimation problem for parabolic DPSs was investigated via a tedious early-lumping design in [13]. However, the observation spillover caused by neglecting the remaining infinite-dimensional dynamics in the design stage affects the estimation performance severely and a large number of in-domain measurements are required to obtain a satisfied performance. In summary, the current research mostly focused on the detection while the localization of fault was not well considered.

To distinguish the actuator or sensor fault discussed in the literature, the *distributed abnormality* term $s(x,t)$ is defined to include all unknown distributed terms with *unknown shape function* in the space without loss of generality. The distributed abnormalities widely exist in industry applications. For example, the distributed thermal faults in Li-ion batteries [1, 2] that can lead to battery degradation and failure belong to the distributed abnormality. The in-domain actuator faults, for example, the actuator faults of diffusion-reaction processes [14] that widely exist in chemical processes are special cases of the defined distributed abnormality.

To achieve the purpose of abnormality localization, typically one needs the *full-state measurement* to construct a distributed residual. However, this requirement is nearly impossible to meet in applications. The question then transforms into constructing a distributed residual using only a *limited* number of measurements and evaluating it.

Motivated by the above considerations, by using a limited number of in-domain measurements plus one boundary measurement, we intend to construct an abnormality detection filter (ADF) and abnormality localization filter (ALF) for the detection and localization of distributed abnormalities, respectively in this chapter. Since backstepping techniques have shown their effectiveness for observer design [15, 16], they are applied for the ADF and ALF design in this chapter. Furthermore, due to the fact that early-lumping techniques are not used, no observation spillover exists in the design stage. The main contributions of this chapter are twofold

- An ADF based on backstepping design with time-varying threshold is proposed to avoid false alarms in the abnormality detection.
- The ALF is motivated by inheriting and further expanding the modal approximation in the early-lumping design and the full-state measurement is not necessary.

The rest of this chapter is concluded as follows. The problem statement is discussed in Sect. 4.2. In Sect. 4.3, the proposed ADF and ALF are studied analytically. Numerical simulations on a heat transfer rod are presented in Sect. 4.4. Finally, we conclude this chapter in Sect. 4.5.

4.2 Problem Statement

Notations: $\mathbb{R}$ and $\mathbb{R}^n$ are the set of all real numbers and n-dimensional Euclidean space with the norm $||\cdot||$, respectively. $|\omega|$ denotes the absolute value of ω for any $\omega \in \mathbb{R}$. $\mathcal{L}^2([0,1]) \triangleq \mathcal{L}^2([0,1];\mathbb{R})$ is a real Hilbert space of square integrable functions $\omega(x) : [0,1] \to \mathbb{R}$ with the spatial $\mathcal{L}^2$ norm $||\omega(\cdot)||_2 \triangleq \sqrt{\int_0^1 \omega^2(x)\mathrm{d}x}$. The spatial $\mathcal{L}^\infty$ norm is defined as $||\omega(\cdot)||_\infty \triangleq \max_{x\in[0,1]} |\omega(x)|$. $u_t \triangleq \partial u/\partial t$, $u_x \triangleq \partial u/\partial x$, $u_{xx} \triangleq \partial^2 u/\partial x^2$.

Consider a class of linear DPSs described by the following parabolic PDE in one spatial dimension:

$$u_t(x,t) = u_{xx}(x,t) + \lambda(x)u(x,t) + Q(x,t) + d(x,t) + s(x,t),$$

$$\boldsymbol{y}(t) = \int_0^1 \boldsymbol{c}(x)u(x,t)\mathrm{d}x, \tag{4.1a}$$

$$\boldsymbol{c}(x) \triangleq [\delta(x-0), \delta(x-x_1), \cdots, \delta(x-x_N)]^{\mathrm{T}}, \tag{4.1b}$$

subject to the following boundary conditions:

$$u_x(0,t) = 0, \tag{4.2a}$$

$$u(1,t) = 0, \tag{4.2b}$$

and with the following initial condition:

$$u(x,0) = u_0(x), \tag{4.3}$$

where $u(\cdot,t) \in \mathcal{L}^2([0,1])$ denotes the state variable; $x \in [0,1]$ is the spatial coordinate; $t \in [0,\infty)$ is the time; $\lambda \in C^1[0,1]$; $Q(x,t)$, $d(x,t)$ and $s(x,t)$ denote a continuous distributed input, an *unknown* continuous distributed disturbance, and an *unknown* continuous distributed abnormality, respectively; $\delta(\cdot)$ is a Dirac distribution [17]; $\boldsymbol{y}(t) \in \mathbb{R}^{N+1} (N \geq 1)$ denotes N *in-domain pointwise measurements* $(0 < x_1 < \cdots < x_N < 1)$ plus one *boundary measurement* $u(0, t)$.

Remark 4.1 For the cases with space and time dependent coefficient $\lambda(x, t)$, backstepping-based design still applies using a time dependent integral kernel [18].

Remark 4.2 The mixed homogeneous Neumann-Dirichlet boundary condition is considered in this chapter as it has been used as a benchmark for demonstration in backstepping-based designs. However, for homogeneous Dirichlet boundary conditions and homogeneous Neumann boundary conditions, the following methodology still applies.

Assumption 4.1 The distributed disturbance $d(x, t)$ is bounded in the sense of spatial $\mathcal{L}^2$ norm, i.e.,

$$||d(\cdot, t)||_2 \leq \bar{d},$$

where $\bar{d} > 0$ is a **known** constant.

The problem investigated in this chapter can be summarized as follows:

Use a *limited number of measurements* (including one boundary measurement) to design an ADF and an ALF for *abnormality detection and localization* in the time and space domain, respectively for systems which subject to the state-space description of the form in (4.1) with the boundary conditions in (4.2) and the initial condition in (4.3).

4.3 PDE Backstepping-Based Abnormality Detection and Localization

4.3.1 Motivation and Proposed Framework

The core of model-based FD lies in the construction of the analytical redundancy [3], which can be used for the residual generation. Motivated by this idea, we consider the following boundary observer for system (4.1):

$$\hat{u}_t(x, t) = \hat{u}_{xx}(x, t) + \lambda(x)\hat{u}(x, t) + Q(x, t) + p_1(x)(u(0, t) - \hat{u}(0, t)),$$

$$\hat{\boldsymbol{y}}(t) = \int_0^1 \boldsymbol{c}(x)\hat{u}(x, t)\mathrm{d}x, \tag{4.4a}$$

$$\hat{u}_x(0, t) = p_{10}(u(0, t) - \hat{u}(0, t)), \tag{4.4b}$$

$$\hat{u}(1, t) = 0, \tag{4.4c}$$

where $p_1(x)$ and p_{10} denote the observer gains to be determined.

It is intuitive to design an ADF for time domain detection first by constructing the following residual $\boldsymbol{r}(t)$:

$$\boldsymbol{r}(t) \triangleq \boldsymbol{y}(t) - \hat{\boldsymbol{y}}(t), \tag{4.5}$$

and evaluate it in a *lumped* manner.

After an abnormality has been detected by the ADF in the time domain, one can design the following *distributed* residual $r_0(x, t)$:

$$r_0(x, t) \triangleq u(x, t) - \hat{u}(x, t), \tag{4.6}$$

for abnormality localization intuitively if the full-state information $u(x, t)$ can be obtained. However, this is quite a restrictive condition since continuous measuring in the space domain [19] is nearly impossible in practice, in certain applications only boundary measurements [20] are available.

To reach a compromise, we aim to design the following *distributed* residual $r(x, t)$ using a limited number of measurements throughout this chapter:

$$r(x, t) \triangleq \boldsymbol{V}(x)(\boldsymbol{y}(t) - \hat{\boldsymbol{y}}(t)), \tag{4.7}$$

where $\boldsymbol{V}(x) \in \mathbb{R}^{1\times(N+1)}$ is defied as a *general post-filter* to be determined. It is motivated by the post-filter $\boldsymbol{V}$ that is commonly used in FD of systems described by ODEs [3]. The difference is that $\boldsymbol{V}$ is a constant matrix while $\boldsymbol{V}(x)$ is a vector function of the space variable x. Furthermore, since backstepping-based observer design calculates the observer gains in a straightforward manner rather than solving several LMIs, it is applied in this chapter for the essential observer design by using only the boundary measurement.

Motivated by the above considerations, a framework which summarizes the main idea is introduced in Fig. 4.1. It mainly consists of four parts: backstepping-based boundary observer design, ADF design, ALF design, and residual evaluation. Note that the ALF is only triggered after the ADF has detected an abnormality in the time domain.

4.3.2 Backstepping-Based Boundary Observer Design and Analysis

By introducing the error variables $\tilde{u}(x, t) \triangleq u(x, t) - \hat{u}(x, t)$, $\tilde{\boldsymbol{y}}(t) \triangleq \boldsymbol{y}(t) - \hat{\boldsymbol{y}}(t)$, we can obtain the following error system:

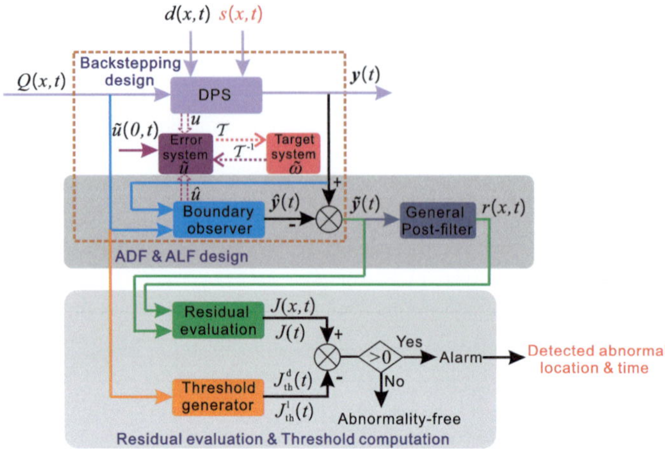

Fig. 4.1 Methodology framework

$$\tilde{u}_t(x,t) = \tilde{u}_{xx}(x,t) + \lambda(x)\tilde{u}(x,t) - p_1(x)\tilde{u}(0,t) + s(x,t) + d(x,t),$$

$$\tilde{\boldsymbol{y}}(t) = \int_0^1 \boldsymbol{c}(x)\tilde{u}(x,t)\mathrm{d}x, \tag{4.8a}$$

$$\tilde{u}_x(0,t) = -p_{10}\tilde{u}(0,t), \tag{4.8b}$$

$$\tilde{u}(1,t) = 0, \tag{4.8c}$$

by combining (4.1), (4.2) with (4.4).

Lemma 4.1 ([20]) *With the following Volterra integral transformation $\mathcal{T}$:*

$$\tilde{u}(x,t) = \tilde{\omega}(x,t) - \int_0^x p(x,\xi)\tilde{\omega}(\xi,t)\mathrm{d}\xi, \tag{4.9}$$

the error system (4.8) is transformed into the following target $\tilde{\omega}$-system:

$$\tilde{\omega}_t(x,t) = \tilde{\omega}_{xx}(x,t) - c\tilde{\omega}(x,t) + s_{\tilde{\omega}}(x,t) + d_{\tilde{\omega}}(x,t), \tag{4.10a}$$

$$\tilde{\omega}_x(0,t) = 0, \tag{4.10b}$$

$$\tilde{\omega}(1,t) = 0, \tag{4.10c}$$

where $c > 1$ is a user defined parameter and $s(x,t) = s_{\tilde{\omega}}(x,t) - \int_0^x p(x,\xi)s_{\tilde{\omega}}(\xi,t)\mathrm{d}\xi$, $d(x,t) = d_{\tilde{\omega}}(x,t) - \int_0^x p(x,\xi)d_{\tilde{\omega}}(\xi,t)\mathrm{d}\xi$ where $s_{\tilde{\omega}}(x,t)$ and $d_{\tilde{\omega}}(x,t)$ denote the transformed abnormality and disturbance in the $\tilde{\omega}$-system, respectively.

Lemma 4.2 ([20]) *The gain kernel $p(x,\xi)$ in (4.9) satisfies the following Klein-Gordon PDEs:*

$$p_{xx}(x,\xi) - p_{\xi\xi}(x,\xi) = -(\lambda(x)+c)p(x,\xi), \tag{4.11a}$$

$$p(1,\xi) = 0, \tag{4.11b}$$

$$p(x,x) = \int_0^x \frac{\lambda) + c}{2} \mathrm{d}y, \tag{4.11c}$$

where $x, \xi \in [0, 1]$, *and the observer gains can be obtained as*

$$p_1(x) = p_\xi(x, 0), \tag{4.12a}$$

$$p_{10} = p(0,0). \tag{4.12b}$$

Lemma 4.3 *The inverse transformation* $\mathcal{T}^{-1}$ *of (4.9) can be obtained as*

$$\tilde{\omega}(x,t) = \tilde{u}(x,t) + \int_0^x k(x,\eta)\tilde{u}(\eta,t)\mathrm{d}\eta.$$

Remark 4.3 Note that with the inverse transformation $\mathcal{T}^{-1}$, it can be derived that

$$d_{\tilde{\omega}}(x,t) = d(x,t) + \int_0^x k(x,\eta)d(\eta,t)\mathrm{d}\eta.$$

Then

$$||d_{\tilde{\omega}}(\cdot,t)||_2 \leq ||d(\cdot,t)||_2 + ||\int_0^x k(x,\eta)d(\eta,t)\mathrm{d}\eta||_2,$$

holds by triangle inequality. Moreover, it can be derived that

$$||\int_0^x k(x,\eta)d(\eta,t)\mathrm{d}\eta||_2 \leq \bar{k}^2\bar{d}^2,$$

where

$$\bar{k} \triangleq \max_{0\leq x\leq 1} ||k(x,\cdot)||_2.$$

Hence it can be obtained that

$$||d_{\tilde{\omega}}(\cdot,t)||_2 \leq (\bar{k}+1)\bar{d}.$$

Lemma 4.4 ([21]) *Let* $V(t)$ *and* $h(t)$ *be regular real functions. Then*

$$\dot{V}(t) \leq -\alpha V(t) + h(t), \forall t \geq 0$$

implies that

$$V(t) \leq e^{-\alpha t}V(0) + \int_0^t e^{-\alpha(t-\tau)}h(\tau)\mathrm{d}\tau, \quad \forall t \geq 0$$

for any finite constant α.

Proof See the proof of Lemma 3.2.4 on page 75 of [21]. □

Definition 4.1 (*Exponential Stability in the Sense of* $||\cdot||_\infty$) The linear PDE (4.10) with $s_{\tilde{\omega}}(x,t)=0$ and $d_{\tilde{\omega}}(x,t)=0$ is said to be exponentially stable in the sense of $||\cdot||_\infty$, if there exist constants $\alpha>0$, $\beta>0$ and $b>0$ such that $||\tilde{\omega}(\cdot,t)||_\infty \le \alpha||\tilde{\omega}_0(\cdot)||_\infty \exp(-\beta t)$ is fulfilled for any $t\ge 0$ and $||\tilde{\omega}_0(\cdot)||_\infty < b$.

Theorem 4.1 *The target $\tilde{\omega}$-system is exponentially stable in $\mathcal{L}^\infty$ under ideal operating conditions ($s_{\tilde{\omega}}(x,t)=0$, $d_{\tilde{\omega}}(x,t)=0$). In addition, when there exists a disturbance, $\tilde{\omega}(x,t)$ is bounded by the following expression:*

$$||\tilde{\omega}(\cdot,t)||_\infty \le \tilde{\omega}_b(t),$$

where

$$\tilde{\omega}_b(t) \triangleq \sqrt{2e^{-2(c-1)t}V(0) + \frac{1}{\gamma(c-1)}(\bar{k}+1)^2\bar{d}^2}, \tag{4.13a}$$

$$V(0) \triangleq \frac{1}{2}\left(\|\tilde{\omega}_0\|_2^2 + \|\tilde{\omega}_{x0}\|_2^2\right), \tag{4.13b}$$

with $0<\gamma\le 2$.

The complete proof is provided in Appendix A.1.

4.3.3 ADF Design and Lumped Residual Evaluation

Following the discussions above, the residual evaluation can provide a feature of the residual signal $\boldsymbol{r}(t)$. Along with the threshold setting, we are able to distinguish the abnormalities from the disturbances exist in the system.

The following *lumped* residual evaluation function and detection threshold are designed:

$$J(t) \triangleq \int_t^{t+T} ||\boldsymbol{r}(\tau)||^2 \mathrm{d}\tau, \tag{4.14a}$$

$$J_{\mathrm{th}}^{\mathrm{d}}(t) \triangleq \sup_{s(x,t)=0} J(t), \tag{4.14b}$$

where $T>0$ is the length of the evaluation time window.

Remark 4.4 There are several residual evaluation functions that are widely used in fault detection, such as the peak value, the average value, the RMS value, and the $\mathcal{L}_2$ norm [3]. The RMS value and $\mathcal{L}_2$ norm are related and they are designed to reduce the FAR at the cost of fault detectability. It is worth noting that $\left(\int_t^{t+T}||\boldsymbol{r}(\tau)||^2\mathrm{d}\tau\right)^{1/2}$ is a general form of the $\mathcal{L}_2$ norm [3, 19] for signal $\boldsymbol{r}(t)$, which measures the energy of a signal over a time interval $[t, t+T]$. Since evaluation over the whole time domain is usually unrealistic, introducing an evaluation window is a practical modification [3]. This kind of evaluation function was widely used in the literature discussing FD issues of dynamic systems, see [22, 23] for reference.

Hence the following decision logic will guarantee reliable abnormality detection performance:

$$\begin{cases} J(t) > J_{\text{th}}^{\text{d}}(t) \Longrightarrow \text{abnormal at time } t \\ J(t) \leq J_{\text{th}}^{\text{d}}(t) \Longrightarrow \text{abnormality-free at time } t. \end{cases} \tag{4.15}$$

Theorem 4.2 (Abnormality detection threshold calculation) *Under the lumped residual evaluation function in (4.14), if the conditions in Theorem 4.1 are satisfied, the time-varying detection threshold $J_{th}^{d}(t)$ for abnormality detection can be computed as*

$$J_{\text{th}}^{\text{d}}(t) = \int_{t}^{t+T} (1 + N(1 + \bar{p}^2))\tilde{\omega}_b^2(\tau)\mathrm{d}\tau, \tag{4.16}$$

where

$$\bar{p} \triangleq \max_{0 \leq x \leq 1} ||p(x, \cdot)||_2. \tag{4.17}$$

Proof First, it can be verified that $\tilde{u}(0, t) = \tilde{\omega}(0, t)$ holds for all $t \geq 0$ by the transformation $\mathcal{T}$ in (4.9). Therefore, we have that

$$\begin{aligned} ||\boldsymbol{r}(t)||^2 &\leq \underbrace{\tilde{\omega}(0, t)^2}_{\text{Boundary measurement}} + \underbrace{N||\tilde{u}(\cdot, t)||_\infty^2}_{N \text{ in-domain measurements}} \leq ||\tilde{\omega}(\cdot, t)||_\infty^2 + N(1 + \bar{p}^2)||\tilde{\omega}(\cdot, t)||_\infty^2 \\ &\leq (1 + N(1 + \bar{p}^2))\tilde{\omega}_b^2(t) \end{aligned} \tag{4.18}$$

by combining (4.9) and the results in Theorem 4.1. Furthermore, since this inequality holds for all $t \geq 0$, we can select the term in (4.16) as the threshold $J_{\text{th}}^{\text{d}}(t)$ for abnormality detection and it can be calculated *off-line*. This completes the proof. □

Remark 4.5 Despite the fact that there exist some fault detection approaches in the literature [2, 8], a new time-varying detection threshold is constructed mainly due to that either the initial error of the observer which may raise the false alarm was not considered or the detection condition was conservative in these works. The "time-varying" characteristic is due to the exponentially decreasing term $2e^{-2(c-1)t}V(0)$ in (4.13). In this manner, the effects of the initial error $\tilde{\omega}(0)$ and $\tilde{\omega}_x(0)$ on the bound of $||\tilde{\omega}(\cdot, t)||_\infty$ are taken into account. Hence false alarm can be avoided. This feature will be demonstrated in the numerical simulations following.

4.3.4 ALF Design and Distributed Residual Evaluation

Once an alarm has been triggered by the ADF, for root tracing and quickly removal of the abnormality, the ALF is activated for abnormality localization.

Remark 4.6 Similar to [24], define a linear unbounded operator $\mathcal{A}$ as

$$\mathcal{A}\bar{u}(x) \triangleq \frac{\mathrm{d}^2\bar{u}(x)}{\mathrm{d}x^2},$$

with the domain

$$\begin{aligned}\mathcal{D}(\mathcal{A}) \triangleq \{&\bar{u} \in \mathcal{L}^2((0,1))|\bar{u}, \mathrm{d}\bar{u}/\mathrm{d}x \text{ are absolutely continuous,}\\ &\mathrm{d}^2\bar{u}/\mathrm{d}x^2 \in \mathcal{L}^2((0,1)), \text{and } \mathrm{d}\bar{u}(x)/\mathrm{d}x|_{x=0} = 0, \bar{u}(1) = 0\}.\end{aligned}$$

Clearly, $\mathcal{A}$ is a linear, symmetric, closed, and self-adjoint operator [25]. Then the eigenvalue problem for the operator $\mathcal{A}$ can be solved with a countable set of real eigenvalues $\{\bar{\lambda}_i\}_{i=1}^{\infty}$ given by $\bar{\lambda}_i = -\frac{(2i-1)^2\pi^2}{4}, i \in \{1, 2, \cdots\}$ with the corresponding eigenfunctions $\{\phi_i(x)\}_{i=1}^{\infty}$ form an orthonormal basis given by: $\phi_i(x) = \sqrt{2}\cos\left(\frac{(2i-1)\pi}{2}x\right), x \in [0, 1], i \in \{1, 2, \ldots\}$. It should be emphasized that the utilization of operator $\mathcal{A}$ is to construct the orthonormal basis $\{\phi_i(x)\}_{i=1}^{\infty}$ for a Galerkin approximation, but is not directly related to the target system (4.10).

Recalling the error system in (4.8), the following expressions can be obtained:

$$\begin{aligned}\boldsymbol{C}_i &\triangleq \int_0^1 \boldsymbol{c}(x)\phi_i(x)\mathrm{d}x \in \mathbb{R}^{(N+1)\times 1},\\ \tilde{u}_i(t) &\triangleq \int_0^1 \phi_i(x)\tilde{u}(x,t)\mathrm{d}x, \tilde{\boldsymbol{y}}(t) = \sum_{i=1}^{\infty} \boldsymbol{C}_i\tilde{u}_i(t),\end{aligned} \tag{4.19}$$

for sufficiently large t under ideal operating conditions.

Remark 4.7 With the aid of series representation [26], the observation error of all available measurements $\tilde{\boldsymbol{y}}(t)$ can be used to derive a distributed residual $r(x,t)$ for abnormality localization. Motivated by the idea of separation of variables, an intuitive selection of $r(x,t)$ is $\sum_{i=1}^{\infty}\phi_i(x)\tilde{u}_i(t)$. In the following text, an alternative representation of $r(x,t)$ is obtained by using the series representation. In this manner, the full-state measurement [19] which is quite restrictive in applications is no longer needed for the distributed residual construction.

By using a finite number of basis functions, the following *virtual* output matrix $\boldsymbol{C}_m$ can be defined:

$$\boldsymbol{C}_m \triangleq \int_0^1 \boldsymbol{c}(x)\boldsymbol{\phi}_m^{\mathrm{T}}(x)\mathrm{d}x \in \mathbb{R}^{(N+1)\times m}, m \le N+1 \tag{4.20}$$

where $\boldsymbol{\phi}_m(x) \triangleq [\phi_1(x), \cdots, \phi_m(x)]^{\mathrm{T}} \in \mathbb{R}^{m\times 1}$ denotes the eigenfunctions corresponding to the largest m eigenvalues. By a proper selection of the in-domain measurements and the number of basis functions, it can be ensured that

$$\operatorname{rank} \boldsymbol{C}_m = m.$$

Then a *distributed* residual $r(x,t)$ is defined using the observer error of all available measurements

$$r(x,t) \triangleq \boldsymbol{V}(x)\tilde{\boldsymbol{y}}(t), \tag{4.21}$$

for abnormality localization, where $\boldsymbol{V}(x) \triangleq \boldsymbol{\phi}_m^{\mathrm{T}}(x)\boldsymbol{C}_m^{+}$, $\boldsymbol{C}_m^{+} \triangleq \left(\boldsymbol{C}_m^{\mathrm{T}}\boldsymbol{C}_m\right)^{-1}\boldsymbol{C}_m^{\mathrm{T}}$ denotes the Moore-Penrose inverse of $\boldsymbol{C}_m$.

Remark 4.8 Note that under ideal operating conditions, for $m \to \infty$, we have that $r(x,t) \to \tilde{u}(x,t)$ by simple calculation. However, it requires an infinite number of measurements to satisfy the compatibility condition, which is unacceptable in real applications. Rewriting $\tilde{\boldsymbol{y}}(t)$ as

$$\tilde{\boldsymbol{y}}(t) = \boldsymbol{C}_m\tilde{\boldsymbol{u}}_m(t) + \sum_{i=m+1}^{\infty} \boldsymbol{C}_i\tilde{u}_i(t).$$

Hence we have

$$\tilde{u}(x,t) = \boldsymbol{\phi}_m^{\mathrm{T}}(x)\tilde{\boldsymbol{u}}_m(t) + \sum_{i=m+1}^{\infty} \phi_i(x)\tilde{u}_i(t),$$

$$r(x,t) = \boldsymbol{\phi}_m^{\mathrm{T}}(x)\tilde{\boldsymbol{u}}_m(t) + \boldsymbol{\phi}_m^{\mathrm{T}}(x)\boldsymbol{C}_m^{+} \sum_{i=m+1}^{\infty} \boldsymbol{C}_i\tilde{u}_i(t),$$

where $\tilde{\boldsymbol{u}}_m(t) \triangleq [\tilde{u}_1(t), \cdots, \tilde{u}_m(t)]^{\mathrm{T}} \in \mathbb{R}^{m\times 1}$. In the spirit of the separation of slow and fast dynamics in [26], denoting $\tilde{\boldsymbol{u}}_f(t) \triangleq [\tilde{u}_{m+1}(t), \cdots, \tilde{u}_{\infty}(t)]^{\mathrm{T}}$, $\boldsymbol{C}_f \triangleq [\boldsymbol{C}_{m+1}, \cdots, \boldsymbol{C}_{\infty}]$, $\boldsymbol{\phi}_f(x) \triangleq [\phi_{m+1}(x), \cdots, \phi_{\infty}(x)]^{\mathrm{T}}$ leads to

$$\begin{aligned}||r(\cdot,t) - \tilde{u}(\cdot,t)||_2 &= ||\boldsymbol{\phi}_m^{\mathrm{T}}(\cdot)\boldsymbol{C}_m^{+}\boldsymbol{C}_f\tilde{\boldsymbol{u}}_f(t) - \boldsymbol{\phi}_f^{\mathrm{T}}(\cdot)\tilde{\boldsymbol{u}}_f(t)||_2 \le ||\boldsymbol{C}_m^{+}\boldsymbol{C}_f\tilde{\boldsymbol{u}}_f(t)|| + ||\tilde{\boldsymbol{u}}_f(t)||\\ &\le (1+\sigma_{\max}(\boldsymbol{C}_m^{+}\boldsymbol{C}_f))||\tilde{\boldsymbol{u}}_f(t)||,\end{aligned}$$

where $\sigma_{\max}(\boldsymbol{C}_m^{+}\boldsymbol{C}_f)$ denotes the maximum singular value of matrix $\boldsymbol{C}_m^{+}\boldsymbol{C}_f$. Hence the error between $r(x,t)$ and $\tilde{u}(x,t)$ is bounded in the sense of $||\cdot||_2$ and the bound is decreasing with the increasing of m according to the singular perturbation theory in [26].

4.3.5 Distributed Residual Evaluation and Threshold Computation

Different from the *lumped* residual evaluation function in (4.14), the following *distributed* residual evaluation function and localization threshold are designed:

$$J(x,t) \triangleq \int_t^{t+T} r^2(x,\tau)\mathrm{d}\tau, \tag{4.22a}$$

$$J_{\mathrm{th}}^{1}(t) \triangleq \sup_{s(x,t)=0} J(x,t). \tag{4.22b}$$

Moreover, the following decision logic will guarantee reliable abnormality localization performance *after* an abnormality has been detected in the time domain by the proposed ADF:

$$\begin{cases} J(x,t) > J_{\mathrm{th}}^{1}(t) \Longrightarrow \text{abnormal for location } x \\ J(x,t) \le J_{\mathrm{th}}^{1}(t) \Longrightarrow \text{abnormality-free for location } x. \end{cases} \tag{4.23}$$

Remark 4.9 The main reason for using the ADF rather ALF for abnormality detection is that the ADF is more sensitive to distributed abnormality since the residual is evaluated in a lumped way. Moreover, this follows the general framework of fault detection and isolation for dynamic systems described by ODEs [3].

Theorem 4.3 (Abnormality localization threshold calculation) *Under the distributed residual evaluation function in (4.22), if the conditions in Theorem 4.1 are satisfied, then the abnormality localization threshold $J_{th}^{l}(t)$ can be calculated as*

$$J_{\mathrm{th}}^{1}(t) = \int_t^{t+T} (2m-1)\pi\sigma_{\max}^2(\boldsymbol{C}_m^{+})(1+N(1+\bar{p}^2))\tilde{\omega}_b^2(\tau)\mathrm{d}\tau, \tag{4.24}$$

where $\sigma_{\max}(\boldsymbol{C}_m^{+})$ denotes the maximum singular value of matrix $\boldsymbol{C}_m^{+}$.

Proof To begin with, some proprieties of the eigenfunctions are illustrated as follows:

$$\begin{aligned} &< \phi_i(\cdot), \phi_j(\cdot) > = 0,\ \|\phi_i(\cdot)\|_2^2 = 1, \\ &< \dot{\phi}_i(\cdot), \dot{\phi}_j(\cdot) > = 0,\ \left\|\dot{\phi}_i(\cdot)\right\|_2^2 = \frac{(2i-1)^2\pi^2}{4}, i, j = 1, \cdots, m, i \ne j, \end{aligned} \tag{4.25}$$

where $< \phi_i(\cdot), \phi_j(\cdot) > \triangleq \int_0^1 \phi_i(x)\phi_j(x)\mathrm{d}x$, $\dot{\phi}_i \triangleq \mathrm{d}\phi_i/\mathrm{d}x$.

To move on, it can be verified that $r(1,t) = 0$. Then Agmon's inequality [20] can be applied such that

$$||r(\cdot,t)||_\infty^2 \le 2||r(\cdot,t)||_2||r_x(\cdot,t)||_2. \tag{4.26}$$

Considering (4.25), $||r(\cdot,t)||_2$ can be further rewritten as

$$||r(\cdot,t)||_2 = ||\phi_m^{\mathrm{T}}(\cdot)\boldsymbol{C}_m^{+}\tilde{\boldsymbol{y}}(t)||_2 = ||\boldsymbol{C}_m^{+}\tilde{\boldsymbol{y}}(t)|| \le \sigma_{\max}(\boldsymbol{C}_m^{+})||\tilde{\boldsymbol{y}}(t)||, \tag{4.27}$$

using the properties discussed in the proof of Theorem 2 in [27].

Moreover, $||r_x(\cdot,t)||_2$ can be rewritten as

$$||r_x(\cdot,t)||_2 = ||\dot{\phi}_m^{\mathrm{T}}(\cdot)\boldsymbol{C}_m^{+}\tilde{\boldsymbol{y}}(t)||_2 \le \frac{(2m-1)\pi}{2}||\boldsymbol{C}_m^{+}\tilde{\boldsymbol{y}}(t)|| \le \frac{(2m-1)\pi}{2}\sigma_{\max}(\boldsymbol{C}_m^{+})||\tilde{\boldsymbol{y}}(t)||. \tag{4.28}$$

Recalling (4.26), we have that

$$||r(\cdot,t)||_\infty^2 \le (2m-1)\pi\sigma_{\max}^2(\boldsymbol{C}_m^{+})||\tilde{\boldsymbol{y}}(t)||^2. \tag{4.29}$$

Recalling (4.18), it can be obtained that

$$||r(\cdot,t)||_\infty^2 \le (2m-1)\pi\sigma_{\max}^2(\boldsymbol{C}_m^{+})(1+N(1+\bar{p}^2))\tilde{\omega}_b^2(t). \tag{4.30}$$

Therefore, one can select the localization threshold $J_{\mathrm{th}}^{1}(t)$ as in (4.24). This completes the proof. □

Remark 4.10 Although the ALF is only activated after the ADF has raised an alarm, the influence of the unknown initial values is still considered in $J_{\mathrm{th}}^{1}(t)$ for accurate fault localization.

Remark 4.11 The main advantages of the proposed backstepping-based design framework over a direct early-lumping design are twofold

- The observation spillover caused by neglecting the remaining infinite-dimensional dynamics in the design stage can introduce extra error into the following abnormality detection and localization.
- For the early-lumping design, the calculation of the detection and localization threshold require that the disturbance is separable and the spatial distribution function is known.

Remark 4.12 It is worth noting that only distributed disturbance is considered in this chapter. However, the proposed method can be applied to boundary disturbances cases. As for the unknown boundary disturbances, one possible solution is to assume that they can be modeled by known finite-dimensional signal processes as in [28]. Then a finite-dimensional reference observer and an infinite-dimensional disturbance observer can be constructed using backstepping techniques. The ALF design procedure then follows. In this sense, the proposed method applies to boundary disturbances.

A summary of the proposed ADF and ALF by pseudo code is shown in Algorithm 4.1.

Algorithm 4.1: ADF & ALF framework .

1: Select $c > 1$ and $\gamma \in (0, 2]$ based on Theorem 4.1.
2: *Off-line* observer gains calculation: $p_1(x)$, $p_{10} \leftarrow$ (4.12).
3: Set $\hat{u}_0(x)$ randomly and select the evaluation time window with length T.
4: *Off-line* detection and localization thresholds calculation: $J_{\text{th}}^{\text{d}}(t) \leftarrow$ (4.16) and (4.17), $J_{\text{th}}^{\text{l}}(t) \leftarrow$ (4.24).
5: *On-line* lumped residual evaluation: $J(t) \leftarrow$ (4.14) .
6: For $t \in [0, t_{max}]$ (t_{max}: considered time length)
  If $J(t) \leq J_{\text{th}}^{\text{d}}(t)$
    Abnormality-free at time t.
  end
  If $J(t) > J_{\text{th}}^{\text{d}}(t)$
    On-line distributed residual evaluation:
    $J(x,t) \leftarrow$ (4.22).
    For $x \in (0, 1]$
      If $J(x,t) \leq J_{\text{th}}^{\text{l}}(t)$
        Abnormality-free for location x.
      end
      If $J(x,t) > J_{\text{th}}^{\text{l}}(t)$
        Abnormal for location x.
      end
    end
  end
end

4.4 Numerical Simulation

Consider a thin rod whose temperature distribution can be modeled by the following parabolic PDE:

$$\begin{aligned} T_t(x,t) &= T_{xx}(x,t) + \lambda_0 T(x,t) + Q(x,t) + d(x,t) + s(x,t), \\ \mathbf{y}(t) &= [T(0,t), T(1/2,t)]^{\mathrm{T}}, && (4.31a) \\ T_x(0,t) &= 0, && (4.31b) \\ T(1,t) &= 0, && (4.31c) \end{aligned}$$

with initial condition $T_0(x) = 2\sin(\pi x)$.

The value of λ_0 is selected as 2. The distributed input $Q(x,t)$ and disturbance $d(x,t)$ adopt the following form:

$$\begin{aligned} Q(x,t) &= e^{(-0.5(x-0.5)^2)}\sin(t), \\ d(x,t) &= 0.005e^{(-0.5(x-0.5)^2)}\sin(t), \end{aligned}$$

for demonstration.

Furthermore, to demonstrate the abnormality localization performance, the following abnormality is injected:

$$s(x,t) = b_s(x)s(t),$$

where

$$b_s(x) = H(x - 1/2) - H(x - 3/4),$$

$$s(t) = \begin{cases} 0, & 0 \leqslant t < 5(\mathrm{s}), \\ 1, & 5 \leqslant t \leqslant 10(\mathrm{s}), \end{cases}$$

with $H(\cdot)$ denoting the standard Heaviside function.

The original PDE system (4.31) is solved numerically by the FDM with the sampling time $\Delta t = 0.001$ s and 19 grid points with respect to the space.

The designed parameters are selected as $c = 10$, $\gamma = 1$ to meet the requirements in Theorem 4.1. Based on the proposed methodology, the abnormality detection results are shown in Fig. 4.2 with the length of the evaluation time window $T = 0.01$ s. It can be observed that the abnormality occurring time $t = 5$ s was well detected and no false alarm raised due to the time-varying threshold setting, as discussed in Remark 4.5. However, if a constant detection threshold (the red dash line in the bottom right corner of Fig. 4.2) is used, false alarm will raise in the initial stage (from 0 to 0.4 s).

Upon the detection of abnormality by ADF in the time domain, the ALF is activated for abnormality localization by constructing the distributed residual $r(x,t)$ as shown in Fig. 4.3b. The state error $\tilde{u}(x,t)$ is presented in Fig. 4.3a only for comparison, note that it is *not available* in the proposed ALF design since full-state measurement is not available in the proposed methodology framework. The results of the distributed residual evaluation $J(x,t)$ and the threshold $J^1_{\mathrm{th}}(t)$ are displayed in Fig. 4.4a from $t = 5$ s and afterward. It can be observed that the value of evaluation function $J(x,t)$ in certain area exceeds the localization threshold. The abnormality localization results at $t = 6$ s is presented in Fig. 4.4b. It can be seen clearly that the abnormality location $[\frac{1}{2}, \frac{3}{4}]$ is detected. Instead of searching the whole area for locations of abnormality, the possible areas of abnormality are widely narrowed down

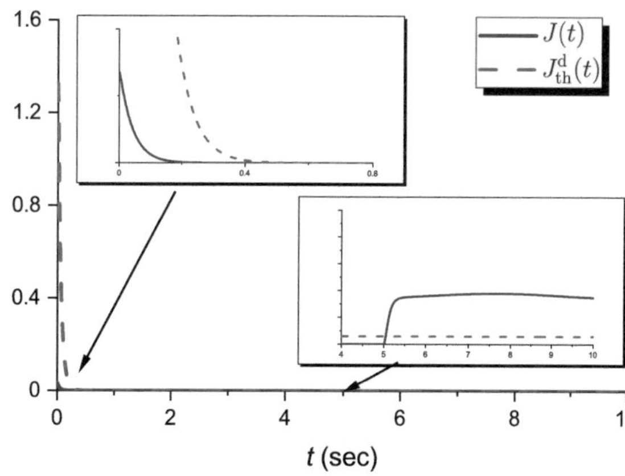

Fig. 4.2 Abnormality detection results in the time domain

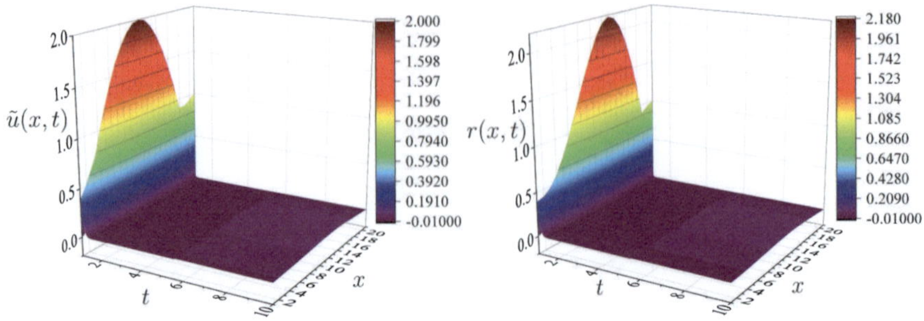

(a) $\tilde{u}(x, t)$. (b) $r(x, t)$.

Fig. 4.3 State error $\tilde{u}(x, t)$ and the distributed residual $r(x, t)$

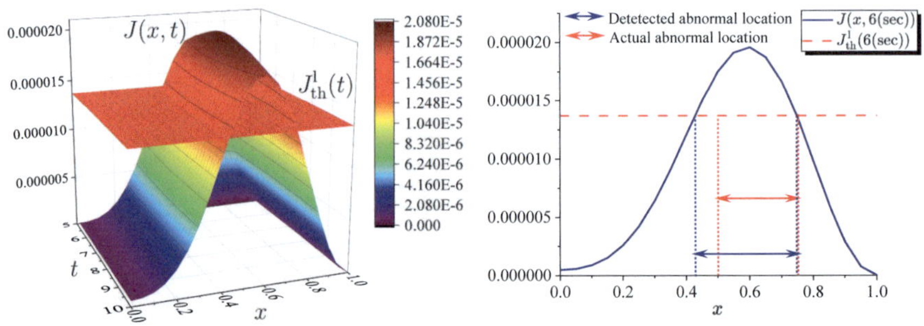

(a) $J(x, t)$ and $J^1_{\rm th}(t)$. (b) Localization results at $t = 6$ sec.

Fig. 4.4 Abnormality localization results

by the proposed framework. In this way, the abnormality can be localized. However, there exists some error between the actual and detected abnormal location due to that full-state measurement is not available and the existence of disturbance.

Remark 4.13 In the ALF design, the number of basis functions is selected as 2 and the compatibility condition can be satisfied. Generally speaking, a larger number of basis functions is beneficial to the residual signal construction since $r(x, t) \to \tilde{u}(x, t)$ when $m \to \infty$. However, this requires a larger number of in-domain measurements N to satisfy the compatibility condition. Moreover, the localization threshold in (4.24) increases with larger N and m, which makes it more conservative and not sensitive to abnormalities with small magnitude. The selection of N and m and the sensors placement is an interesting optimization problem in the sense of optimal abnormality detection and localization which shall be discussed in future work.

4.5 Conclusion

In this chapter, the abnormality localization problem for a class of parabolic DPSs is studied via the backstepping techniques and eigenspectrum. The proposed methodology framework consists of two steps: first, activate the ADF for detection in the time domain; once an alarm raised for the ADF, activate the ALF for localization in the space domain by constructing a distributed residual using a limited number of measurements. For the detection purpose, the residual evaluation is conducted in the traditional "lumped" way. However, for the abnormality localization, the residual is evaluated in a "distributed" way.

Appendix

A.1 Proof of Theorem 4.1

Proof Consider the following Lyapunov function:

$$V(t) = \frac{1}{2}\int_0^1 (\tilde{\omega}^2(x,t) + \tilde{\omega}_x^2(x,t))\mathrm{d}x. \tag{4.32}$$

Using integration by parts and considering (4.11), it can be derived that

$$\dot{V}(t) = -(1+c)||\tilde{\omega}_x(\cdot,t)||_2^2 - c||\tilde{\omega}(\cdot,t)||_2^2 - ||\tilde{\omega}_{xx}(\cdot,t)||_2^2 + \int_0^1 d_{\tilde{\omega}}\tilde{\omega}\mathrm{d}x - \int_0^1 d_{\tilde{\omega}}\tilde{\omega}_{xx}\mathrm{d}x. \tag{4.33}$$

Part I. First let us consider the case when no disturbance exists, i.e., $d_{\tilde{\omega}}(x,t) = 0$. Applying the Poincaré inequality [20] to (4.33) leads to

$$V(t) \le e^{-2(c+\frac{\pi^2}{4})t}V(0), \tag{4.34}$$

by referring to Lemma 4.4.

Applying Agmon's inequality and Young's inequality [20] and considering (4.34) yields

$$||\tilde{\omega}(\cdot,t)||_\infty^2 \le 2||\tilde{\omega}(\cdot,t)||_2||\tilde{\omega}_x(\cdot,t)||_2 \le ||\tilde{\omega}(\cdot,t)||_2^2 + ||\tilde{\omega}_x(\cdot,t)||_2^2 \le e^{-2(c+\frac{\pi^2}{4})t}(||\tilde{\omega}_0||_2^2 + ||\tilde{\omega}_{x0}||_2^2), \tag{4.35}$$

which ensures the exponentially convergence of $\tilde{\omega}(x,t)$ in the $\mathcal{L}^\infty$ norm.

Part II. When there exist disturbances, applying the Cauchy-Schwarz inequality on the fourth and fifth term of (4.33) and further using Young's inequality leads to

$$\begin{aligned}\dot{V}(t) \le& -(1+c)||\tilde{\omega}_x(\cdot,t)||_2^2 - c||\tilde{\omega}(\cdot,t)||_2^2 - ||\tilde{\omega}_{xx}(\cdot,t)||_2^2 \\ &+ \frac{1}{2\gamma}||d_{\tilde{\omega}}(\cdot,t)||_2^2 + \frac{\gamma}{2}||\tilde{\omega}(\cdot,t)||_2^2 + \frac{1}{2\gamma}||d_{\tilde{\omega}}(\cdot,t)||_2^2 + \frac{\gamma}{2}||\tilde{\omega}_{xx}(\cdot,t)||_2^2,\end{aligned} \tag{4.36}$$

where $\gamma > 0$ is a positive constant.

Selecting γ which satisfies $0 < \gamma \le 2$, then it can be further derived that

$$\dot{V}(t) \le -(1+c)||\tilde{\omega}_x(\cdot,t)||_2^2 + (1-c)||\tilde{\omega}(\cdot,t)||_2^2 + \frac{1}{\gamma}||d_{\tilde{\omega}}(\cdot,t)||_2^2 \le 2(1-c)V(t) + \frac{1}{\gamma}||d_{\tilde{\omega}}(\cdot,t)||_2^2. \tag{4.37}$$

Since $c > 1$, combining Remark 4.3 with Lemma 4.4, we have that

$$V(t) \le e^{-2(c-1)t}V(0) + \frac{1}{2\gamma(c-1)}(\bar{k}+1)^2\bar{d}^2. \tag{4.38}$$

Applying Agmon's inequality and Young's inequality [20] yields

$$\begin{aligned}||\tilde{\omega}(\cdot,t)||_\infty^2 &\le 2||\tilde{\omega}(\cdot,t)||_2||\tilde{\omega}_x(\cdot,t)||_2 \le ||\tilde{\omega}(\cdot,t)||_2^2 + ||\tilde{\omega}_x(\cdot,t)||_2^2 \\ &\le 2e^{-2(c-1)t}V(0) + \frac{1}{\gamma(c-1)}(\bar{k}+1)^2\bar{d}^2.\end{aligned} \tag{4.39}$$

Hence

$$||\tilde{\omega}(\cdot,t)||_\infty \le \sqrt{2e^{-2(c-1)t}V(0) + \frac{1}{\gamma(c-1)}(\bar{k}+1)^2\bar{d}^2}. \tag{4.40}$$

This completes the proof. □

References

1. Dey S, Perez HE, Moura SJ (2017) Model-based battery thermal fault diagnostics: algorithms, analysis, and experiments. IEEE Trans Control Syst Technol 27(2):576–587
2. Dey S, Perez HE, Moura SJ (2019) Robust fault detection of a class of uncertain linear parabolic PDEs. Automatica 107:502–510
3. Ding SX (2013) Model-based fault diagnosis techniques: design schemes, algorithms, and tools, 2nd edn. Springer, London
4. Ding SX (2014) Data-driven design of fault diagnosis and fault-tolerant control systems. Springer, London
5. Ghantasala S, El-Farra NH (2009) Robust actuator fault isolation and management in constrained uncertain parabolic PDE systems. Automatica 45(10):2368–2373
6. Deutscher J, Harkort C (2010) A parametric approach to finite-dimensional control of linear distributed-parameter systems. Int J Control 83(8):1674–1685
7. Deutscher J (2016) Fault detection for linear distributed-parameter systems using finite-dimensional functional observers. Int J Control 89(3):550–563

8. Cai J, Ferdowsi H, Sarangapani J (2016) Model-based fault detection, estimation, and prediction for a class of linear distributed parameter systems. Automatica 66:122–131
9. Shinbrot M (1954) On the analysis of linear and nonlinear dynamical systems from transient-response data. NACA Technical Note No. 3288
10. Fischer F, Deutscher J (2020) Flatness-based algebraic fault diagnosis for distributed-parameter systems. Automatica 117:108987
11. Asiri S, Laleg-Kirati T-M (2017) Modulating functions-based method for parameters and source estimation in one-dimensional partial differential equations. Inverse Probl Sci Eng 25(8):1191–1215
12. Asiri S, Liu D-Y, Laleg-Kirati T-M (2020) Selection of modulating functions' design parameters for estimation problems. IEEE Control Syst Lett 5(1):277–282
13. Feng Y, Li H-X, Yang H-D (2019) Abnormal source identification for parabolic distributed parameter systems. IEEE Trans Syst Man Cybern Syst 51(9):5698–5707
14. El-Farra NH, Ghantasala S (2007) Actuator fault isolation and reconfiguration in transport-reaction processes. AIChE J 53(6):1518–1537
15. Smyshlyaev A, Krstic M (2004) Closed-form boundary state feedbacks for a class of 1-D partial integro-differential equations. IEEE Trans Autom Control 49(12):2185–2202
16. Smyshlyaev A, Krstic M (2005) Backstepping observers for a class of parabolic PDEs. Syst Control Lett 54(7):613–625
17. Balakrishnan V (2003) All about the Dirac delta function (?). Resonance 8(8):48–58
18. Kerschbaum S, Deutscher J (2020) Backstepping control of coupled linear parabolic PDEs with space and time dependent coefficients. IEEE Trans Autom Control 65(7):3060–3067
19. Li L, Ding SX, Peng K et al (2019) Fuzzy fault detection filter design for nonlinear distributed parameter systems. IEEE Access 7:11105–11113
20. Krstic M, Smyshlyaev A (2008) Boundary control of PDEs: a course on backstepping designs. SIAM, Philadelphia
21. Ioannou PA, Sun J (2012) Robust adaptive control. Courier Corporation, Chelmsford
22. Li L, Ding SX, Qiu J et al (2016) Fuzzy observer-based fault detection design approach for nonlinear processes. IEEE Trans Syst Man Cybern Syst 47(8):1941–1952
23. Li L, Ding SX (2020) Gap metric techniques and their application to fault detection performance analysis and fault isolation schemes. Automatica 118:109029
24. Wang J-W, Liu Y-Q, Sun C-Y (2018) Pointwise exponential stabilization of a linear parabolic PDE system using non-collocated pointwise observation. Automatica 93:197–210
25. Curtain RF, Zwart H (1995) An introduction to infinite-dimensional linear systems theory. Springer, New York
26. Christofides PD (2012) Nonlinear and robust control of PDE systems: methods and applications to transport-reaction processes. Springer, Berlin
27. Wu H-N, Li H-X (2011) Robust adaptive neural observer design for a class of nonlinear parabolic PDE systems. J Process Control 21(8):1172–1182
28. Deutscher J (2015) A backstepping approach to the output regulation of boundary controlled parabolic PDEs. Automatica 57:56–64

Chapter 5
Spatial Decomposition-Based Fault Detection Framework for Parabolic-Distributed Parameter Processes

Abstract Fault detection in distributed parameter systems (such as thermal or fluid processes) plays a crucial role in ensuring both safety and efficiency. Current data-driven strategies often disregard the dynamic evolution of these processes, making them unsuitable for rapidly changing or transient conditions. Meanwhile, most reported model-based approaches rely heavily on the backstepping technique, which lacks sufficient redundancy for reliable fault diagnosis since only boundary measurements are utilized. To address these limitations, this work investigates the robust fault detection problem from a model-based perspective, incorporating both boundary and in-domain sensing. A real-time fault detection filter is proposed, eliminating the dependence on extensive data acquisition and offline training. Comprehensive theoretical analysis is provided to guide parameter tuning and threshold design. In particular, a time-varying threshold is constructed to suppress false alarms during transient stages. Experimental validation on a hot strip mill cooling system demonstrates the approach's feasibility and potential for industrial deployment.

5.1 Introduction

Distributed parameter processes are prevalent in various industrial applications, such as transport–reaction processes [1], snap curing processes [2–4], and battery thermal processes [5–7]. These processes are typically described by PDEs, which inherently capture spatial–temporal dynamics [8–12]. With the growing demand for high production quality and economic operation, ensuring system safety has become a critical concern. Undetected faults in such processes may result in severe failures or even irreversible damage. For instance, distributed thermal faults in Li-ion batteries [5, 13] can trigger battery degradation, malfunction, or thermal runaway, while in-domain actuator faults in diffusion–reaction processes [1, 14], which are common in chemical industries, may also lead to significant consequences. Although advanced control strategies for DPSs have been extensively studied in recent years [15, 16], including sampled-data [17, 18] and fuzzy control [19], research on FD for distributed parameter processes remains relatively limited.

© The Author(s), under exclusive license to Springer Nature Singapore Pte Ltd. 2026

Y. Feng et al., *Spatio-Temporal Abnormality Diagnosis for Industrial Distributed Parameter Systems*, https://doi.org/10.1007/978-981-95-3750-1_5

On one side, fault detection and fault-tolerant control of technical processes represented by ODEs have been extensively studied over the past decades, including both model-based [20–22] and data-driven strategies [23]. On the other side, investigations concerning FD of systems governed by PDEs remain comparatively limited. Similar to the ODE case, existing studies can generally be grouped into these two main categories.

In the context of data-driven fault detection for distributed parameter processes, several studies have appeared in recent years (see [24–27]). For instance, a process monitoring scheme for parabolic DPSs was developed using the Galerkin method [24]. Another work introduced a new data-driven strategy for fault detection and localization in parabolic DPSs [25]. Nevertheless, these approaches present limitations: some rely on modal approximation [24], which may result in the so-called *observation spillover*, while others overlook the dynamic evolution of the processes [25, 27], making them unsuitable for highly dynamic or transient scenarios.

Existing model-based fault detection strategies can be broadly classified into two groups. The first group corresponds to the so-called –early-lumping–approach [28], where the PDE model is approximated by a finite set of ODEs that are then analyzed; representative works include [1, 14, 29]. The second group refers to the –late-lumping–approach [30, 31], which investigates the PDE system directly without performing such approximations. Most of the research in this category relies on backstepping design techniques [32], see [13, 33] for examples. Early-lumping approaches have been criticized because the modal approximation used often introduces the problem of *observation spillover*. In addition, backstepping-based detection methods may lack redundancy in certain applications since they only exploit boundary measurements. Consequently, a general framework capable of handling both boundary and in-domain sensing scenarios is still absent.

Building upon the above considerations, this work develops a methodological framework for robust fault detection of distributed parameter processes, adopting a model-based viewpoint and incorporating both boundary and in-domain sensing scenarios. Specifically, a FDF is constructed, inspired by the design principles of FDFs in LPSs [20], to generate the diagnostic signals. In addition, the residual evaluation strategy and the procedure for threshold determination are carried out in a comparable fashion.

The organization of this chapter is arranged as follows. Section 5.2 specifies the problem formulation in detail. Section 5.3 introduces the Luenberger-type PDE observer together with the design of the FDF. Application examples on a hot strip mill cooling system are provided in Sect. 5.4. Concluding discussions are given in Sect. 5.5.

5.2 Preliminaries and Problem Statement

Notations: $\mathbb{R}^n$ stands for the n-dimensional Euclidean space, with the norm denoted by $|\cdot|_2$. The space $\mathcal{L}^2([0, L]) \triangleq \mathcal{L}^2((0, L); \mathbb{R})$ is a real Hilbert space consisting of square-integrable functions $\omega(x) : [0, L] \to \mathbb{R}$, where the $\mathcal{L}^2$ spatial norm is defined as $||\omega(\cdot)||_2 \triangleq \sqrt{\int_0^L \omega^2(x)\mathrm{d}x}$. For a given natural number $\bar{l}$, the Sobolev space $\mathcal{H}^{\bar{l}}(0, 1) \triangleq \mathcal{W}^{\bar{l},2}((0, 1); \mathbb{R})$ consists of absolutely continuous functions $\bar{\omega}(x)$: $(0, 1) \to \mathbb{R}$ whose derivatives $\mathrm{d}^i\bar{\omega}(x)/\mathrm{d}x^i$ up to order $\bar{l} \geq 1$ are square integrable, and its norm is given by $||\bar{\omega}(\cdot)||_{\mathcal{H}^{\bar{l}}} \triangleq \sqrt{\int_0^1 \sum_{i=0}^{\bar{l}} \left(\frac{\mathrm{d}^i\bar{\omega}(x)}{\mathrm{d}x^i}\right)^2 \mathrm{d}x}$. For any function $\omega(\cdot) \in \mathcal{H}^1(0, 1)$, the $\mathcal{L}^\infty$ spatial norm is defined as $||\omega(\cdot)||_\infty \triangleq \max_{x\in[0,1]} |\omega(x)|$. Moreover, partial derivatives are denoted as $T_t \triangleq \partial T/\partial t$, $T_x \triangleq \partial T/\partial x$, $T_{xx} \triangleq \partial^2 T/\partial x^2$.

The class of DPSs under study is described by the following PDE:

$$T_t(x, t) = T_{xx}(x, t) + d(x, t) + f(x, t), \tag{5.1a}$$

$$T_x(0, t) = \eta_1 T(0, t), \tag{5.1b}$$

$$T_x(1, t) = -\eta_2 T(1, t), \tag{5.1c}$$

$$T(x, 0) = T_0(x), \tag{5.1d}$$

$$\boldsymbol{y}(t) = \int_0^1 \boldsymbol{c}(x)T(x, t)\mathrm{d}x, \tag{5.1e}$$

where $T(\cdot, t) \in \mathcal{L}^2([0, 1])$ represents the system state; $x \in [0, 1]$ is the spatial variable and $t \in [0, \infty)$ denotes time; $d(x, t)$ is an *unknown* disturbance in the system, including process noise and model uncertainties; $f(x, t)$ is an *unknown* fault; $\boldsymbol{y}(t) \triangleq [y_1(t) \;\; y_2(t) \;\; \cdots \;\; y_n(t)]^{\mathrm{T}} \in \mathbb{R}^n$ represents the n-dimensional sensor output, while $\boldsymbol{c}(x) \triangleq [c_1(x) \;\; c_2(x) \;\; \cdots \;\; c_n(x)]^{\mathrm{T}} \in \mathbb{R}^n$ characterizes the measurement configuration. Without loss of generality, the most commonly employed **pointwise measurement** [34, 35] is adopted as follows:

$$c_i(x) = \delta(x - \bar{x}_i), i \in \mathcal{N} \triangleq \{1, 2, \cdots, n\}$$

this measurement corresponds to pointwise sensors located at $\bar{x}_i, i \in \mathcal{N}$ within the spatial domain (0, 1), where $\delta(\cdot)$ represents the Dirac delta function.

Assumption 5.1 The distributed disturbance $d(x, t)$ satisfies that

$$||d(\cdot, t)||_2 \leq \bar{d} < \infty,$$

where $\bar{d} > 0$ is a **known** constant.

Remark 5.1 It should be emphasized that both the fault $f(x,t)$ and the disturbance $d(x,t)$ are present in the system. Such a configuration is commonly adopted in the literature for both ODE-based [20] and PDE-based [13, 30, 36] systems. In practice, the disturbance $d(x,t)$, considered as an unknown input, includes model uncertainties and process noise. The term $f(x,t)$ accounts for all potential faults and takes the value zero in a fault-free scenario. This modeling choice is crucial for robust fault detection, as neglecting it may lead to false alarms.

The problem under consideration can be stated as follows:

Employ the pointwise measurements $\mathbf{y}(t)$ to construct a FDF and produce the associated residual signal for *robust fault detection.* for DPSs governed by the state-space representation given in (5.1a)–(5.1e).

5.3 Infinite-Dimensional FDF Design

5.3.1 Motivation and Proposed Framework

The fundamental principle of model-based FD for ODE systems lies in the construction of analytical redundancy [20], which forms the basis for residual generation. Following this principle, a FDF can be designed in a similar manner.

It is worth noting that in ODE systems, both the state and output vectors are finite-dimensional. In contrast, for systems governed by (5.1), the state $T(x,t)$ resides in an *infinite-dimensional* space, whereas the output $\mathbf{y}(t)$ remains finite-dimensional. A central difficulty in designing the FDF lies in mapping the finite-dimensional output onto the infinite-dimensional state in order to carry out the *output injection* process.

Although backstepping-based boundary observer designs offer advantages due to their relatively simple construction (see [5, 13, 32, 33, 37] for reference), extending these methods to incorporate in-domain measurements (both pointwise and piecewise [34]) remains a challenging research problem. This extension is particularly important for fault detection in DPSs, as in-domain sensing enhances the redundancy of the FDF.

Accordingly, a FDF for DPSs is developed to accommodate boundary, pointwise, and piecewise measurements, drawing on the *unified* Lyapunov-based compensator approach [38]. To facilitate the FDF construction, certain results from [38] are further extended from the spatial $\mathcal{L}^2$ norm to the spatial $\mathcal{L}^\infty$ norm.

Motivated by the preceding discussion, Fig. 5.1 presents a framework comprising a FDF along with the associated residual evaluation and threshold computation procedures.

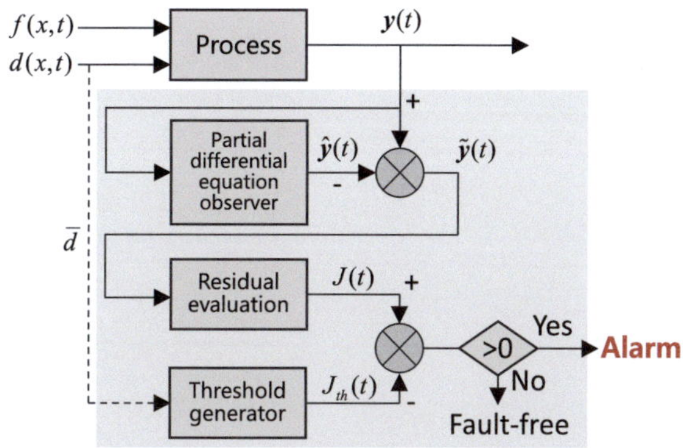

Fig. 5.1 Methodology framework

5.3.2 *Spatial Domain Decomposition-Based PDE Observer Design and Analysis*

Inspired by the spatial domain decomposition approach in [34, 38], the following PDE observer is designed for the system described in (5.1a)–(5.1e):

$$\hat{T}_t(x,t) = \hat{T}_{xx}(x,t) + \boldsymbol{g}^{\mathrm{T}}(x)\boldsymbol{L}(\boldsymbol{y}(t) - \hat{\boldsymbol{y}}(t)), \tag{5.2a}$$

$$\hat{T}_x(0,t) = \eta_1 \hat{T}(0,t), \tag{5.2b}$$

$$\hat{T}_x(1,t) = -\eta_2 \hat{T}(1,t), \tag{5.2c}$$

$$\hat{\boldsymbol{y}}(t) = \int_0^1 \boldsymbol{c}(x)\hat{T}(x,t)\mathrm{d}x, \tag{5.2d}$$

where $\boldsymbol{g}(x) \triangleq [g_1(x) \quad g_2(x) \quad \cdots \quad g_n(x)]^{\mathrm{T}}$ and the elements are defined as

$$g_i(x) \triangleq \begin{cases} 1, \ x \in [x_i, x_{i+1}] \\ 0, \ x \notin [x_i, x_{i+1}] \end{cases} \quad i \in \mathcal{N} \tag{5.3}$$

where $\bar{x}_i \in (x_i, x_{i+1})$ for $i \in \mathcal{N}$, with $0 = x_1 < x_2 < \cdots < x_n < x_{n+1} = 1$, and $\boldsymbol{L} \triangleq \mathrm{diag}\{l_1, l_2, \cdots, l_n\}$ denotes the observer gain matrix. This represents the core concept of the spatial domain decomposition approach [34, 39], as illustrated in Fig. 5.2.

By defining the state error $e(x,t) \triangleq T(x,t) - \hat{T}(x,t)$ and the output error $\tilde{\boldsymbol{y}}(t) \triangleq \boldsymbol{y}(t) - \hat{\boldsymbol{y}}(t)$, the resulting error system can be expressed as:

$$e_t(x,t) = e_{xx}(x,t) + f(x,t) + d(x,t) - \boldsymbol{g}^{\mathrm{T}}(x)\boldsymbol{L}\int_0^1 \boldsymbol{c}(x)e(x,t)\mathrm{d}x, \tag{5.4a}$$

$$e_x(0,t) = \eta_1 e(0,t), \tag{5.4b}$$

$$e_x(1,t) = -\eta_2 e(1,t), \tag{5.4c}$$

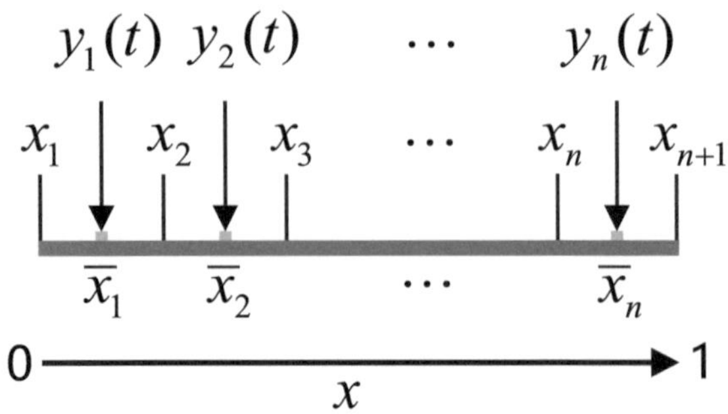

Fig. 5.2 Illustration of the spatial decomposition approach with pointwise measurement

by combining (5.1) with (5.2).

Lemma 5.1 *[38] Given a scalar function $z \in \mathcal{H}^1(0, 1)$, the following inequality holds:*

$$\int_0^1 (z(y) - z(\bar{x}))^2 \mathrm{d}s \leq 4\phi\pi^{-2} \int_0^1 (\mathrm{d}z(y)/\mathrm{d}y)^2 \mathrm{d}y,$$

where $\bar{x} \in [0, 1]$, $\phi \triangleq \max\left\{\bar{x}^2, (1-\bar{x})^2\right\}$.

Lemma 5.2 *[40] Given regular real functions $V(t)$ and $h(t)$,*

$$\dot{V}(t) \leq -\beta V(t) + h(t), \forall t \geq 0$$

implies that

$$V(t) \leq e^{-\alpha t} V(0) + \int_0^t e^{-\beta(t-\upsilon)} h(\upsilon) \mathrm{d}\upsilon, \quad \forall t \geq 0$$

for any finite positive constant α.

Lemma 5.3 (Agmon's inequality [32]) *Given a scalar function $z \in \mathcal{H}^1(0, 1)$, the following inequalities hold:*

$$||z(\cdot, t)||_\infty^2 \leq z^2(0, t) + 2||z(\cdot, t)||_2 ||z_y(\cdot, t)||_2,$$
$$||z(\cdot, t)||_\infty^2 \leq z^2(1, t) + 2||z(\cdot, t)||_2 ||z_y(\cdot, t)||_2.$$

Theorem 5.1 (Fault detectability) *For positive constant γ, if there exists positive scalars $p, q \in (0, \frac{8}{\pi^2})$ and $\bar{l}_i, i \in \mathcal{N}$ such that the following LMIs can be satisfied:*

$$\boldsymbol{\Psi}_i + \frac{p\eta_1 q}{2}\boldsymbol{I} \leq 0, i \in \mathcal{N} \tag{5.5}$$

where

$$\mathbf{\Psi}_i \triangleq \begin{bmatrix} -\frac{p\pi^2}{4\phi_i} & \star & \star & \star \\ (\frac{p\pi^2}{4\phi_i} - \frac{\bar{l}_i}{2}) & -\frac{p\pi^2}{4\phi_i} & \star & \star \\ 0 & \frac{\bar{l}_i}{2} & -p & \star \\ \frac{p}{2} & 0 & -\frac{p}{2} & -\gamma^2 \end{bmatrix}, i \in \mathcal{N},$$

and $\phi_i \triangleq \max\{(\bar{x}_i - x_i)^2, (x_{i+1} - \bar{x}_i)^2\}$*, then* $e(x, t)$ *is UUB in the sense of* $|| \cdot ||_\infty$ *with ultimate bound* $\bar{e}$ *where*

$$\bar{e} \triangleq \sqrt{\frac{2}{p \min\{\eta_1, 1\}\bar{\rho}}}\gamma \bar{d}, \tag{5.6}$$

with

$$\bar{\rho} \triangleq \min\{\eta_1 q, 2 - \frac{q\pi^2}{4}\}. \tag{5.7}$$

The elements l_i *of the observer gain matrix can be calculated according to*

$$l_i = p^{-1}\bar{l}_i, i \in \mathcal{N}. \tag{5.8}$$

Proof Consider the following Lyapunov function:

$$V(t) \triangleq V_1(t) + V_2(t), \tag{5.9}$$

where

$$\begin{aligned} V_1(t) &\triangleq \frac{p}{2}\int_0^1 (e^2(x, t) + e_x^2(x, t))\mathrm{d}x, \\ V_2(t) &\triangleq \frac{p\eta_1}{2} e^2(0, t) + \frac{p\eta_2}{2} e^2(1, t). \end{aligned} \tag{5.10}$$

The time derivative of $V_1(t)$ can be calculated as

$$\dot{V}_1(t) = \int_0^1 pe(x, t)e_t(x, t)\mathrm{d}x + \int_0^1 pe_x(x, t)e_{xt}(x, t)\mathrm{d}x. \tag{5.11}$$

Set

$$\bar{l}_i = pl_i, i \in \mathcal{N}.$$

Using integration by parts, one gets

$$\begin{aligned}\int_0^1 pe(x,t)e_t(x,t)\mathrm{d}x =& \underbrace{\int_0^1 pe(x,t)e_{xx}(x,t)\mathrm{d}x}_{\text{Intergration by parts}+(5.4b)-(5.5c)} - \int_0^1 pe(x,t)\mathbf{g}^{\mathrm{T}}(x)\mathbf{L}\mathrm{d}x\int_0^1 \mathbf{c}(x)e(x,t)\mathrm{d}x \\ &+ \int_0^1 pe(x,t)d(x,t)\mathrm{d}x \\ =& -p\eta_2 e^2(1,t) - p\eta_1 e^2(0,t) - \int_0^1 pe_x^2(x,t)\mathrm{d}x \\ &- \sum_{i=1}^n \int_{x_i}^{x_{i+1}} \bar{l}_i e(x,t)e(\bar{x}_i,t)\mathrm{d}x + \int_0^1 pe(x,t)d(x,t)\mathrm{d}x,\end{aligned} \tag{5.12}$$

in conjunction with (5.4a) and the boundary conditions specified in (5.4b)–(5.4c).

Furthermore, the final term on the right-hand side of (5.11) can be reformulated as

$$\begin{aligned}\underbrace{\int_0^1 pe_x(x,t)e_{xt}(x,t)\mathrm{d}x}_{\text{Intergration by parts}+(5.4b)-(5.4c)} =& -p\eta_2 e(1,t)e_t(1,t) - p\eta_1 e(0,t)e_t(0,t) - \int_0^1 pe_{xx}(x,t)e_t(x,t)\mathrm{d}x \\ =& -p\eta_2 e(1,t)e_t(1,t) - p\eta_1 e(0,t)e_t(0,t) - \int_0^1 pe_{xx}^2(x,t)\mathrm{d}x \\ &\sum_{i=1}^n \int_{x_i}^{x_{i+1}} \bar{l}_i e_{xx}(x,t)e(\bar{x}_i,t)\mathrm{d}x - \int_0^1 pe_{xx}(x,t)d(x,t)\mathrm{d}x,\end{aligned} \tag{5.13}$$

taking into account (5.4a) along with the boundary conditions given in (5.4b)–(5.4c).

By combining (5.12) and (5.13), it follows that

$$\begin{aligned}\dot{V}_1(t) =& -p\eta_2 e^2(1,t) - p\eta_1 e^2(0,t) - p\eta_2 e(1,t)e_t(1,t) - p\eta_1 e(0,t)e_t(0,t) - \int_0^1 pe_x^2(x,t)\mathrm{d}x \\ &- \int_0^1 pe_{xx}^2(x,t)\mathrm{d}x - \sum_{i=1}^n \int_{x_i}^{x_{i+1}} \bar{l}_i e(x,t)e(\bar{x}_i,t)\mathrm{d}x + \sum_{i=1}^n \int_{x_i}^{x_{i+1}} \bar{l}_i e_{xx}(x,t)e(\bar{x}_i,t)\mathrm{d}x \\ &+ \int_0^1 pe(x,t)d(x,t)\mathrm{d}x - \int_0^1 pe_{xx}(x,t)d(x,t)\mathrm{d}x.\end{aligned} \tag{5.14}$$

In addition, it can be determined that

$$\dot{V}_2(t) = p\eta_1 e(0,t)e_t(0,t) + p\eta_2 e(1,t)e_t(1,t). \tag{5.15}$$

By combining (5.14) and (5.15), it can be inferred that

$$\begin{aligned}\dot{V}(t) =& -p\eta_2 e^2(1,t) - p\eta_1 e^2(0,t) - \int_0^1 pe_x^2(x,t)\mathrm{d}x - \int_0^1 pe_{xx}^2(x,t)\mathrm{d}x - \sum_{i=1}^n \int_{x_i}^{x_{i+1}} \bar{l}_i e(x,t)e(\bar{x}_i,t)\mathrm{d}x \\ &+ \sum_{i=1}^n \int_{x_i}^{x_{i+1}} \bar{l}_i e_{xx}(x,t)e(\bar{x}_i,t)\mathrm{d}x + \int_0^1 pe(x,t)d(x,t)\mathrm{d}x - \int_0^1 pe_{xx}(x,t)d(x,t)\mathrm{d}x.\end{aligned} \tag{5.16}$$

Applying Lemma 5.1 to each interval $[x_i, x_{i+1}]$, $i \in \mathcal{N}$, it follows that

$$\int_{x_i}^{x_{i+1}} e_x^2(x,t)\mathrm{d}x \geq \frac{\pi^2}{4\phi_i} \int_{x_i}^{x_{i+1}} (e(x,t) - e(\bar{x}_i, t))^2 \,\mathrm{d}x. \tag{5.17}$$

By substituting (5.17) into (5.16) and taking into account that $0 = x_1 < x_2 < \cdots < x_n < x_{n+1} = 1$ with $\eta_1 > 0$, $\eta_2 > 0$, it follows that

$$\begin{aligned}
\dot{V}(t) \leq & -p\eta_2 e^2(1,t) - p\eta_1 e^2(0,t) - \sum_{i=1}^{n} \int_{x_i}^{x_{i+1}} \frac{p\pi^2}{4\phi_i} e^2(x,t)\mathrm{d}x - \sum_{i=1}^{n} \int_{x_i}^{x_{i+1}} \frac{p\pi^2}{4\phi_i} e^2(\bar{x}_i,t)\mathrm{d}x \\
& - \sum_{i=1}^{n} \int_{x_i}^{x_{i+1}} p e_{xx}^2(x,t)\mathrm{d}x + \sum_{i=1}^{n} \int_{x_i}^{x_{i+1}} (\frac{p\pi^2}{2\phi_i} - \bar{l}_i) e(x,t) e(\bar{x}_i,t)\mathrm{d}x \\
& + \sum_{i=1}^{n} \int_{x_i}^{x_{i+1}} \bar{l}_i e_{xx}(x,t) e(\bar{x}_i,t)\mathrm{d}x + \sum_{i=1}^{n} \int_{x_i}^{x_{i+1}} p e(x,t) d(x,t)\mathrm{d}x \\
& - \sum_{i=1}^{n} \int_{x_i}^{x_{i+1}} p e_{xx}(x,t) d(x,t)\mathrm{d}x \\
= & \sum_{i=1}^{n} \int_{x_i}^{x_{i+1}} \boldsymbol{\xi}_i^T(x,t) \boldsymbol{\Psi}_i \boldsymbol{\xi}_i(x,t)\mathrm{d}x - p\eta_2 e^2(1,t) - p\eta_1 e^2(0,t) + \gamma^2 ||d(\cdot,t)||_2^2,
\end{aligned} \tag{5.18}$$

where

$$\boldsymbol{\xi}_i(x,t) \triangleq [e(x,t) \quad e(\bar{x}_i,t) \quad e_{xx}(x,t) \quad d(x,t)]^{\mathrm{T}}.$$

By inserting (5.5) into (5.18) and taking Assumption 5.1 into account, it follows that

$$\begin{aligned}
\dot{V}(t) \leq & -\frac{p\eta_1 q}{2} \sum_{i=1}^{n} \int_{x_i}^{x_{i+1}} \boldsymbol{\xi}_i^T(x,t) \boldsymbol{\Psi}_i \boldsymbol{\xi}_i(x,t)\mathrm{d}x - p\eta_2 e^2(1,t) - p\eta_1 e^2(0,t) + \gamma^2 \bar{d}^2 \\
\leq & -\frac{p\eta_1 q}{2} (||e(\cdot,t)||_2^2 + ||e_{xx}(\cdot,t)||_2^2) - p\eta_2 e^2(1,t) - p\eta_1 e^2(0,t) + \gamma^2 \bar{d}^2 \\
\leq & -(\eta_1 q \frac{p}{2} ||e(\cdot,t)||_2^2 + \frac{\pi^2 \eta_1 q}{4} \frac{p}{2} ||e_x(\cdot,t)||_2^2 + 2\frac{p\eta_2}{2} e^2(1,t) + (2 - \frac{q\pi^2}{4}) \frac{p\eta_1}{2} e^2(0,t)) + \gamma^2 \bar{d}^2 \\
\leq & -\bar{\rho} V(t) + \gamma^2 \bar{d}^2,
\end{aligned} \tag{5.19}$$

by employing Lemma 5.1 and referring to the definition of $\bar{\rho}$ in (5.7).

Hence one gets

$$V(t) \leq V(0)\exp(-\bar{\rho}t) + \frac{\gamma^2 \bar{d}^2}{\bar{\rho}}, \tag{5.20}$$

by Lemma 5.2.

By referring to Lemma 5.3 and Young's inequality [32], we obtain

$$\begin{aligned}
||e(\cdot,t)||_\infty^2 & \leq e^2(0,t) + 2||e(\cdot,t)||_2 ||e_x(\cdot,t)||_2 \leq e^2(0,t) + ||e(\cdot,t)||_2^2 + ||e_x(\cdot,t)||_2^2 \\
& \leq e_B^2(t) \triangleq \frac{2V(0)}{p\min\{\eta_1, 1\}} \exp(-\bar{\rho}t) + \bar{e}^2,
\end{aligned} \tag{5.21}$$

using (5.20).

From Assumption 5.1, it follows that $0 < \bar{e} < \infty$. Consequently, the state error $e(x, t)$ meets the UUB condition with respect to the $|\cdot|_\infty$ norm, with $\bar{e}$ as the ultimate bound. □

Remark 5.2 It should be noted that when deriving (5.19) and (5.21) using variants of the Poincaré-Wirtinger and Agmon inequalities, respectively, $e(0, t)$ is employed in both cases. One may replace it with $e(1, t)$, which leads to modifications in the LMIs in (5.5).

$$\boldsymbol{\Psi}_i + \frac{p\eta_2 q}{2}\boldsymbol{I} \leq 0, i \in \mathcal{N}$$

the corresponding result holds under this condition. Furthermore, an ultimate bound can be derived using the same procedure; the details are omitted for brevity.

Remark 5.3 The proposed spatial domain decomposition method incorporates boundary measurements at $x = 0$ or $x = 1$ when $\bar{x}_1 = 0$ or $\bar{x}_{n+1} = 1$. In addition, as demonstrated in [38], the approach can be readily extended to handle piecewise measurements. Moreover, it is applicable to homogeneous Dirichlet, homogeneous Neumann, mixed homogeneous Neumann-Dirichlet, and mixed homogeneous Dirichlet-Neumann boundary conditions. These extensions are straightforward and thus not elaborated here.

5.3.3 *FDF Design and Analysis*

For the purpose of fault detection, the residual signal $\boldsymbol{r}(t)$ is intuitively chosen as follows:

$$\boldsymbol{r}(t) = \tilde{\boldsymbol{y}}(t). \tag{5.22}$$

Next, the *lumped* residual evaluation method and its corresponding threshold are presented as follows:

$$J(t) = \int_t^{t+\Delta} |\boldsymbol{r}(\tau)|_2^2 \mathrm{d}\tau, \tag{5.23a}$$

$$J_{\mathrm{th}}(t) = \sup_{f(x,t)=0} J(t), \tag{5.23b}$$

where $\Delta > 0$ is the length of the evaluation time window.

Remark 5.4 Several residual evaluation metrics are commonly employed in fault detection, including the peak value, mean value, root-mean-square (RMS), and the $\mathcal{L}_2$ norm [20]. The RMS and $\mathcal{L}_2$ norms are related and aim to lower the FAR at the expense of fault sensitivity. Notably, $\left(\int t^{t+T} |\boldsymbol{r}(\tau)|^2 \mathrm{d}\tau\right)^{1/2}$ represents a general form of the $\mathcal{L}_2$ norm [20, 41] for the signal $\boldsymbol{r}t)$, measuring its energy over the interval $[t, t+T]$. Since evaluating over the entire time domain is often impractical, introducing a finite evaluation window is a convenient modification [20]. This type of evaluation function has been widely adopted in studies addressing fault detection for dynamic systems, see [42, 43] for reference.

Therefore, the following decision logic ensures reliable fault detection performance:

$$\begin{cases} J(t) > J_{\text{th}}(t) \Longrightarrow \text{faulty at time } t \\ \text{Otherwise} \Longrightarrow \text{fault-free at time } t. \end{cases} \tag{5.24}$$

Theorem 5.2 (Fault detection threshold calculation) *When employing the lumped residual evaluation function in (5.23) and the conditions in Theorem 5.1 are met, the time-dependent detection threshold $J_{th}(t)$ for fault detection can be obtained as*

$$J_{th}(t) = \int_t^{t+\Delta} n e_B^2(\tau)\mathrm{d}\tau. \tag{5.25}$$

Proof The result follows directly from (5.21) and the properties of $||\cdot||_\infty$, and is omitted here for brevity.

Remark 5.5 It should be noted that the detection threshold $J_{\text{th}}(t)$ is designed to vary over time instead of using the commonly adopted constant value [29, 33]. The underlying idea is that the initial error $e_0(x)$ affects $V(0)$, while the time-dependent component in (5.25) decays exponentially. If the threshold were set solely based on the ultimate bound $\bar{e}$, false alarms would inevitably occur during the initial transient phase [29]. To mitigate this, a compensating time-varying term is included in the threshold to reduce the FAR.

5.4 Illustrative Example: Hot Strip Mill Cooling System

5.4.1 System Description

Under the assumptions of *fault-free* and *disturbance-free* conditions, the thermal model of a hot strip mill can be described as follows [24, 44–46]:

$$\rho c \bar{T}_\tau(\xi, \tau) = k \bar{T}_{\xi\xi}(\xi, \tau), \tag{5.26a}$$

$$\bar{T}_\xi(0, \tau) = \frac{h}{k}(\bar{T}(0, \tau) - T_w), \tag{5.26b}$$

$$\bar{T}_\xi(L, \tau) = -\frac{h}{k}(\bar{T}(L, \tau) - T_w), \tag{5.26c}$$

$$\bar{T}(\xi, 0) = \bar{T}_0(\xi), \tag{5.26d}$$

where $\bar{T}(\xi, \tau) \in \mathcal{L}^2([0, L])$ represents the temperature distribution along the thickness of the strip, with L being the strip thickness; $\xi \in [0, L]$ denotes the spatial coordinate, and $\tau \in [0, \infty)$ indicates time. The parameters ρ, k, and c represent the density, average thermal conductivity, and average specific heat, respectively; h is the

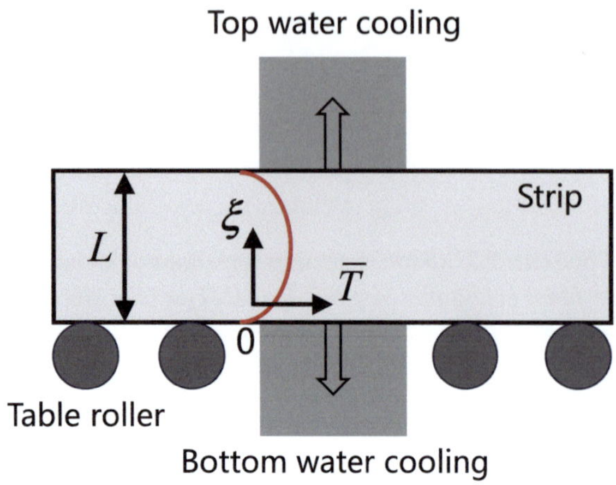

Fig. 5.3 Hot strip mill cooling system

heat transfer coefficient, and T_w denotes the water temperature. Figure 5.3 illustrates the arrangement of the strip on the runout table.

To simplify the analysis, the disturbances and faults present in the process are taken into account, and a scaling transformation is applied [32, 44]

$$\begin{aligned} x &= \frac{\xi}{L}, \\ t &= \frac{k\tau}{\rho c L^2}, \\ T &= \frac{\bar{T} - T_w}{\bar{T}_0 - T_w}, \end{aligned}$$

to system (5.26) leads to the following normalized equation:

$$\begin{aligned} T_t(x,t) &= T_{xx}(x,t) + d(x,t) + f(x,t), \\ T_x(0,t) &= \eta T(0,t), \\ T_x(1,t) &= -\eta T(1,t), \\ T(x,0) &= T_0(x) = 1, \\ y(t) &= T(0,t), \end{aligned}$$

where $\eta \triangleq \frac{hL}{k} > 0$. For practical purposes, the *boundary* temperature measurement $T(0,t)$ at the bottom of the strip is employed for fault detection.

Table 5.1 Process parameter setting

	Definition (Unit)	Value
L	Thickness ([m])	0.01
ρ	Density ([Kg/m^3])	7850
k	Average thermal conductivity ([W/m °C])	28
c	Average specific heat ([J/Kg °C])	932
h	Heat transfer coefficients([W/m^2 °C])	2000
T_w	Temperature of the cooling water ([°C])	30

5.4.2 Parameter Setting

Typically, the temperature of the strip decreases from approximately 900 to 650 °C [44], so the initial condition in (5.26d) is set as $\bar{T}_0(\xi) = 900$ without affecting generality. Additionally, as reported in [44, 47], the parameters employed in the simulation along with their practical values are summarized in Table 5.1.

5.4.3 Simulation Results

The simulation results of system (5.26) under normal operating conditions are shown in Fig. 5.4, with the following disturbance taken into account:

$$d(x,t) = 0.05 \exp\left(-0.5(x-0.5)^2\right) \sin(t).$$

In addition, the following fault is introduced:

$$f(x,t) = b_f(x) f(t),$$

where

$$b_f(x) = H(x-0.5) - H(x-0.75),$$

represents the fault shape function, with $H(\cdot)$ denoting the standard Heaviside function. Two types of faults are considered.

$$\text{Abrupt fault}: f(t) = \begin{cases} 0, & t \in [0, 60) \\ 1, & t \in [60, \infty) \end{cases}$$

$$\text{Periodic fault}: f(t) = \begin{cases} 0, & t \in [0, 60) \\ 1 + \sin(5t), & t \in [60, \infty) \end{cases}$$

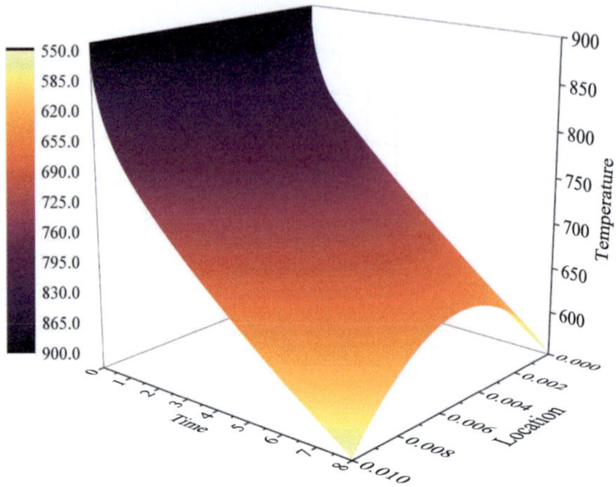

Fig. 5.4 Temperature distribution profile $\bar{T}(\xi, \tau)$ under normal operating condition

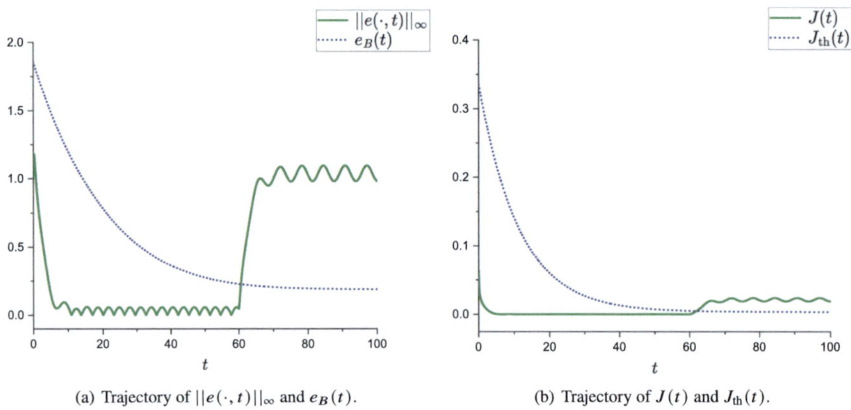

Fig. 5.5 Simulation results with abrupt fault

With the evaluation time window set to $\Delta = 0.1$, the simulation results for both abrupt and periodic faults are shown in Figs. 5.5 and 5.6, respectively. It can be observed that both faults are successfully detected at $t = 60$. Moreover, the results in Figs. 5.5a and 5.6a indicate that $||e(\cdot, t)||_\infty$ remains UUB under fault-free conditions, confirming the validity of Theorem 5.1.

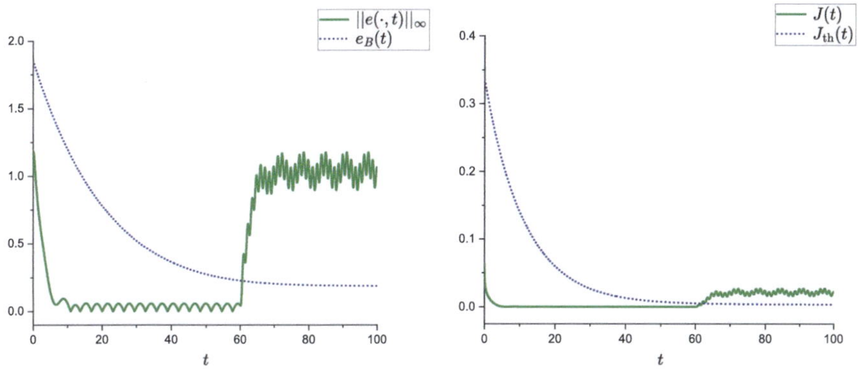

(a) Trajectory of $||e(\cdot,t)||_\infty$ and $e_B(t)$.

(b) Trajectory of $J(t)$ and $J_{\text{th}}(t)$.

Fig. 5.6 Simulation results with periodic fault

5.5 Conclusion

This chapter presents a unified fault detection framework that incorporates both boundary and in-domain sensors, without relying on modal approximation or interpolation. A PDE observer-based FDF is developed, inspired by the FDF design for LPSs. Rigorous theoretical analysis is provided to ensure the reliability of the proposed detection approach.

References

1. El-Farra NH, Ghantasala S (2007) Actuator fault isolation and reconfiguration in transport-reaction processes. AIChE J 53(6):1518–1537
2. Deng H, Li H-X, Chen G (2005) Spectral-approximation-based intelligent modeling for distributed thermal processes. IEEE Trans Control Syst Technol 13(5):686–700
3. Chen L, Li H-X, Yang H-D (2019) Dimension embedded basis function for spatiotemporal modeling of distributed parameter system. IEEE Trans Ind Inform 16(9):5846–5854
4. Chen L, Li H-X, Yang H-D (2020) Spatiotemporal modeling for distributed parameter system under sparse sensing. Ind Eng Chem Res 59(37):16321–16329
5. Dey S, Perez HE, Moura SJ (2017) Model-based battery thermal fault diagnostics: algorithms, analysis, and experiments. IEEE Trans Control Syst Technol 27(2):576–587
6. Wang B-C, Li H-X, Yang H-D (2019) Spatial correlation-based incremental learning for spatiotemporal modeling of battery thermal process. IEEE Trans Ind Electron 67(4):2885–2893
7. Meng X-B, Li H-X, Yang H-D (2018) Evolutionary design of spatio-temporal learning model for thermal distribution in lithium-ion batteries. IEEE Trans Ind Inform 15(5):2838–2848
8. Li H-X, Qi C (2010) Modeling of distributed parameter systems for applications—a synthesized review from time-space separation. J Process Control 20(8):891–901
9. Wang Z, Li H-X, Chen C (2019) Reinforcement learning-based optimal sensor placement for spatiotemporal modeling. IEEE Trans Cybern 50(6):2861–2871
10. Tsai S-H, Wang J-W, Song E-S et al (2019) Robust $\mathcal{H}_\infty$ control for nonlinear hyperbolic PDE systems based on the polynomial fuzzy model. IEEE Trans Cybern 51(7):3789–3801

11. Wang J-W (2018) Dynamic boundary fuzzy control design of semilinear parabolic PDE systems with spatially noncollocated discrete observation. IEEE Trans Cybern 49(8):3041–3051
12. Wang J-W, Wu H-N (2019) Design of suboptimal local piecewise fuzzy controller with multiple constraints for quasi-linear spatiotemporal dynamic systems. IEEE Trans Cybern 51(5):2433–2445
13. Dey S, Perez HE, Moura SJ (2019) Robust fault detection of a class of uncertain linear parabolic PDEs. Automatica 107:502–510
14. El-Farra NH (2006) Integrated fault detection and fault-tolerant control architectures for distributed processes. Ind Eng Chem Res 45(25):8338–8351
15. Liu Z, Liu J, He W (2017) Modeling and vibration control of a flexible aerial refueling hose with variable lengths and input constraint. Automatica 77:302–310
16. Liu Z, He X, Zhao Z et al (2020) Vibration control for spatial aerial refueling hoses with bounded actuators. IEEE Trans Ind Electron 68(5):4209–4217
17. Wang Z-P, Wu H-N (2019) Sampled-data fuzzy control with guaranteed cost for nonlinear parabolic PDE systems via static output feedback. IEEE Trans Fuzzy Syst 28(10):2452–2465
18. Wang Z-P, Wu H-N, Huang T (2020) Sampled-data fuzzy control for nonlinear delayed distributed parameter systems. IEEE Trans Fuzzy Syst 29(10):3054–3066
19. Wang Z-P, Wu H-N, Li H-X (2019) Fuzzy control under spatially local averaged measurements for nonlinear distributed parameter systems with time-varying delay. IEEE Trans Cybern 51(3):1359–1369
20. Ding SX (2013) Model-based fault diagnosis techniques: design schemes, algorithms, and tools, 2nd edn. Springer, London
21. Zhang C, Yang H, Jiang B (2020) Fault estimation and accommodation of fractional-order nonlinear, switched, and interconnected systems. IEEE Trans Cybern 52(3):1443–1453
22. Yang H, Zhang C, An Z et al (2020) Exponential small-gain theorem and fault tolerant safe control of interconnected nonlinear systems. Automatica 115:108866
23. Ding SX (2014) Data-driven design of fault diagnosis and fault-tolerant control systems. Springer, London
24. Dong J, Wang Q, Wang M et al (2018) Data-driven quality monitoring techniques for distributed parameter systems with application to hot-rolled strip laminar cooling process. IEEE Access 6:16646–16654
25. Feng Y, Li H-X (2018) Detection and spatial identification of fault for parabolic distributed parameter systems. IEEE Trans Ind Electron 66(9):7300–7309
26. Feng Y, Li H-X, Yang H-D (2019) Abnormal source identification for parabolic distributed parameter systems. IEEE Trans Syst Man Cybern Syst 51(9):5698–5707
27. Feng Y, Li H-X (2020) Dynamic spatial independent component analysis based abnormality localization for distributed parameter systems. IEEE Trans Ind Inform 16(5):2929–2936
28. Ray WH (1981) Advanced process control. McGraw-Hill, New York
29. Ghantasala S, El-Farra NH (2009) Robust actuator fault isolation and management in constrained uncertain parabolic PDE systems. Automatica 45(10):2368–2373
30. Fischer F, Deutscher J (2020) Flatness-based algebraic fault diagnosis for distributed-parameter systems. Automatica 117:108987
31. Feng Y, Wang Y, Wang JW et al (2020) Abnormal spatio-temporal source estimation for a linear unstable parabolic distributed parameter system, an adaptive PDE observer perspective. J Franklin Inst 358(2):1656–1672
32. Krstic M, Smyshlyaev A (2008) Boundary control of PDEs: a course on backstepping designs. SIAM, Philadelphia
33. Cai J, Ferdowsi H, Sarangapani J (2016) Model-based fault detection, estimation, and prediction for a class of linear distributed parameter systems. Automatica 66:122–131
34. Wang J-W, Liu Y-Q, Hu Y-Y et al (2017) A spatial domain decomposition approach to distributed $\mathcal{H}_\infty$ observer design of a linear unstable parabolic distributed parameter system with spatially discrete sensors. Int J Control 90(12):2772–2785
35. Wang J-W, Liu Y-Q, Sun C-Y (2018) Pointwise exponential stabilization of a linear parabolic PDE system using non-collocated pointwise observation. Automatica 93:197–210

36. Deutscher J (2016) Fault detection for linear distributed-parameter systems using finite-dimensional functional observers. Int J Control 89(3):550–563
37. Smyshlyaev A, Krstic M (2005) Backstepping observers for a class of parabolic PDEs. Syst Control Lett 54(7):613–625
38. Wang J-W (2021) A unified Lyapunov-based design for a dynamic compensator of linear parabolic MIMO PDEs. Int J Control 94(7):1804–1811
39. Wang J, Wang J (2020) Dynamic compensator design of linear parabolic MIMO PDEs in N-dimensional spatial domain. IEEE Trans Autom Control 66(3):1399–1406
40. Ioannou PA, Sun J (2012) Robust adaptive control. Courier Corporation, Chelmsford
41. Li L, Ding SX, Peng K et al (2019) Fuzzy fault detection filter design for nonlinear distributed parameter systems. IEEE Access 7:11105–11113
42. Li L, Ding SX, Qiu J et al (2016) Fuzzy observer-based fault detection design approach for nonlinear processes. IEEE Trans Syst Man Cybern Syst 47(8):1941–1952
43. Li L, Ding SX (2020) Gap metric techniques and their application to fault detection performance analysis and fault isolation schemes. Automatica 118:109029
44. Kumar RK, Sinha SK, Lahiri AK (1997) Modeling of the cooling process on the runout table of a hot strip mill—a parallel approach. IEEE Trans Ind Appl 33(3):807–814
45. Hashimoto T, Yoshioka Y, Ohtsuka T (2012) Receding horizon control for hot strip mill cooling systems. IEEE/ASME Trans Mechatron 18(3):998–1005
46. Yang Z, Luo X (2019) Parallel numerical calculation on GPU for the 3-dimensional mathematical model in the walking beam reheating furnace. IEEE Access 7:44583–44595
47. Prieto MM, Ruiz LS, Menendez JA (2001) Thermal performance of numerical model of hot strip mill runout table. Ironmaking Steelmaking 28(6):474–480

Chapter 6
Inverse S-T Model-Based Abnormal Source Identification for Parabolic DPSs

Abstract Detecting abnormal sources within distributed parameter systems is a type of inverse source problem. Despite its industrial importance, studies on this topic remain scarce. This chapter presents the first effort to identify spatio-temporal anomalies in a particular class of distributed parameter systems. A new inverse spatio-temporal model is introduced, including an adaptive state observer for source detection and an adaptive estimation algorithm. One main advantage is that the method requires only system outputs, without direct measurement of system states. Theoretical analysis confirms that the estimation error converges. Finally, the approach is validated on a heat transfer rod containing an abnormal spatio-temporal source.

6.1 Introduction

Industrial systems such as thermal processes and transport-reaction operations can be represented by DPSs, where inputs, outputs, and parameters vary across time and space, exhibiting S-T dynamics [1–8]. Unexpected events in DPSs may lead to controller malfunctions or undesired system responses, posing risks to safe and reliable operation. These anomalies can generally be interpreted as arising from an unknown S-T source $f(z, t)$ within the system, equivalently considered as a S-T fault. The abnormal source term $f(z, t)$ represents unknown factors that may trigger adverse behaviors, including actuator or sensor faults and coupled disturbances or noise within system dynamics. Identifying such sources has potential applications in chemical process monitoring [9, 10], Li-ion battery FD [11], and control of vibrating single-link flexible manipulators [12]. On one hand, detecting and estimating DPSs' abnormal sources is critical for industrial use, yet remains insufficiently explored. On the other hand, disturbance observer-based control (DOBC) for nonlinear parabolic PDEs has been studied [13–15], where disturbances were generated by known ODE [13, 14] or PDE [15] exosystems with unknown initial states.

The abnormal S-T source resembles the source term in inverse source problems for the wave equation, where a modulating functions-based approach has been applied [16]. Nevertheless, this technique relies on complete state measurements and

© The Author(s), under exclusive license to Springer Nature Singapore Pte Ltd. 2026
Y. Feng et al., *Spatio-Temporal Abnormality Diagnosis for Industrial Distributed Parameter Systems*, https://doi.org/10.1007/978-981-95-3750-1_6

the computation of time derivatives of the observed signals, which is challenging to implement in practical industrial settings [17].

Over recent decades, actuator and sensor fault detection and diagnosis in DPSs have received increasing attention, with several studies reported [9, 10, 18–21]. However, identifying and detecting faults in DPSs is inherently less challenging than locating abnormal S-T sources, mainly because the spatial distribution of actuator or sensor faults is typically not taken into account.

- In the actuator fault case, the spatial distribution function of the fault was assumed to be the same as that of the actuator and was known a prior;
- In the sensor fault case, the spatial distribution characteristic of the fault was not considered for the most commonly used point-wise measurement sensors.

Hence, actuator and sensor FD in DPSs is generally restricted to the temporal domain, whereas detecting abnormal S-T sources requires analysis across both time and space. Existing strategies can be grouped into two main categories: one employs a finite-dimensional ODE representation of the DPS, and the other works directly with the original PDE formulation. For instance, a finite-dimensional residual generator was proposed in [22] for fault detection in linear DPSs. A model-based fault detection scheme was developed in [23], utilizing a state observer derived from the original PDE system. By applying the modulating functions technique [24] to the PDE system, algebraic expressions for identifying actuator or sensor faults were obtained in [25–27]. Despite these advances, research on detecting and identifying abnormal S-T sources in DPSs remains limited, even though it is highly relevant for practical applications.

Recently, data-driven techniques have been applied to detect abnormal S-T sources in DPSs for engineering purposes [28, 29]. However, the problem of identifying these sources–estimating the unknown term $f(z, t)$–has not yet been addressed, even though it is essential for removing anomalies and restoring the process. Furthermore, source identification is more challenging than detection because it requires dynamically tracking the abnormal S-T source rather than merely determining detection thresholds based on residuals [28, 29].

Adaptive observers [30–33] are widely used to estimate unknown terms in systems described by ODEs. Nevertheless, due to the infinite-dimensional nature of DPSs, design methods developed for ODE-based observers cannot be directly applied to PDE-modeled DPSs. Parabolic DPSs, a representative subclass, have spatial differential operators that can be decomposed into a finite-dimensional slow subsystem and an infinite-dimensional fast subsystem [34]. This property offers a promising avenue for employing adaptive observers to identify abnormal S-T sources.

The benefits of employing adaptive observers for identifying abnormal S-T sources, compared with modulating-functions techniques [16], include:

- First, no state measurement is needed, so fewer sensors are required in industrial applications;
- Second, neither the time or space derivative of the measurement is required, which will improve the identification performance.

Motivated by the considerations discussed above, this chapter presents the first study on identifying abnormal S-T sources in a class of linear parabolic DPSs. An inverse S-T model is developed based on the separation of variables principle. The model incorporates an adaptive state observer for source detection along with an adaptive algorithm for source estimation. The feasibility and effectiveness of the proposed approach are verified through both theoretical analysis and numerical simulations.

The remainder of this chapter is organized as follows: Sect. 6.2 outlines the problem formulation. Section 6.3.1 presents the design of the inverse S-T model for abnormal source identification. Theoretical analysis and numerical simulations are provided in Sect. 6.3.2 and Sect. 6.4, respectively. Finally, Sect. 6.5 concludes the chapter

6.2 Preliminaries

Consider a class of linear DPSs modeled by the following parabolic PDE:

$$\begin{aligned} \frac{\partial x(z,t)}{\partial t} &= a_1\frac{\partial x(z,t)}{\partial z} + a_2\frac{\partial^2 x(z,t)}{\partial z^2} + a_3 x(z,t) + k_u \boldsymbol{b}_u^{\mathrm{T}}(z)\boldsymbol{u}(t) + f(z,t), \\ \boldsymbol{y}(t) &= \int_{\alpha_1}^{\alpha_2} \boldsymbol{c}(z) k_y x(z,t)\mathrm{d}z, \end{aligned} \tag{6.1}$$

subject to the following boundary conditions:

$$\begin{aligned} c_1 x(\alpha_1,t) + d_1\frac{\partial x}{\partial z}(\alpha_1,t) &= r_1, \\ c_2 x(\alpha_2,t) + d_2\frac{\partial x}{\partial z}(\alpha_2,t) &= r_2, \end{aligned} \tag{6.2}$$

and with the following initial condition:

$$x(z,0) = x_0(z), \tag{6.3}$$

where $x(z,t)$ denotes the system state, with the spatial domain defined as $[\alpha_1, \alpha_2]$, z representing the spatial coordinate, and t the temporal variable. The vector $\boldsymbol{u}(t) \in \mathbb{R}^{n_u}$ corresponds to the control input, while $f(z,t) \in \mathbb{R}$ represents the unknown abnormal S-T source in the DPS, which is responsible for abnormal behaviors or events. To distinguish it from the manipulated input $\boldsymbol{u}(t)$, $f(z,t)$ is defined as the **"abnormal source"**, whereas $\boldsymbol{u}(t)$ is considered the **"normal source"**. The abnormal source $f(z,t)$ is independent of the state $x(z,t)$ and cannot be directly measured. The system output is $\boldsymbol{y}(t) \in \mathbb{R}^{n_y}$, and $\partial x/\partial z$ and $\partial^2 x/\partial z^2$ denote the first- and second-order spatial derivatives of the state, respectively. Constant coefficients include

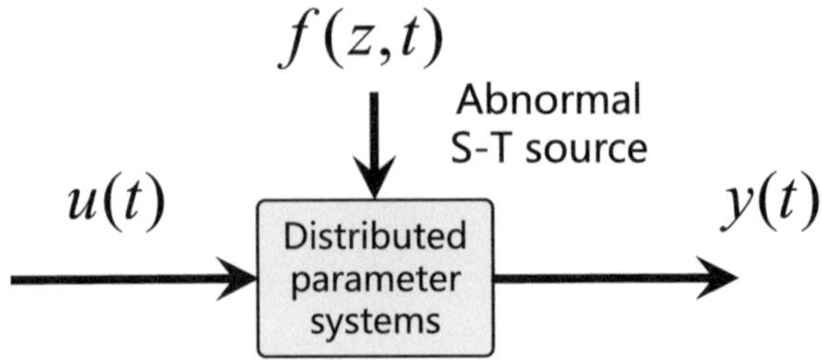

Fig. 6.1 System description

$a_1, a_2, a_3, k_u, k_y, c_1, c_2, d_1, d_2, r_1$, and r_2. The i-th component of the known smooth function $\boldsymbol{b}_u(z) \in \mathbb{R}^{n_u}$ describes how the i-th control input is distributed over the spatial domain, and the i-th component of $\boldsymbol{c}(z) \in \mathbb{R}^{n_y}$ is determined by the corresponding sensor type (point or distributed). The initial condition is given by $x_0(z)$. A schematic of the system is shown in Fig. 6.1.

The following Hilbert Space is defined throughout this chapter:

$$\mathcal{H} \triangleq \mathcal{L}_2([\alpha_1, \alpha_2]; \mathbb{R}).$$

In this Hilbert space, the corresponding inner product and norm are defined as follows:

$$< x_1(\cdot), x_2(\cdot) > \triangleq \int_{\alpha_1}^{\alpha_2} x_1(z)x_2(z)\mathrm{d}z,$$

$$\|x_1(\cdot)\|_2 \triangleq < x_1(\cdot), x_1(\cdot) >^{1/2},$$

where $x_1(\cdot)$ and $x_2(\cdot)$ are two elements of $\mathcal{H}$.

The problem investigated in this chapter can be summarized as follows:

Utilize the system output $\boldsymbol{y}(t)$ to design an inverse S-T model to identify the abnormal S-T source $f(z, t)$ of systems,

which follows the system dynamics in (6.1) while satisfying the boundary conditions in (6.2) and the initial condition in (6.3).

6.3 Inverse S-T Modeling

6.3.1 *Inverse S-T Model Design for Abnormal S-T Source Identification*

In this section, an inverse S-T model is proposed to estimate the unknown source term $f(z, t)$ using only the system output $\boldsymbol{y}(t)$, without requiring direct measurement of the state.

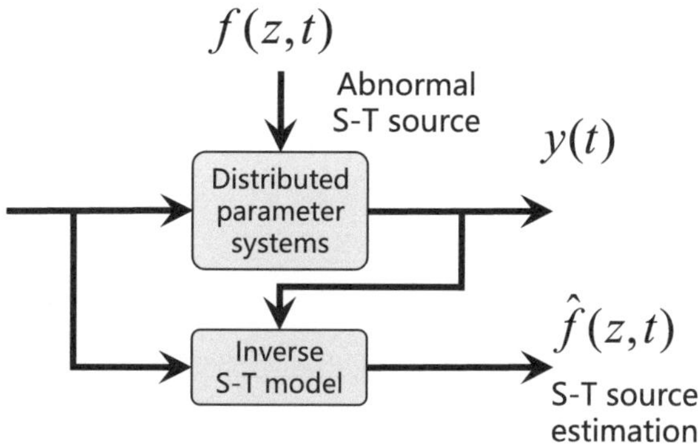

Fig. 6.2 Framework of the inverse S-T model

Inspired by adaptive observer theory [30–33], the inverse S-T model depicted in Fig. 6.2 is constructed using an adaptive state observer for the purpose of source identification.

Taking into account the infinite-dimensional nature of the linear DPSs described in (6.1)–(6.3), an approximate finite-dimensional model capturing the dominant dynamics of the original system is first derived to facilitate the design of the adaptive state observer as follows:

$$\begin{aligned} \dot{\boldsymbol{x}}_s(t) &= \boldsymbol{A}_s\boldsymbol{x}_s(t) + \boldsymbol{B}_{u,s}\boldsymbol{u}(t) + \boldsymbol{f}_s(t), \\ \boldsymbol{y}_s(t) &= \boldsymbol{C}_s\boldsymbol{x}_s(t), \end{aligned} \tag{6.4}$$

and the advantages include:

- This allows the use of finite-dimensional observers for abnormal S-T source identification, avoiding the need for complex infinite-dimensional observer design.

In industrial applications, using the dominant dynamics model is sufficient to achieve satisfactory performance without additional design effort. Further details on the model reduction are provided in the Appendix.

Remark 6.1 (Time/space decoupled form of the abnormal S-T source)
Based on the definitions of $\boldsymbol{f}_s(t)$ and $\boldsymbol{f}_f(t)$ in (6.30), one can readily derive:

$$f(z,t) = \boldsymbol{\phi}^{\mathrm{T}}(z)\boldsymbol{f}(t) = \boldsymbol{\phi}_s^{\mathrm{T}}(z)\boldsymbol{f}_s(t) + \boldsymbol{\phi}_f^{\mathrm{T}}(z)\boldsymbol{f}_f(t).$$

In this approach, the unknown abnormal S-T source can be represented as a combination of the basis functions $\boldsymbol{\phi}(z)$ and the associated temporal coefficients $\boldsymbol{f}(t)$. Consequently, identifying the abnormal S-T source is reduced to determining the temporal coefficients $\boldsymbol{f}(t)$.

Inspired by the adaptive FD observers presented in [30–33], an adaptive state observer for the system in (6.4) is formulated as follows:

$$\begin{aligned} \dot{\hat{\boldsymbol{x}}}_s(t) &= \boldsymbol{A}_s\hat{\boldsymbol{x}}_s(t) + \boldsymbol{B}_{u,s}\boldsymbol{u}(t) + \hat{\boldsymbol{f}}_s(t) - \boldsymbol{L}(\hat{\boldsymbol{y}}_s(t) - \boldsymbol{y}(t)), \\ \hat{\boldsymbol{y}}_s(t) &= \boldsymbol{C}_s\hat{\boldsymbol{x}}_s(t), \end{aligned} \tag{6.5}$$

where $\hat{\boldsymbol{x}}_s(t) \in \mathbb{R}^m$, $\hat{\boldsymbol{y}}_s(t) \in \mathbb{R}^{n_y}$, and $\hat{\boldsymbol{f}}_s(t) \in \mathbb{R}^m$ are the estimates of $\boldsymbol{x}_s(t)$, $\boldsymbol{y}_s(t)$, and $\boldsymbol{f}_s(t)$, respectively. $\boldsymbol{L} \in \mathbb{R}^{m\times n_y}$ is the observer gain matrix.

Define $\boldsymbol{e}_x(t) = \hat{\boldsymbol{x}}_s(t) - \boldsymbol{x}_s(t)$, $\boldsymbol{e}_y(t) = \hat{\boldsymbol{y}}_s(t) - \boldsymbol{y}(t)$, and $\boldsymbol{e}_f(t) = \hat{\boldsymbol{f}}_s(t) - \boldsymbol{f}_s(t)$, then the error dynamics are obtained by combining (6.4) and (6.5):

$$\begin{aligned} \dot{\boldsymbol{e}}_x(t) &= (\boldsymbol{A}_s - \boldsymbol{L}\boldsymbol{C}_s)\boldsymbol{e}_x(t) + \boldsymbol{e}_f(t) + \boldsymbol{L}\boldsymbol{y}_f(t), \\ \boldsymbol{e}_y(t) &= \boldsymbol{C}_s\boldsymbol{e}_x(t) - \boldsymbol{y}_f(t). \end{aligned} \tag{6.6}$$

Remark 6.2 It should be noted that the output error $\boldsymbol{e}_y(t)$, which serves in the design of the adaptive state observer, is defined as:

$$\boldsymbol{e}_y(t) = \hat{\boldsymbol{y}}_s(t) - \boldsymbol{y}(t),$$

instead of using the error between the slow-system output estimate $\hat{\boldsymbol{y}}_s(t)$ and the actual slow-system output $\boldsymbol{y}_s(t)$. The reason is that $\boldsymbol{y}_s(t)$ in (6.4) cannot be directly measured, whereas the original PDE system output $\boldsymbol{y}(t)$ in (6.1) is accessible via measurement sensors. This leads to $\boldsymbol{e}_y(t) \neq \boldsymbol{C}_s\boldsymbol{e}_x(t)$, which differs from the equality assumed in [30–33]. Therefore, as indicated in (6.6), the fast subsystem output $\boldsymbol{y}_f(t)$ must be incorporated in the adaptive state observer design, introducing additional errors in source identification. In practice, a common remedy is to select a sufficiently high order m for the slow subsystem.

Then the following adaptive algorithm for source estimation is introduced:

$$\dot{\hat{\boldsymbol{f}}}_s(t) = -\boldsymbol{\Gamma}\boldsymbol{F}(\dot{\boldsymbol{e}}_y(t) + \sigma\boldsymbol{e}_y(t)), \tag{6.7}$$

where $\boldsymbol{F} \in \mathbb{R}^{m\times n_y}$, and the symmetric positive definite matrix $\boldsymbol{\Gamma} \in \mathbb{R}^{m\times m}$ represents the learning rate.

Remark 6.3 It can be observed that the adaptive source estimation algorithm in (6.7) comprises both a proportional and an integral component of $\boldsymbol{e}_y(t)$, and it can be reformulated as:

$$\hat{\boldsymbol{f}}_s(t) = -\boldsymbol{\Gamma}\boldsymbol{F}(\boldsymbol{e}_y(t) + \sigma\int_0^t \boldsymbol{e}_y(\tau)\mathrm{d}\tau).$$

The proportional component is incorporated to enhance the speed of abnormal source estimation.

Finally, the estimation of the abnormal S-T source can be achieved through a **time-space synthesis**:

$$\hat{f}(z,t) = \boldsymbol{\phi}_s{}^{\mathrm{T}}(z)\hat{\boldsymbol{f}}_s(t). \tag{6.8}$$

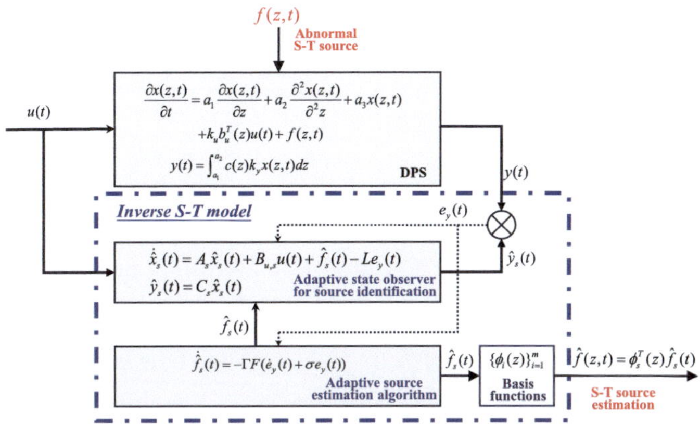

Fig. 6.3 Schematic graph of the proposed inverse S-T model-based method for source identification

Remark 6.4 It should be noted that a common assumption for the abnormal S-T source $f(z,t)$ is that it can be expressed using the following basis function expansion:

$$f(z,t) = \sum_{i=1}^{n} f_i(t)\upsilon_i(z), \ 1 \leqslant n < \infty$$

and a finite set of basis functions $\{\upsilon_i(z)\}_{i=1}^{n}$ is considered, where n remains unknown for abnormal S-T source identification. Unlike the polynomial-type basis functions employed in [16], eigenfunctions $\{\phi_i(z)\}_{i=1}^{n}$ are adopted as the basis, i.e. $\upsilon_i(z) = \phi_i(z), i = 1, \cdots, n$. When the slow subsystem order satisfies $m \geq n$, it follows that $f_f(t) = 0$. Since the inverse S-T model relies on an approximate finite-dimensional representation, $f_f(t)$ is neglected in the source identification. To further enhance identification performance, a sufficiently large m can be chosen, which also reduces the effects of disregarding the fast subsystem.

A schematic of the proposed inverse S-T model-based approach for abnormal source identification is presented in Fig. 6.3. The model comprises the adaptive state observer for source identification in (6.5) and the adaptive source estimation algorithm in (6.7). Notably, no state measurements are required in the proposed inverse S-T model.

6.3.2 *Theoretical Analysis*

The following assumptions and lemma are considered necessary for the design of the proposed inverse S-T model.

Assumption 6.1 The output of the $\boldsymbol{x}_f$-subsystem $\boldsymbol{y}_f(t)$ and its derivative with respect to time $\dot{\boldsymbol{y}}_f(t)$ satisfy that:

$$\left\|\boldsymbol{y}_f(t)\right\|_{\text{peak}} \triangleq \sup_t ||\boldsymbol{y}_f(t)|| < \infty,$$
$$\left\|\dot{\boldsymbol{y}}_f(t)\right\|_{\text{peak}} \triangleq \sup_t ||\dot{\boldsymbol{y}}_f(t)|| < \infty, \ \forall t \geqslant 0$$

where $\|\cdot\|_{\text{peak}}$ and $\|\cdot\|$ denote the so-called peak-norm [35] and Euclidean norm, respectively.

Assumption 6.2 The derivative of $\boldsymbol{f}_s(t)$ with respect to time is norm bounded, i.e.

$$\left\|\dot{\boldsymbol{f}}_s(t)\right\|^2 \leqslant f_1, \ \forall t \geq 0$$

where $f_1 \in [0, \infty)$ is a constant.

Assumption 6.3 The abnormal S-T source $f(z, t)$ satisfies that:

$$\|f(z,t)\|_2^2 = \left\|\boldsymbol{f}_s(t)\right\|^2 + \left\|\boldsymbol{f}_f(t)\right\|^2 \leqslant f_2, \ \forall t \geq 0$$

where $f_2 \in [0, \infty)$ is a constant.

Assumption 6.4 $(\boldsymbol{A}_s, \boldsymbol{C}_s)$ is observable and $\boldsymbol{C}_s$ is of full column rank.

Remark 6.5 In Assumption 6.4, it is typically required that $\boldsymbol{C}_s$ has full column rank, a condition also referred to as output separability in fault isolation [36, 37]. This requirement can be fulfilled by an appropriate selection of measurement sensors and dominant modes.

Lemma 6.1 *[31] For a given positive scalar $\mu > 0$ and a symmetric positive definite matrix $\boldsymbol{P}$, the following inequality holds:*

$$2\boldsymbol{x}^{\mathrm{T}}\boldsymbol{y} \leqslant \frac{1}{\mu}\boldsymbol{x}^{\mathrm{T}}\boldsymbol{P}\boldsymbol{x} + \mu\boldsymbol{y}^{\mathrm{T}}\boldsymbol{P}^{-1}\boldsymbol{y}, \ \boldsymbol{x}, \boldsymbol{y} \in \mathbb{R}^n.$$

Lemma 6.2 *[38] Let $V(t)$ and $g(t)$ be real functions. Then*

$$\dot{V}(t) \leq -\alpha V(t) + g(t), \forall t \geq 0$$

implies that

$$V(t) \leq e^{-\alpha t}V(0) + \int_0^t e^{-\alpha(t-\tau)}g(\tau)d\tau, \quad \forall t \geq 0$$

for any finite constant α.

Theorem 6.1 *Under Assumptions 6.1, 6.2, 6.4, and 6.5, given scalars* $\mu_1, \mu_2, \sigma > 0$*, if there exist symmetric positive definite matrices* $\boldsymbol{P} \in \mathbb{R}^{m\times m}$*,* $\boldsymbol{G}_1 \in \mathbb{R}^{m\times m}$*,* $\boldsymbol{G}_2 \in \mathbb{R}^{m\times m}$*, matrices* $\boldsymbol{X} \in \mathbb{R}^{m\times n_y}$*,* $\boldsymbol{F} \in \mathbb{R}^{m\times n_y}$*, and a positive constant* ε_1 *such that the following LMI is satisfied:*

$$\boldsymbol{\Xi} \triangleq \begin{bmatrix} \boldsymbol{\Xi}_{11} & * & * \\ \frac{1}{\sigma}(\boldsymbol{X}\boldsymbol{C}_s - \boldsymbol{P}\boldsymbol{A}_s) & \boldsymbol{\Xi}_{22} & * \\ \boldsymbol{X}^{\mathrm{T}} & \boldsymbol{F}^{\mathrm{T}} - \frac{1}{\sigma}\boldsymbol{X}^{\mathrm{T}} & -\varepsilon_1 \boldsymbol{I} \end{bmatrix} < 0, \tag{6.9}$$

where

$$\boldsymbol{\Xi}_{11} \triangleq \boldsymbol{P}\boldsymbol{A}_s + \boldsymbol{A}_s^{\mathrm{T}}\boldsymbol{P} - \boldsymbol{X}\boldsymbol{C}_s - \boldsymbol{C}_s^{\mathrm{T}}\boldsymbol{X}^{\mathrm{T}},$$

$$\boldsymbol{\Xi}_{22} \triangleq -\frac{2}{\sigma}\boldsymbol{P} + \frac{1}{\sigma\mu_1}\boldsymbol{G}_1 + \frac{1}{\sigma\mu_2}\boldsymbol{G}_2,$$

$$\boldsymbol{X} = \boldsymbol{P}\boldsymbol{L}, \tag{6.10}$$

and the following condition holds:

$$\boldsymbol{P} = \boldsymbol{F}\boldsymbol{C}_s, \tag{6.11}$$

then the adaptive source estimation algorithm in (6.7) can realize $\boldsymbol{e}_x(t)$ *and* $\boldsymbol{e}_f(t)$ *UUB where the symmetric positive definite matrix* $\boldsymbol{\Gamma} \in \mathbb{R}^{m\times m}$ *denotes the learning rate.*

Proof Choose the Lyapunov candidate as follows:

$$V(t) = \boldsymbol{e}_x^{\mathrm{T}}(t)\boldsymbol{P}\boldsymbol{e}_x(t) + \frac{1}{\sigma}\boldsymbol{e}_f^{\mathrm{T}}(t)\boldsymbol{\Gamma}^{-1}\boldsymbol{e}_f(t). \tag{6.12}$$

Combining (6.6), (6.7), (6.10), and (6.11), its derivative with respect to time is:

$$\begin{aligned} \dot{V}(t) &= \boldsymbol{e}_x^{\mathrm{T}}(t)(\boldsymbol{P}\boldsymbol{A}_s + \boldsymbol{A}_s^{\mathrm{T}}\boldsymbol{P} - \boldsymbol{X}\boldsymbol{C}_s - \boldsymbol{C}_s^{\mathrm{T}}\boldsymbol{X}^{\mathrm{T}})\boldsymbol{e}_x(t) \\ &\quad + \frac{2}{\sigma}\boldsymbol{e}_f^{\mathrm{T}}(t)(\boldsymbol{X}\boldsymbol{C}_s - \boldsymbol{P}\boldsymbol{A}_s)\boldsymbol{e}_x(t) + 2\boldsymbol{y}_f^{\mathrm{T}}(t)\boldsymbol{X}^{\mathrm{T}}\boldsymbol{e}_x(t) \\ &\quad + 2\boldsymbol{y}_f^{\mathrm{T}}(t)(\boldsymbol{F}^{\mathrm{T}} - \frac{1}{\sigma}\boldsymbol{X}^{\mathrm{T}})\boldsymbol{e}_f(t) - \frac{2}{\sigma}\boldsymbol{e}_f^{\mathrm{T}}(t)\boldsymbol{\Gamma}^{-1}\dot{\boldsymbol{f}}_s(t) \\ &\quad - \frac{2}{\sigma}\boldsymbol{e}_f^{\mathrm{T}}(t)\boldsymbol{P}\boldsymbol{e}_f(t) + \frac{2}{\sigma}\boldsymbol{e}_f^{\mathrm{T}}(t)\boldsymbol{F}\dot{\boldsymbol{y}}_f(t). \end{aligned} \tag{6.13}$$

Combining Lemma 6.1, Assumptions 6.1, and 6.2, it can be obtained that:

$$\begin{aligned} -\frac{2}{\sigma}\boldsymbol{e}_f^{\mathrm{T}}(t)\boldsymbol{\Gamma}^{-1}\dot{\boldsymbol{f}}_s(t) &\leqslant \frac{1}{\sigma\mu_1}\boldsymbol{e}_f^{\mathrm{T}}(t)\boldsymbol{G}_1\boldsymbol{e}_f(t) + \frac{\mu_1}{\sigma}\dot{\boldsymbol{f}}_s^{\mathrm{T}}(t)\boldsymbol{\Gamma}^{-1}\boldsymbol{G}_1^{-1}\boldsymbol{\Gamma}^{-1}\dot{\boldsymbol{f}}_s(t) \\ &\leqslant \frac{1}{\sigma\mu_1}\boldsymbol{e}_f^{\mathrm{T}}(t)\boldsymbol{G}_1\boldsymbol{e}_f(t) + \frac{\mu_1}{\sigma}f_1\lambda_{\max}(\boldsymbol{\Gamma}^{-1}\boldsymbol{G}_1^{-1}\boldsymbol{\Gamma}^{-1}). \end{aligned} \tag{6.14}$$

$$\begin{aligned}\frac{2}{\sigma}\boldsymbol{e}_f^{\mathrm{T}}(t)\boldsymbol{F}\dot{\boldsymbol{y}}_f(t) &\leqslant \frac{1}{\sigma\mu_2}\boldsymbol{e}_f^{\mathrm{T}}(t)\boldsymbol{G}_2\boldsymbol{e}_f(t)+\frac{\mu_2}{\sigma}\dot{\boldsymbol{y}}_f^{\mathrm{T}}(t)\boldsymbol{F}^{\mathrm{T}}\boldsymbol{G}_2^{-1}\boldsymbol{F}\dot{\boldsymbol{y}}_f(t)\\ &\leqslant \frac{1}{\sigma\mu_2}\boldsymbol{e}_f^{\mathrm{T}}(t)\boldsymbol{G}_2\boldsymbol{e}_f(t)+\frac{\mu_2}{\sigma}\lambda_{\max}(\boldsymbol{F}^{\mathrm{T}}\boldsymbol{G}_2^{-1}\boldsymbol{F})\left\|\dot{\boldsymbol{y}}_f(t)\right\|_{\text{peak}}^2.\end{aligned} \tag{6.15}$$

Substituting (6.14) and (6.15) into (6.13) and considering Assumption 6.1 yields:

$$\dot{V}(t)\leqslant \boldsymbol{\xi}^{\mathrm{T}}(t)\boldsymbol{\Xi}\boldsymbol{\xi}(t)+\beta+\varepsilon_1\left\|\boldsymbol{y}_f(t)\right\|_{\text{peak}}^2+\varepsilon_2\left\|\dot{\boldsymbol{y}}_f(t)\right\|_{\text{peak}}^2, \tag{6.16}$$

where

$$\begin{aligned}\boldsymbol{\xi}(t) &\triangleq \begin{bmatrix}\boldsymbol{e}_x(t)\\ \boldsymbol{e}_f(t)\\ \boldsymbol{y}_f(t)\end{bmatrix},\\ \boldsymbol{\Xi} &\triangleq \begin{bmatrix}\boldsymbol{\Xi}_{11} & * & *\\ \frac{1}{\sigma}(\boldsymbol{X}\boldsymbol{C}_s-\boldsymbol{P}\boldsymbol{A}_s) & \boldsymbol{\Xi}_{22} & *\\ \boldsymbol{X}^{\mathrm{T}} & \boldsymbol{F}^{\mathrm{T}}-\frac{1}{\sigma}\boldsymbol{X}^{\mathrm{T}} & -\varepsilon_1\boldsymbol{I}\end{bmatrix},\\ \boldsymbol{\Xi}_{11} &\triangleq \boldsymbol{P}\boldsymbol{A}_s+\boldsymbol{A}_s^{\mathrm{T}}\boldsymbol{P}-\boldsymbol{X}\boldsymbol{C}_s-\boldsymbol{C}_s^{\mathrm{T}}\boldsymbol{X}^{\mathrm{T}},\\ \boldsymbol{\Xi}_{22} &\triangleq -\frac{2}{\sigma}\boldsymbol{P}+\frac{1}{\sigma\mu_1}\boldsymbol{G}_1+\frac{1}{\sigma\mu_2}\boldsymbol{G}_2,\\ \beta &\triangleq \frac{\mu_1}{\sigma}f_1\lambda_{\max}(\boldsymbol{\Gamma}^{-1}\boldsymbol{G}_1^{-1}\boldsymbol{\Gamma}^{-1}),\ \varepsilon_2\triangleq\frac{\mu_2}{\sigma}\lambda_{\max}(\boldsymbol{F}^{\mathrm{T}}\boldsymbol{G}_2^{-1}\boldsymbol{F}).\end{aligned}$$

Hence, when $\boldsymbol{\Xi}<0$, one can obtain that:

$$\begin{aligned}\dot{V}(t) &\leqslant -\lambda_{\min}(-\boldsymbol{\Xi})\|\boldsymbol{\xi}(t)\|^2+\beta+\varepsilon_1\left\|\boldsymbol{y}_f(t)\right\|_{peak}^2+\varepsilon_2\left\|\dot{\boldsymbol{y}}_f(t)\right\|_{\text{peak}}^2\\ &= -\lambda_{\min}(-\boldsymbol{\Xi})(\|\boldsymbol{e}_x(t)\|^2+\left\|\boldsymbol{e}_f(t)\right\|^2+\left\|\boldsymbol{y}_f(t)\right\|^2)+\beta+\varepsilon_1\left\|\boldsymbol{y}_f(t)\right\|_{\text{peak}}^2+\varepsilon_2\left\|\dot{\boldsymbol{y}}_f(t)\right\|_{\text{peak}}^2\\ &\leqslant -\lambda_{\min}(-\boldsymbol{\Xi})(\|\boldsymbol{e}_x(t)\|^2+\left\|\boldsymbol{e}_f(t)\right\|^2)+\beta+\varepsilon_1\left\|\boldsymbol{y}_f(t)\right\|_{\text{peak}}^2+\varepsilon_2\left\|\dot{\boldsymbol{y}}_f(t)\right\|_{\text{peak}}^2.\end{aligned} \tag{6.17}$$

According to the definition of $V(t)$ in (6.12), it can be derived that:

$$\begin{aligned}V(t) &\leqslant \lambda_{\max}(\boldsymbol{P})\|\boldsymbol{e}_x(t)\|^2+\frac{1}{\sigma}\lambda_{\max}(\boldsymbol{\Gamma}^{-1})\left\|\boldsymbol{e}_f(t)\right\|^2\\ &\leqslant \max\{\lambda_{\max}(\boldsymbol{P}),\frac{1}{\sigma}\lambda_{\max}(\boldsymbol{\Gamma}^{-1})\}(\|\boldsymbol{e}_x(t)\|^2+\left\|\boldsymbol{e}_f(t)\right\|^2).\end{aligned} \tag{6.18}$$

Combining (6.17) and (6.18), it can be obtained that:

$$\dot{V}(t)\leqslant -\alpha V(t)+\beta+\varepsilon_1\left\|\boldsymbol{y}_f(t)\right\|_{\text{peak}}^2+\varepsilon_2\left\|\dot{\boldsymbol{y}}_f(t)\right\|_{\text{peak}}^2, \tag{6.19}$$

where

$$\alpha = \frac{\lambda_{\min}(-\boldsymbol{\Xi})}{\max\{\lambda_{\max}(\boldsymbol{P}), \frac{1}{\sigma}\lambda_{\max}(\boldsymbol{\Gamma}^{-1})\}}.$$

By Lemma 6.2, it can be obtained that:

$$\begin{aligned}
V(t) &\leqslant e^{-\alpha t}V(0) + (\beta + \varepsilon_1\left\|\boldsymbol{y}_f(t)\right\|_{\text{peak}}^2 + \varepsilon_2\left\|\dot{\boldsymbol{y}}_f(t)\right\|_{\text{peak}}^2) \\
\int_0^t e^{-\alpha(t-\tau)}\mathrm{d}\tau &= e^{-\alpha t}V(0) + (\beta + \varepsilon_1\left\|\boldsymbol{y}_f(t)\right\|_{\text{peak}}^2 + \varepsilon_2\left\|\dot{\boldsymbol{y}}_f(t)\right\|_{\text{peak}}^2) \\
\frac{1}{\alpha}\left(1 - e^{-\alpha t}\right) &\leqslant e^{-\alpha t}V(0) + (\frac{\beta}{\alpha} + \frac{\varepsilon_1}{\alpha}\left\|\boldsymbol{y}_f(t)\right\|_{\text{peak}}^2 + \frac{\varepsilon_2}{\alpha}\left\|\dot{\boldsymbol{y}}_f(t)\right\|_{\text{peak}}^2) \\
\sup_{t\in[0,\infty)}\left\{1 - e^{-\alpha t}\right\} &\leqslant e^{-\alpha t}V(0) + (\frac{\beta}{\alpha} + \frac{\varepsilon_1}{\alpha}\left\|\boldsymbol{y}_f(t)\right\|_{\text{peak}}^2 + \frac{\varepsilon_2}{\alpha}\left\|\dot{\boldsymbol{y}}_f(t)\right\|_{\text{peak}}^2) \\
&\leqslant e^{-\alpha t}V(0) + (\sqrt{\frac{\beta}{\alpha}} + \sqrt{\frac{\varepsilon_1}{\alpha}}\left\|\boldsymbol{y}_f(t)\right\|_{\text{peak}} + \sqrt{\frac{\varepsilon_2}{\alpha}}\left\|\dot{\boldsymbol{y}}_f(t)\right\|_{\text{peak}})^2.
\end{aligned} \tag{6.20}$$

Meanwhile, considering (6.12), it can be derived that:

$$\begin{aligned}
V(t) &\geq \lambda_{\min}(\boldsymbol{P})\|\boldsymbol{e}_x(t)\|^2 + \frac{1}{\sigma}\lambda_{\min}(\boldsymbol{\Gamma}^{-1})\left\|\boldsymbol{e}_f(t)\right\|^2 \\
&\geq \min\{\lambda_{\min}(\boldsymbol{P}), \frac{1}{\sigma}\lambda_{\min}(\boldsymbol{\Gamma}^{-1})\}(\|\boldsymbol{e}_x(t)\|^2 + \left\|\boldsymbol{e}_f(t)\right\|^2).
\end{aligned} \tag{6.21}$$

Combining (6.20) with (6.21), it can be derived that:

$$\begin{aligned}
\|\boldsymbol{e}_x(t)\|^2 + \left\|\boldsymbol{e}_f(t)\right\|^2 &\leqslant \frac{e^{-\alpha t}V(0)}{\min\{\lambda_{\min}(\boldsymbol{P}), \frac{1}{\sigma}\lambda_{\min}(\boldsymbol{\Gamma}^{-1})\}} + \rho^2 \\
&\leqslant \left(\sqrt{\frac{e^{-\alpha t}V(0)}{\min\{\lambda_{\min}(\boldsymbol{P}), \frac{1}{\sigma}\lambda_{\min}(\boldsymbol{\Gamma}^{-1})\}}} + \rho\right)^2,
\end{aligned} \tag{6.22}$$

where

$$\rho = \sqrt{\frac{1}{\min\{\lambda_{\min}(\boldsymbol{P}), \frac{1}{\sigma}\lambda_{\min}(\boldsymbol{\Gamma}^{-1})\}}}(\sqrt{\frac{\beta}{\alpha}} + \sqrt{\frac{\varepsilon_1}{\alpha}}\left\|\boldsymbol{y}_f(t)\right\|_{\text{peak}} + \sqrt{\frac{\varepsilon_2}{\alpha}}\left\|\dot{\boldsymbol{y}}_f(t)\right\|_{\text{peak}}). \tag{6.23}$$

Hence it can be further derived that

$$\begin{aligned}
\|\boldsymbol{e}_x(t)\| &\leqslant \sqrt{\frac{e^{-\alpha t}V(0)}{\min\{\lambda_{\min}(\boldsymbol{P}), \frac{1}{\sigma}\lambda_{\min}(\boldsymbol{\Gamma}^{-1})\}}} + \rho, \\
\left\|\boldsymbol{e}_f(t)\right\| &\leqslant \sqrt{\frac{e^{-\alpha t}V(0)}{\min\{\lambda_{\min}(\boldsymbol{P}), \frac{1}{\sigma}\lambda_{\min}(\boldsymbol{\Gamma}^{-1})\}}} + \rho.
\end{aligned} \tag{6.24}$$

Referring to Assumption 6.1, it follows that $0 \leq \rho < \infty$. Consequently, from (6.24), there exists a time $T \to \infty$ such that for all $t > T$, $|\boldsymbol{e}_x(t)| \leq \rho$ and $|\boldsymbol{e}_f(t)| \leq \rho$. In other words, $\boldsymbol{e}_x(t)$ and $\boldsymbol{e}_f(t)$ are UUB with an ultimate bound of ρ. This completes the proof.

Corollary 6.1 *The abnormal S-T source estimation error of the original PDE system in (6.1)–(6.3) can be obtained as*

$$e_f(z, t) = \hat{f}(z, t) - f(z, t),$$

where $\hat{f}(z, t) = \boldsymbol{\phi}_s^{\mathrm{T}}(z)\hat{\boldsymbol{f}}_s(t)$, *as shown in (6.8). Hence the square of the norm for the abnormal S-T source estimation error in* $\mathcal{H}$ *can be obtained as*

$$\begin{aligned}
\|e_f(z,t)\|_2^2 &= \left\|\hat{f}(z,t) - f(z,t)\right\|_2^2 \\
&= \left\|\boldsymbol{\phi}_s^{\mathrm{T}}(z)\hat{\boldsymbol{f}}_s(t) - \boldsymbol{\phi}_s^{\mathrm{T}}(z)\boldsymbol{f}_s(t) - \boldsymbol{\phi}_f^{\mathrm{T}}(z)\boldsymbol{f}_f(t)\right\|_2^2 \\
&= \left\|\boldsymbol{\phi}_s^{\mathrm{T}}(z)\boldsymbol{e}_f(t) - \boldsymbol{\phi}_f^{\mathrm{T}}(z)\boldsymbol{f}_f(t)\right\|_2^2 \\
&= \left\|\boldsymbol{e}_f(t)\right\|^2 + \left\|\boldsymbol{f}_f(t)\right\|^2.
\end{aligned}$$

Combining Theorem 6.1 with Assumption 6.3, it follows that the abnormal S-T source estimation error $e_f(z, t)$ *is uniformly UUB under the conditions of Theorem 6.1, with an ultimate bound of* $\rho + \sqrt{f_2}$.

Remark 6.6 Inequality (6.9) can be addressed using the MATLAB LMI toolbox. Nevertheless, the equality constraint in (6.11) renders the problem challenging. As a result, this condition is reformulated as the following optimization problem: minimize η subject to the constraints (6.9), (6.10) and

$$\begin{bmatrix} \eta \boldsymbol{I} & * \\ (\boldsymbol{P} - \boldsymbol{F}\boldsymbol{C}_s)^{\mathrm{T}} & \eta \boldsymbol{I} \end{bmatrix} > 0. \tag{6.25}$$

6.4 Numerical Simulation

Consider a slender rod whose temperature profile can be described by the following parabolic PDE:

$$\frac{\partial x(z,t)}{\partial t} = \frac{\partial^2 x(z,t)}{\partial z^2} + \beta_U(b_u(z)u(t) - x(z,t)) + f(z,t), \tag{6.26}$$

$$\boldsymbol{y}(t) = \begin{bmatrix} \int_0^{\pi} \delta(z - \frac{\pi}{4})x(z,t)\mathrm{d}z \\ \int_0^{\pi} \delta(z - \frac{3\pi}{4})x(z,t)\mathrm{d}z \end{bmatrix} = \begin{bmatrix} x(\frac{\pi}{4}, t) \\ x(\frac{3\pi}{4}, t) \end{bmatrix}, \tag{6.27}$$

subject to the Dirichlet boundary conditions:

$$x(0,t) = 0,\ \ x(\pi,t) = 0, \tag{6.28}$$

where $x(z,t)$ denotes the dimensionless temperature of the rod; β_U represents the dimensionless heat transfer coefficient; $u(t)$ is the controlled input (temperature of the cooling medium); and $\boldsymbol{y}(t)$ corresponds to the thermocouple measurements at $z = \pi/4$ and $z = 3\pi/4$. The typical value of β_U is set to 2. The actuator distribution function is defined as:

$$b_u(z) = \sqrt{\frac{2}{\pi}}\sin(z).$$

And the control input is selected as:

$$u(t) = 1.$$

The eigenvalue problem for the spatial differential operator:

$$\begin{aligned} \mathcal{A}x &= \frac{\partial^2 x}{\partial z^2} - \beta_U x, \\ x \in \mathcal{S}(\mathcal{A}) &= \{x \in \mathcal{L}_2([0,\pi];\mathbb{R}); x(0,t) = 0, x(\pi,t) = 0\} \end{aligned}$$

can be directly solved as:

$$\lambda_j = -j^2 - 2,\ \ \phi_j(z) = \sqrt{\frac{2}{\pi}}\sin(jz),\ \ j = 1,\ldots,\infty.$$

The solution procedures for this type of eigenvalue problem can be found in Chap. 4 of [39]. The first two eigenvalues are selected as the dominant ones, yielding $\varepsilon = |\lambda_1| / |\lambda_3| \approx 0.273$. Following the method outlined in Sect. 6.3.1, the resulting 2-D slow subsystem is derived as:

$$\begin{aligned} \dot{\boldsymbol{x}}_s(t) &= \boldsymbol{A}_s\boldsymbol{x}_s(t) + \boldsymbol{B}_{u,s}\boldsymbol{u}(t) + \boldsymbol{f}_s(t), \\ \boldsymbol{y}_s(t) &= \boldsymbol{C}_s\boldsymbol{x}_s(t), \end{aligned} \tag{6.29}$$

where

$$\boldsymbol{x}_s(t) = \begin{bmatrix} x_1(t) \\ x_2(t) \end{bmatrix},\ \boldsymbol{A}_s = \begin{bmatrix} -3 & 0 \\ 0 & -6 \end{bmatrix},\ \boldsymbol{B}_{u,s} = \begin{bmatrix} 2 \\ 0 \end{bmatrix},$$

and

$$\boldsymbol{C}_s = \int_0^{\pi} \begin{bmatrix} \delta(z - \frac{\pi}{4}) \\ \delta(z - \frac{3\pi}{4}) \end{bmatrix} [\phi_1(z)\ \phi_2(z)]\,\mathrm{d}z = \begin{bmatrix} \sqrt{\frac{1}{\pi}} & \sqrt{\frac{2}{\pi}} \\ \sqrt{\frac{1}{\pi}} & -\sqrt{\frac{2}{\pi}} \end{bmatrix}.$$

It can be observed that Assumption 6.4 holds. In this numerical simulation, the following abnormal S-T source is considered in accordance with Remark 6.4:

$$f(z,t) = [\phi_1(z)\ \phi_2(z)] \begin{bmatrix} f_1(t) \\ f_2(t) \end{bmatrix}.$$

Given that $b_u(z) = \phi_1(z)$ is specified, the abnormal source $f_1(t)$ can be interpreted as an actuator fault.

To assess the effectiveness of the proposed abnormal S-T source identification method, the following types of abnormal sources are considered:

$$f_1(t) = \begin{cases} 0, & 0 \leqslant t < 10\ (\mathrm{s}) \\ 2, & 10 \leqslant t \leqslant 80\ (\mathrm{s}) \end{cases}$$

$$f_2(t) = \begin{cases} 0, & 0 \leqslant t < 40\ (\mathrm{s}) \\ 3, & 40 \leqslant t \leqslant 80\ (\mathrm{s}) \end{cases}$$

and

$$f_1(t) = \begin{cases} 0, & 0 \leqslant t < 10\ (\mathrm{s}) \\ 2 - e^{-0.01(t-10)}, & 10 \leqslant t \leqslant 80\ (\mathrm{s}) \end{cases}$$

$$f_2(t) = \begin{cases} 0, & 0 \leqslant t < 40\ (\mathrm{s}) \\ 3 - e^{-0.02(t-40)}. & 40 \leqslant t \leqslant 80\ (\mathrm{s}) \end{cases}$$

The first type of abnormal source can be regarded as a sudden (abrupt) source, whereas the second represents an incipient source. Based on the definition of $\boldsymbol{f}_s(t)$ in (6.30), it follows that:

$$\boldsymbol{f}_s(t) = \begin{bmatrix} f_{s1}(t) \\ f_{s2}(t) \end{bmatrix} = \begin{bmatrix} f_1(t) \\ f_2(t) \end{bmatrix}.$$

According to Theorem 6.1, by selecting $\mu_1 = 1$, $\mu_2 = 1$, and $\sigma = 1$, and solving (6.9), (6.10), and (6.25), it can be determined that:

$$\begin{aligned}
\eta &= 9.4277 \times 10^{-12}, \\
\boldsymbol{P} &= \begin{bmatrix} 0.1774 & 0 \\ 0 & 0.0609 \end{bmatrix}, \boldsymbol{G}_1 = \begin{bmatrix} 0.0193 & 0 \\ 0 & 0.0102 \end{bmatrix}, \\
\boldsymbol{G}_2 &= \begin{bmatrix} 0.0193 & 0 \\ 0 & 0.0102 \end{bmatrix}, \boldsymbol{X} = \begin{bmatrix} -0.1106 & -0.1106 \\ -0.1588 & 0.1588 \end{bmatrix}, \\
\boldsymbol{F} &= \begin{bmatrix} 0.1572 & 0.1572 \\ 0.0382 & -0.0382 \end{bmatrix}, \boldsymbol{L} = \begin{bmatrix} -0.6231 & -0.6231 \\ -2.6069 & 2.6069 \end{bmatrix},
\end{aligned}$$

by the MATLAB LMI toolbox.

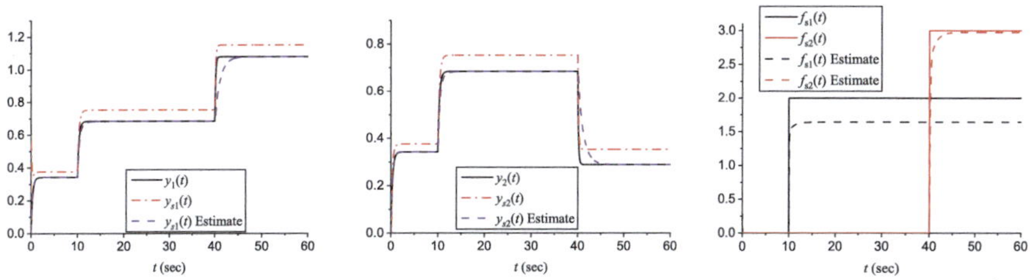

(a) The first output of the original PDE system $y_1(t)$, the slow subsystem $y_{s1}(t)$, and its estimation $\hat{y}_{s1}(t)$

(b) The second output of the original PDE system $y_2(t)$, the slow subsystem $y_{s2}(t)$, and its estimation $\hat{y}_{s2}(t)$

(c) $\boldsymbol{f}_s(t)$ and its estimation $\hat{\boldsymbol{f}}_s(t)$

Fig. 6.4 Estimation results of abnormal source 1. (Abrupt source)

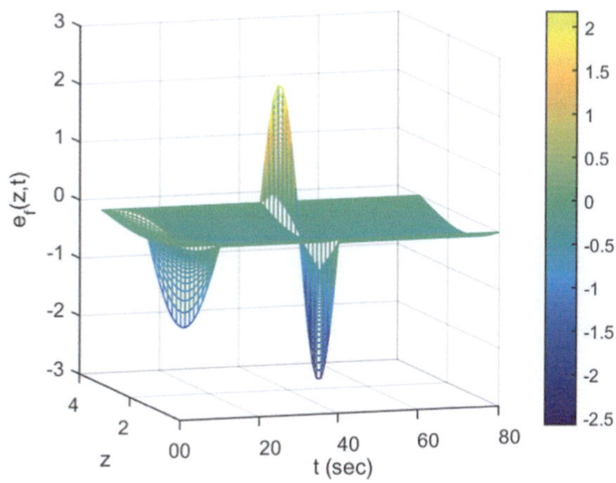

Fig. 6.5 $e_f(z,t)$ of abnormal source 1. (Abrupt source)

Furthermore, the learning rate is set as $\boldsymbol{\Gamma} = diag(100, 100)$. The original PDE system in (6.26) is numerically solved using the FDM [40] with a sampling interval of $\Delta t = 0.01$ sec. By employing the adaptive state observer (6.5) along with the adaptive source estimation algorithm (6.7), the source estimate of the slow subsystem, $\hat{\boldsymbol{f}}_s(t) = [\hat{f}_{s1}(t), \hat{f}_{s2}(t)]^{\mathrm{T}}$, can be obtained, and the abnormal S-T source for the original PDE system (6.26) is subsequently derived by:

$$\hat{f}(z,t) = [\phi_1(z)\ \phi_2(z)] \begin{bmatrix} \hat{f}_{s1}(t) \\ \hat{f}_{s2}(t) \end{bmatrix},$$

as shown in (6.8).

The simulation results for the two types of abnormal sources are presented in Figs. 6.4a, 6.5 and 6.6a, 6.7, respectively. It can be observed from Figs. 6.4a, b, 6.6a, b that minor discrepancies exist between the slow subsystem output $\boldsymbol{y}_s(t)$ and the original PDE system output $\boldsymbol{y}(t)$, which arise from neglecting the fast subsystem. Meanwhile, the estimated output $\hat{\boldsymbol{y}}_s(t)$ rapidly tracks $\boldsymbol{y}(t)$ for both types of abnormal

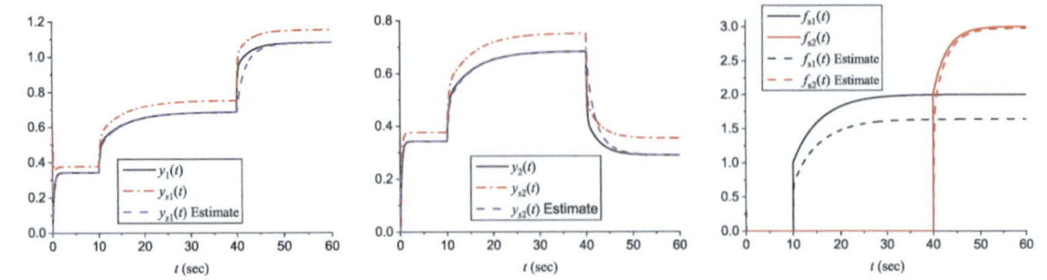

(a) The first output of the original PDE system $y_1(t)$, the slow subsystem $y_{s1}(t)$, and its estimation $\hat{y}_{s1}(t)$

(b) The second output of the original PDE system $y_2(t)$, the slow subsystem $y_{s2}(t)$, and its estimation $\hat{y}_{s2}(t)$

(c) $\boldsymbol{f}_s(t)$ and its estimation $\hat{\boldsymbol{f}}_s(t)$

Fig. 6.6 Estimation results of abnormal source 2. (Incipient source)

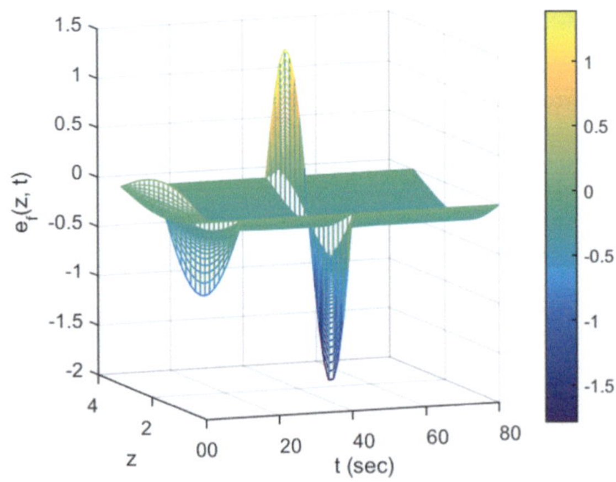

Fig. 6.7 $e_f(z,t)$ of abnormal source 2. (Incipient source)

sources. Small errors also exist between $\boldsymbol{f}_s(t)$ and its estimate $\hat{\boldsymbol{f}}_s(t)$, mainly due to truncation errors caused by ignoring the fast subsystem. Moreover, the abnormal S-T source estimate $\hat{f}(z,t)$ tracks $f(z,t)$ with minor deviations, as shown in Figs. 6.5 and 6.7, and the occurrence times of the abnormal sources ($t = 10$ sec and $t = 40$ sec) can be successfully detected. To better illustrate the performance of the proposed abnormal S-T source identification approach, the RMSE is defined as follows:

$$\text{RMSE} = \left(\int \sum e_f(z,t)^2 \mathrm{d}z \Big/ \int \mathrm{d}z \sum \Delta t\right)^{1/2}.$$

The computed RMSE values for the abrupt and incipient abnormal source estimations in Figs. 6.5 and 6.7 are 0.2007 and 0.1919, respectively. As noted in Remark 6.2, the primary source of the estimation error $e_f(z,t)$ is the omission of the fast subsystem.

For simplicity, the above S-T source is generated using the eigenfunctions of the spatial operator $\mathcal{A}$ as basis functions. To further evaluate the effectiveness of the proposed approach, a more general S-T source $f(z,t)$ is considered as follows:

Table 6.1 Identification results of a general S-T source

(m, n_y)	Γ	RMSE	Ideal RMSE
(2, 2)	$100 * I_2$	0.7709	0.7497
(2, 3)	$100 * I_2$	0.7517	0.7497
(3, 3)	$100 * I_3$	0.6377	0.5901
(2, 4)	$100 * I_2$	0.7518	0.7497
(3, 4)	$100 * I_3$	0.6454	0.5901
(4, 4)	$100 * I_4$	0.5102	0.4286

$$f(z,t) = f(t)b_f(z),$$

where

$$b_f(z) = H(z) - H(z - \pi/4),$$
$$f(t) = \begin{cases} 0, & 0 \leqslant t < 10\,(\mathrm{s}) \\ 2, & 10 \leqslant t \leqslant 80\,(\mathrm{s}) \end{cases}$$

$H(\cdot)$ denotes the standard Heaviside function.

To assess the performance of the proposed method, n_y point-wise measurements are evenly distributed across the spatial domain $[0, \pi]$. In addition, the first m eigenvalues are chosen as the dominant modes. The abnormal source estimation results are presented below, evaluated using the RMSE performance metric.

As indicated in Table 6.1, the performance of the proposed method for this S-T source is slightly inferior to that for the source generated using the same basis functions. This is mainly due to the relatively large truncation error for this type of source, which requires a higher number of basis functions for accurate representation. The table also shows that RMSE decreases as the number of dominant modes m increases, which is expected. To provide a clearer evaluation of the method's performance, another metric called "Ideal RMSE" is introduced as follows:

$$e_{\mathrm{ideal}\,f}(z,t) = \boldsymbol{\phi}_s^{\mathrm{T}}(z)\boldsymbol{f}_s(t) - f(z,t),$$
$$\boldsymbol{f}_s(t) =< \boldsymbol{\phi}_s^{\mathrm{T}}(z), b_f(z) > f(t),$$
$$\text{Ideal RMSE} = \left(\int \sum e_{\mathrm{ideal}\,f}(z,t)^2 \mathrm{d}z \Big/ \int \mathrm{d}z \sum \Delta t\right)^{1/2},$$

this metric provides a lower bound for the RMSE. By comparing the RMSE with the Ideal RMSE, it is evident that the proposed method achieves satisfactory performance for this S-T source. For sources generated using the same basis functions, the Ideal RMSE is clearly zero.

In practice, for a fixed number of sensors, determining their optimal placement along with selecting the learning rate $\boldsymbol{\Gamma}$ to enhance estimation accuracy constitutes an interesting multi-objective optimization problem. In this work, the emphasis is placed on developing a general framework for addressing such inverse source estimation problems rather than achieving optimal estimation performance, which is inherently challenging.

Remark 6.7 Given the stringent requirements of the modulating functions-based approach, it is neither fair nor informative to compare its performance with the proposed method on the same problem. Specifically, if the conditions for the modulating functions-based method are not satisfied, it will fail to operate. Conversely, since this chapter aims to relax such constraints for industrial applications, the practical significance of the proposed approach would be substantially reduced even if these conditions are met, regardless of the comparative outcome.

6.5 Conclusion

This chapter presents a first study on the identification of abnormal S-T sources in a class of linear parabolic DPSs. An inverse S-T model is developed for this purpose, comprising an adaptive state observer for source estimation and an adaptive source estimation algorithm. Theoretical analysis ensures the convergence of the abnormal S-T source estimation error. Finally, numerical simulations on a heat rod with an abnormal S-T source demonstrate the effectiveness of the proposed approach.

Appendix

A.1 Model Reduction for Parabolic DPSs

For the spatial operator $\mathcal{A}$ described above, the corresponding eigenvalue problem is formulated as follows:

$$\mathcal{A}x = a_1\frac{\partial x}{\partial z} + a_2\frac{\partial^2 x}{\partial z^2} + a_3 x, x \in S(\mathcal{A}) \triangleq \{x \in \mathcal{H}, \text{ and the conditions in (6.2) hold}\}.$$

For the spatial operator $\mathcal{A}$ defined above, the associated eigenvalue problem is given by:

$$\mathcal{A}\phi_j(z) = \lambda_j\phi_j(z), \ j = 1, \cdots, \infty,$$

where ϕ_j denotes the eigenfunction and λ_j the corresponding eigenvalue. Let $\sigma(\mathcal{A})$ represent the set of all eigenvalues of the operator $\mathcal{A}$, i.e., $\sigma(\mathcal{A}) = \{\lambda_1, \cdots, \}$. It is assumed that $\sigma(\mathcal{A})$ can be separated into a finite-dimensional subset containing m slow eigenvalues and a stable infinite-dimensional complement consisting of the remaining fast eigenvalues, with a significant gap between the slow and fast eigenvalues.

Assumption 6.5
1. $Re\{\lambda_1\} \geqslant Re\{\lambda_2\} \geqslant \cdots \geqslant Re\{\lambda_j\} \geqslant \cdots$, where $Re\{\lambda_j\}$ denotes the real part of λ_j.
2. $\sigma(\mathcal{A})$ can be partitioned as $\sigma(\mathcal{A}) = \sigma_1(\mathcal{A}) + \sigma_2(\mathcal{A})$, where $\sigma_1(\mathcal{A})$ consists of the first m (with m finite) eigenvalues, that is $\sigma_1(\mathcal{A}) = \{\lambda_1, \cdots, \lambda_m\}$, and $|Re\{\lambda_1\}|/|Re\{\lambda_m\}| = O(1)$.
3. $Re\{\lambda_{m+1}\} < 0$ and $|Re\{\lambda_m\}|/|Re\{\lambda_{m+1}\}| = O(\varepsilon)$, where $\varepsilon < 1$ is a small positive number.

By employing the separation of variables theorem [41] and applying Galerkin's method [34], the system can be represented as the following infinite-dimensional model in terms of ODEs:

$$\begin{aligned} \dot{\boldsymbol{x}}_s(t) &= \boldsymbol{A}_s \boldsymbol{x}_s(t) + \boldsymbol{B}_{u,s}\boldsymbol{u}(t) + \boldsymbol{f}_s(t), \\ \dot{\boldsymbol{x}}_f(t) &= \boldsymbol{A}_f \boldsymbol{x}_f(t) + \boldsymbol{B}_{u,f}\boldsymbol{u}(t) + \boldsymbol{f}_f(t), \\ \boldsymbol{y}(t) &= \boldsymbol{C}_s \boldsymbol{x}_s(t) + \boldsymbol{C}_f \boldsymbol{x}_f(t) = \boldsymbol{y}_s(t) + \boldsymbol{y}_f(t), \end{aligned} \tag{6.30}$$

with the initial conditions:

$$\boldsymbol{x}_s(0) = \mathcal{P}_s \boldsymbol{x}_0(z), \boldsymbol{x}_f(0) = \mathcal{P}_f \boldsymbol{x}_0(z),$$

where

$$\begin{aligned} \boldsymbol{A}_s &= \mathcal{P}_s \mathcal{A} \boldsymbol{\phi}^{\mathrm{T}}(z) = diag\{\lambda_1, \cdots, \lambda_m\}, \\ \boldsymbol{A}_f &= \mathcal{P}_f \mathcal{A} \boldsymbol{\phi}^{\mathrm{T}}(z) = diag\{\lambda_{m+1}, \cdots, \lambda_\infty\}, \\ \boldsymbol{B}_{u,s} &= \mathcal{P}_s k_u \boldsymbol{b}_u^{\mathrm{T}}(z), \ \ \boldsymbol{B}_{u,f} = \mathcal{P}_f k_u \boldsymbol{b}_u^{\mathrm{T}}(z), \\ \boldsymbol{f}_s(t) &= \mathcal{P}_s f(z,t), \ \ \boldsymbol{f}_f(t) = \mathcal{P}_f f(z,t), \\ \boldsymbol{C}_s &= \int_{\alpha_1}^{\alpha_2} \boldsymbol{c}(z) k_y \boldsymbol{\phi}_s^{\mathrm{T}}(z) \mathrm{d}z, \boldsymbol{C}_f = \int_{\alpha_1}^{\alpha_2} \boldsymbol{c}(z) k_y \boldsymbol{\phi}_f^{\mathrm{T}}(z) \mathrm{d}z. \end{aligned}$$

The derivation of this infinite-dimensional ODE system is straightforward and thus omitted here. Interested readers are referred to [34, 42] for further details.

By defining $\varepsilon = \frac{|\mathrm{Re}\,\lambda_1|}{|\mathrm{Re}\,\lambda_{m+1}|}$ and scaling the $\boldsymbol{x}_f$ subsystem with ε, the resulting singularly perturbed model is obtained as follows:

$$\begin{aligned} \dot{\boldsymbol{x}}_s(t) &= \boldsymbol{A}_s \boldsymbol{x}_s(t) + \boldsymbol{B}_{u,s}\boldsymbol{u}(t) + \boldsymbol{f}_s(t), \\ \varepsilon \dot{\boldsymbol{x}}_f(t) &= \boldsymbol{A}_{f\varepsilon} \boldsymbol{x}_f(t) + \varepsilon \boldsymbol{B}_{u,f}\boldsymbol{u}(t) + \varepsilon \boldsymbol{f}_f(t), \end{aligned} \tag{6.31}$$

where $\boldsymbol{A}_{f\varepsilon} = \varepsilon \boldsymbol{A}_f$.

The singular perturbation theory [43] can then be employed. By introducing the fast time scale $\tau = t/\varepsilon$ and setting $\varepsilon = 0$, the following infinite-dimensional fast subsystem can be directly derived from model (6.31):

$$\frac{d\boldsymbol{x}_f(\tau)}{d\tau} = \boldsymbol{A}_{f\varepsilon} \boldsymbol{x}_f(\tau). \tag{6.32}$$

Since $Re\{\lambda_{m+1}\} < 0$ and based on the definition of ε, it follows that (6.32) is globally exponentially stable. By setting $\varepsilon = 0$ in (6.31) and considering the nonsingularity of $\boldsymbol{A}_{f\varepsilon}$, one obtains $\boldsymbol{x}_f(t) = 0$. Substituting $\boldsymbol{x}_f(t) = 0$ into (6.31) leads to:

$$\begin{aligned}\dot{\boldsymbol{x}}_s(t) &= \boldsymbol{A}_s\boldsymbol{x}_s(t) + \boldsymbol{B}_{u,s}\boldsymbol{u}(t) + \boldsymbol{f}_s(t),\\ \boldsymbol{y}_s(t) &= \boldsymbol{C}_s\boldsymbol{x}_s(t).\end{aligned}$$

References

1. Li H-X, Qi C (2010) Modeling of distributed parameter systems for applications–A synthesized review from time-space separation. J Process Control 20(8):891–901
2. Wang Z, Li H-X (2019) Incremental spatio-temporal learning for online modeling of distributed parameter systems. IEEE Trans Syst Man Cybern Syst 49(12):2612–2622
3. Wang Z, Li H-X, Chen C (2019) Reinforcement learning-based optimal sensor placement for spatiotemporal modeling. IEEE Trans Cybern 50(6):2861–2871
4. Wang Z, Li H-X (2019) Dissimilarity analysis-based multimode modeling for complex distributed parameter systems. IEEE Trans Syst Man Cybern Syst 51(5):2789–2797
5. Xu K-K, Li H-X, Yang H-D (2019) Kernel-based random vector functional-link network for fast learning of spatiotemporal dynamic processes. IEEE Trans Syst Man Cybern Syst 49(5):1016–1026
6. Wang B-C, Li H-X (2019) A sliding window based dynamic spatiotemporal modeling for distributed parameter systems with time-dependent boundary conditions. IEEE Trans Ind Inform 15(4):2044–2053
7. Wang B-C, Li H-X, Yang H-D (2019) Spatial correlation based incremental learning for spatiotemporal modeling of battery thermal process. IEEE Trans Ind Electron 67(4):2885–2893
8. Meng X-B, Li H-X, Yang H-D (2018) Evolutionary design of spatio-temporal leaning model for thermal distribution in lithium-ion batteries. IEEE Trans Ind Inform 15(5):2838–2848
9. El-Farra NH (2006) Integrated fault detection and fault-tolerant control architectures for distributed processes. Ind Eng Chem Res 45(25):8338–8351
10. El-Farra NH, Ghantasala S (2007) Actuator fault isolation and reconfiguration in transport-reaction processes. AIChE J 53(6):1518–1537
11. Wei J, Dong G, Chen Z (2019) Lyapunov-based thermal fault diagnosis of cylindrical lithium-ion batteries. IEEE Trans Ind Electron 67(6):4670–4679
12. Zhao Z, He X, Ahn CK (2019) Boundary disturbance observer-based control of a vibrating single-link flexible manipulator. IEEE Trans Syst Man Cybern Syst 51(4):2382–2390
13. Wu H-N, Wang H-D, Guo L (2016) Disturbance rejection fuzzy control for nonlinear parabolic PDE systems via multiple observers. IEEE Trans Fuzzy Syst 24(6):1334–1348
14. Wu H-N, Wang H-D, Guo L (2016) Finite dimensional disturbance observer based control for nonlinear parabolic PDE systems via output feedback. J Process Control 48:25–40
15. Wang H-D, Wu H-N, Guo L (2016) Low dimensional disturbance observer-based control for nonlinear parabolic PDE systems with spatio-temporal disturbances. Int J Robust Nonlinear Control 26(12):2686–2707
16. Asiri S, Laleg-Kirati T-M (2017) Modulating functions-based method for parameters and source estimation in one-dimensional partial differential equations. Inverse Probl Sci Eng 25(8):1191–1215
17. Fischer F, Deutscher J, Laleg-Kirati T-M (2018) Source estimation for first order time-varying hyperbolic systems. In: Proceedings of the 23rd international symposium on mathematical theory of networks and systems, Hong Kong University of Science and Technology, Hong Kong, pp 78–84

18. Demetriou MA (2002) A model-based fault detection and diagnosis scheme for distributed parameter systems: a learning systems approach. ESAIM Control Optim Calc Var 7:43–67
19. Demetriou MA, Ito K, Smith RC (2007) Adaptive monitoring and accommodation of nonlinear actuator faults in positive real infinite dimensional systems. IEEE Trans Autom Control 52(12):2332–2338
20. Armaou A, Demetriou MA (2008) Robust detection and accommodation of incipient component and actuator faults in nonlinear distributed processes. AIChE J 54(10):2651–2662
21. Ghantasala S, El-Farra NH (2009) Robust actuator fault isolation and management in constrained uncertain parabolic PDE systems. Automatica 45(10):2368–2373
22. Deutscher J (2016) Fault detection for linear distributed-parameter systems using finite-dimensional functional observers. Int J Control 89(3):550–563
23. Cai J, Ferdowsi H, Sarangapani J (2016) Model-based fault detection, estimation, and prediction for a class of linear distributed parameter systems. Automatica 66:122–131
24. Shinbrot M (1954) On the analysis of linear and nonlinear dynamical systems from transient-response data. NACA Technical Note No. 3288
25. Fischer F, Deutscher J (2016) Algebraic fault detection and isolation for parabolic distributed-parameter systems using modulation functions. IFAC-PapersOnLine 49(8):162–167
26. Fischer F, Deutscher J (2017) Fault detection for parabolic systems with distributed inputs and outputs using the modulation function approach. IFAC-PapersOnLine 50(1):6774–6779
27. Fischer F, Deutscher J (2018) Modulating function based fault detection for parabolic systems with polynomial faults. IFAC-PapersOnLine 51(24):359–366
28. Feng Y, Li H-X (2019) Detection and spatial identification of fault for parabolic distributed parameter systems. IEEE Trans Ind Electron 66(9):7300–7309
29. Feng Y, Li H-X (2019) Dynamic spatial independent component analysis based abnormality localization for distributed parameter systems. IEEE Trans Ind Inform 16(5):2929–2936
30. Wang H, Daley S (1996) Actuator fault diagnosis: an adaptive observer-based technique. IEEE Trans Autom Control 41(7):1073–1078
31. Jiang B, Wang JL, Soh YC (2002) An adaptive technique for robust diagnosis of faults with independent effects on system outputs. Int J Control 75(11):792–802
32. Jiang B, Staroswiecki M, Cocquempot V (2006) Fault accommodation for nonlinear dynamic systems. IEEE Trans Autom Control 51(9):1578–1583
33. Zhang K, Jiang B, Cocquempot V (2008) Adaptive observer-based fast fault estimation. Int J Control Autom Syst 6(3):320–326
34. Christofides PD (2012) Nonlinear and robust control of PDE systems: methods and applications to transport-reaction processes. Springer, Berlin
35. Ding SX (2008) Model-based fault diagnosis techniques: design schemes, algorithms, and tools. Springer, Berlin
36. White J, Speyer JL (1987) Detection filter design: spectral theory and algorithms. IEEE Trans Autom Control 32(7):593–603
37. Liu B, Si J (1997) Fault isolation filter design for linear time-invariant systems. IEEE Trans Autom Control 42(5):704–707
38. Ioannou PA, Sun J (2012) Robust adaptive control. Courier Corporation, North Chelmsford
39. Ray WH (1981) Advanced process control. McGraw-Hill, Columbus
40. Strikwerda JC (2004) Finite difference schemes and partial differential equations. SIAM, Philadelphia
41. MacCluer CR (1994) Boundary value problems and orthogonal expansions: physical problems from a Sobolev viewpoint. IEEE Press, Piscataway
42. Wu H-N, Li H-X (2011) Robust adaptive neural observer design for a class of nonlinear parabolic PDE systems. J Process Control 21(8):1172–1182
43. Khalil HK, Grizzle JW (2002) Nonlinear systems, 3rd edn. Prentice Hall, Upper Saddle River

Chapter 7
Adaptive PDE Observer-Based Abnormal Source Estimation for A Linear Unstable Parabolic DPS

Abstract Detection and estimation of abnormalities for distributed parameter system have wide applications in industry, e.g., battery thermal fault detection, quality monitoring of hot-rolled strip laminar cooling process. In this chapter, the abnormal spatio-temporal source detection and estimation problem for a linear unstable distributed parameter system is first studied. The proposed methodology consists of two steps: first, an abnormality detection filter which generates a residual signal for abnormality detection in the time domain is constructed using pointwise measurement; Then, an adaptive Luenberger-type partial differential equation observer including an adaptive estimation algorithm is designed and triggered only when an alarm raises from the abnormality detection filter. Theoretic analysis based on the spatial domain decomposition approach is presented to show the convergence of the estimation errors. Finally, an illustrative example is presented to show the performance of the proposed method.

7.1 Introduction

Many industrial processes whose dynamic behaviors are dominated by PDEs belong to the category of DPSs, such as heat processes, fluid processes, etc. One key feature of DPSs is the so-called S-T dynamics [1], which indicates that the system input, output and even parameters can vary both in space and time domain. With the rapidity increasing demand for the high production quality with economic operations for these industrial processes, the requirement of system safety is a critical issue that needs to be concerned. S-T abnormalities [2–8] hidden in the system may cause undesired system response or even system failures, which are harmful to the safe operations. For example, as discussed in [9], abnormal heat generation from electrochemical side reactions or internal failure caused by mechanical or thermal abuse [10, 11] may lead to thermal runaway and eventually a fire. The abnormality detection and estimation mainly focus on the heat processes in industry applications, with a particular interest in the thermal fault detection of Li-ion batteries [12, 13].

Without loss of generality, these abnormalities can all be modeled by an *unknown* abnormal S-T source $s(x, t)$ with *unknown dynamic characteristic*. This distributed

© The Author(s), under exclusive license to Springer Nature Singapore Pte Ltd. 2026
Y. Feng et al., *Spatio-Temporal Abnormality Diagnosis for Industrial Distributed Parameter Systems*, https://doi.org/10.1007/978-981-95-3750-1_7

abnormality can be viewed as a special type of "fault" in the process. For decades, a number of results have been reported on the FD of DPSs. These works can be roughly classified into two categories: One was the so-called "early-lumping" [14] approach which transforms the original PDE system into an ODE system in the first place and then analyze the ODE model, representative works can be referred to [15, 16]; the other was the so-called "late-lumping" approach which directly investigates the PDE system without approximation. Researches belong to this category are mainly based on the backstepping design techniques [17], see [18, 19] for reference. "Early-lumping"-based approaches have been criticized for the *observation spillover* induced by the modal approximation used.

To the best of our knowledge, almost all existing works reported so far only concentrates on fault detection in the time-domain, while the fault localization and estimation problem is not well studied. The main reason is that the spatial distribution characteristic of the most studied actuator or sensor fault was not taken into consideration since the locations of both actuator and sensor are known *a prior*. However, for the distributed fault such as the abnormal heat generation in batteries [19], fault detection in the time domain alone can not provide sufficient information for root tracing and fault removal.

Motivated by the above considerations, we intend to study the abnormal source estimation problem for a class of linear unstable DPS. Considering the fact that adaptive observers [20, 21] have been widely used for fault estimation in ODE systems, we attempt to design an adaptive PDE observer for abnormal source estimation. Moreover, an ADF is designed first for abnormality detection in the time domain. The adaptive PDE observer begins to work once an alarm raises by the ADF.

The main contribution of this chapter are twofold:

- An adaptive PDE observer along with an adaptive estimation algorithm are first introduced in the spirit of the well-known adaptive ODE observer.
- A general methodology framework consisting of an ADF and adaptive PDE observer is proposed to achieve the purpose of abnormality detection and estimation sequentially, which have not been well discussed in the literature.

The remainder of this chapter is organized as follows: In Sect. 7.2, the detailed system and problem description are introduced. The main design procedures along with theoretical analysis are given in Sect. 7.3. Numerical simulations for validating the proposed method are presented in Sect. 7.4. Finally, we conclude this chapter in Sect. 7.5.

7.2 Preliminaries

Notations: $\mathbb{R}$ and $\mathbb{R}^n$ are the set of all real numbers and n-dimensional Euclidean space with the norm $|\cdot|_2$, respectively. $|\omega|$ denotes the absolute value of ω for any

$\omega \in \mathbb{R}$. For a given constant $L > 0$, $\mathcal{L}^2([0, L]) \triangleq \mathcal{L}^2([0, L]; \mathbb{R})$ is a real Hilbert space of square integrable functions $\omega(x) : [0, L] \to \mathbb{R}$ with the spatial $\mathcal{L}^2$ norm $||\omega(\cdot)||_2 \triangleq \sqrt{\int_0^L \omega^2(x)\mathrm{d}x}$. $u_t \triangleq \partial u/\partial t$, $u_x \triangleq \partial u/\partial x$, $u_{xx} \triangleq \partial^2 u/\partial x^2$.

Consider a linear parabolic DPS of the following form:

$$u_t(x,t) = u_{xx}(x,t) + \lambda u(x,t) + s(x,t), x \in (0, L), t > 0 \tag{7.1}$$

subject to the following homogeneous Dirichlet boundary conditions:

$$u(0,t) = u(L,t) = 0, t > 0 \tag{7.2}$$

and with the following initial condition:

$$u(x,0) = u_0(x), x \in [0, L] \tag{7.3}$$

where $u(x,t) \in \mathcal{L}^2([0, L])$ denotes the state variable; $x \in [0, L]$ is the spatial coordinate; $t \in [0, \infty)$ is the time; $\lambda \in \mathbb{R}$ is a constant; $s(x,t)$ denotes an **unknown** abnormal S-T source.

The measurement equation for the PDE system (7.1)–(7.3) is defined as:

$$\boldsymbol{y}(t) = \int_0^L \boldsymbol{c}(x)u(x,t)\mathrm{d}x, \tag{7.4}$$

where $\boldsymbol{y}(t) \triangleq [y_1(t) \quad y_2(t) \quad \cdots \quad y_n(t)]^T \in \mathbb{R}^n$ denotes the measurement output for the system, it is a vector which consists of n number of continuous elements with respect to the time variable t. And $\boldsymbol{c}(x) \triangleq [c_1(x) \quad c_2(x) \quad \cdots \quad c_n(x)]^T \in \mathbb{R}^n$ is a known integrable vector function with respect to the spatial position x, in which the i-th element $c_i(x)$ describes the distribution of the i-th sensor over the spatial domain $[0, L]$.

Without loss of generality, the most frequently used **pointwise measurement** [22, 23] is selected with

$$c_i(x) = \delta\left(x - \bar{x}_i\right), i \in \mathcal{N} \triangleq \{1, 2, \cdots, n\}$$

corresponding to pointwise measurement at the positions $\bar{x}_i, i \in \mathcal{N}$ in the spatial domain $(0, L)$ and $\delta(\cdot)$ denotes the Dirac delta function.

Remark 7.1 When $s(x,t) = 0$, the PDE system (7.1)–(7.3) is an abnormality-free system. Using the idea of separation of variables [17], it can be concluded that the abnormality-free system is **unstable** when $\lambda > L^{-2}\pi^2$.

Definition 7.1 [22] (Exponential Stability in the Sense of $|| \cdot ||_2$). The abnormality-free linear PDE system (7.1)–(7.3) is said to be exponential stable in the sense of $|| \cdot ||_2$, if there exists constants $\alpha \geq 1$ and $\beta > 0$ such that $||u(\cdot,t)||_2 \leq \alpha ||u_0(\cdot)||_2 \exp(-\beta t)$ is fulfilled for any $t \geq 0$.

The problem investigated in this chapter can be summarized as follows: detect and estimate the unknown abnormal S-T source $s(x,t)$ for the unstable parabolic PDE system (7.1)–(7.3).

7.3 Adaptive PDE Observer Design

7.3.1 ADF Design and Analysis

To facilitate the adaptive abnormal S-T source estimation, an ADF is constructed first such that the source estimation is only activated after an abnormal source has been detected in the time domain.

Intuitively, a Luenberger-type PDE observer [24] using only pointwise measurement is constructed for system (7.1)–(7.3) as follows:

$$\begin{aligned} \bar{u}_t(x,t) &= \bar{u}_{xx}(x,t) + \lambda\bar{u}(x,t) + \boldsymbol{c}^{\mathrm{T}}(x)\boldsymbol{H}(\boldsymbol{y}(t) - \bar{\boldsymbol{y}}(t)), \\ \bar{u}(0,t) &= \bar{u}(L,t) = 0, \\ \bar{\boldsymbol{y}}(t) &= \int_0^L \boldsymbol{c}(x)\bar{u}(x,t)\mathrm{d}x, \end{aligned} \tag{7.5}$$

with $\boldsymbol{H} \triangleq \operatorname{diag}\{h_1, h_2, \cdots, h_n\}$ denoting the observer gain matrix. The following estimation error system can be derived:

$$\begin{aligned} \tilde{u}_t(x,t) &= \tilde{u}_{xx}(x,t) + \lambda\tilde{u}(x,t) + s(x,t) - \boldsymbol{c}^{\mathrm{T}}(x)\boldsymbol{H}(\boldsymbol{y}(t) - \bar{\boldsymbol{y}}(t)), \\ \tilde{u}(0,t) &= \tilde{u}(L,t) = 0, \end{aligned} \tag{7.6}$$

where $\tilde{u}(x,t) \triangleq u(x,t) - \bar{u}(x,t)$.

Lemma 7.1 *(Theorem 3.1 of [22])*
Given a constant $\gamma > 0$, if there exists scalar $q > 0$ and $\bar{h}_i, i \in \mathcal{N}$ satisfying the following LMIs:

$$\boldsymbol{\Xi}_i \triangleq \begin{bmatrix} q\lambda + 1 - \frac{q\pi^2}{4\phi_i} & \star & \star & \star \\ \frac{q\pi^2}{4\phi_i} & -\frac{\bar{h}_i}{x_{i+1}-x_i} - \frac{q\pi^2}{4\phi_i} & \star & \star \\ \frac{q}{2} & 0 & -\gamma^2 & \star \\ 0 & -\frac{\bar{h}_i}{2(x_{i+1}-x_i)} & 0 & -\frac{\gamma^2}{(x_{i+1}-x_i)} \end{bmatrix} < 0, i \in \mathcal{N} \tag{7.7}$$

where $\phi_i \triangleq \max\{(\bar{x}_i - x_i)^2, (x_{i+1} - \bar{x}_i)^2\}, 0 = x_1 < x_2 < \cdots < x_{n+1} = L$ and $\bar{x}_i \in [x_i, x_{i+1}], i \in \mathcal{N}$. Then the estimation error system (7.6) is exponential stable in the sense of $\|\cdot\|_2$ when $s(x,t) = 0$. The observer gains h_i are given as:

$$h_i = q^{-1}\bar{h}_i, i \in \mathcal{N}. \tag{7.8}$$

Thanks to Lemma 7.1, we can select a residual signal $r(t) = |\boldsymbol{y}(t) - \bar{\boldsymbol{y}}(t)|_2^2$ for abnormality detection in the time domain.

Corollary 7.1 *If conditions in Lemma 7.1 are satisfied, then the residual signal $r(t)$ can guarantee reliable abnormality detection performance with the following decision logic:*

$$\begin{cases} r(t) > r_{th} = 0 \Longrightarrow \textit{Abnormal at time } t \\ \textit{otherwise} \Longrightarrow \textit{Abnormality-free at time } t \end{cases} \tag{7.9}$$

for $t > t_0$, where t_0 denotes the first time that $r(t) \to 0$.

7.3.2 Motivation: Adaptive Observers for ODEs

Before introducing the proposed adaptive PDE observer, a quick review of adaptive observers for ODEs is first presented. For simple illustration, let us start with the following class of linear time invariant (LTI) systems:

$$\begin{aligned} \dot{z}(t) &= \boldsymbol{A}z(t) + \boldsymbol{B}\boldsymbol{u}(t) + \boldsymbol{E}\boldsymbol{f}(t), \\ \xi(t) &= \boldsymbol{C}z(t), \end{aligned}$$

where $z(t) \in \mathbb{R}^p$, $\boldsymbol{u}(t) \in \mathbb{R}^q$ and $\xi(t) \in \mathbb{R}^r$ denote the state, control, and output, respectively. $\boldsymbol{f}(t) \in \mathbb{R}^m$ denotes an unknown term (fault, disturbance, etc.) to be estimated. $\boldsymbol{A}$, $\boldsymbol{B}$, $\boldsymbol{C}$, and $\boldsymbol{E}$ are constant matrices of appropriate dimensions.

Intuitively, the following Luenberger-type observer can be designed using the format of "copy of the plant plus output injection":

$$\begin{aligned} \dot{\hat{z}}(t) &= \boldsymbol{A}\hat{z}(t) + \boldsymbol{B}\boldsymbol{u}(t) + \boldsymbol{E}\hat{\boldsymbol{f}}(t) + \boldsymbol{L}_o(\xi(t) - \hat{\xi}(t)), \\ \hat{\xi}(t) &= \boldsymbol{C}\hat{z}(t), \end{aligned}$$

where $\hat{z}(t)$, $\hat{\xi}(t)$, and $\hat{\boldsymbol{f}}(t)$ denote the estimation of state, output and unknown term, respectively. $\boldsymbol{L}_o$ denotes the observer gain matrix. The main difficulty lies in the adaptive estimation algorithm design for $\boldsymbol{f}(t)$ since it does not have a known dynamic as the state variable $z(t)$, which makes the problem interesting yet challenging.

For decades, a number of works have been reported on the design of the adaptive estimation algorithms (see [20, 25–29] for reference). The fundamental idea of these works can be concluded as making full use of the output estimation error $\xi(t) - \hat{\xi}(t)$ to update the estimation of $\hat{\boldsymbol{f}}(t)$. The motivations are twofold:

- On the one hand, $\xi(t) - \hat{\xi}(t)$ can be obtained directly since $\xi(t)$ is measured, this is the same reason for using it in the state observer design;
- On the other hand, as long as $\boldsymbol{f}(t) - \hat{\boldsymbol{f}}(t) \neq 0$, we have that $\xi(t) - \hat{\xi}(t) \neq 0$, hence the adaptive estimation algorithm is triggered to update $\hat{\boldsymbol{f}}(t)$.

One representative adaptive estimation algorithm motivated by [20] for this linear system is:

$$\frac{d\hat{\boldsymbol{f}}(t)}{dt} = \boldsymbol{\Gamma}(\boldsymbol{\xi}(t) - \hat{\boldsymbol{\xi}}(t)), \tag{7.10}$$

where $\boldsymbol{\Gamma} = \boldsymbol{\Gamma}^{\mathrm{T}} > 0$ of appropriate dimension defines the learning rate. Under this setting, the algorithm (7.10) stops updating the value of $\hat{\boldsymbol{f}}(t)$ when $\boldsymbol{\xi}(t) \to \hat{\boldsymbol{\xi}}(t)$.

In summary, the intuitive idea of the adaptive observers is to make full use of the output estimation error $\boldsymbol{\xi}(t) - \hat{\boldsymbol{\xi}}(t)$ for updating $\hat{\boldsymbol{f}}(t)$ adaptively. Bearing this in mind, an adaptive Luenberger-type PDE observer is initiated.

7.3.3 Adaptive Luenberger-Type Observer Design: from ODEs to PDEs

Intuitively, an adaptive Luenberger-type PDE observer for the PDE system (7.1)–(7.3) can be designed as follows:

$$\begin{aligned} \hat{u}_t(x,t) &= \hat{u}_{xx}(x,t) + \lambda\hat{u}(x,t) + \hat{s}(x,t) + \boldsymbol{l}(x)(\boldsymbol{y}(t) - \hat{\boldsymbol{y}}(t)), x \in (0, L), t > 0 \\ \hat{\boldsymbol{y}}(t) &= \int_0^L \boldsymbol{c}(x)\hat{u}(x,t)\mathrm{d}x, \end{aligned} \tag{7.11}$$

subject to the following homogeneous Dirichlet boundary conditions:

$$\hat{u}(0,t) = \hat{u}(L,t) = 0, t > 0$$

where $\hat{u}(x,t)$, $\hat{s}(x,t)$, and $\hat{\boldsymbol{y}}(t)$ denote the state, abnormal source, and output estimation, respectively. $\boldsymbol{l}(x) \in \mathbb{R}^{1\times n}$ denotes the observer gain to be determined.

Comparing this adaptive Luenberger-type PDE observer with the adaptive ODE observer introduced in the previous section, it is obvious that the observer gain is a vector function of the spatial variable x rather than a constant matrix, which makes the problem more complicated.

Fortunately, knowledge of the measurement equation $\boldsymbol{c}(x)$ can be used to reduce the design effort. In fact, we can select the observer gain $\boldsymbol{l}(x)$ in the following form:

$$\boldsymbol{l}(x) = \boldsymbol{c}^{\mathrm{T}}(x)\boldsymbol{L}, \tag{7.12}$$

where $\boldsymbol{L} \in \mathbb{R}^{n\times n}$ is a constant matrix to be determined. In this manner, the problem transfers from the determination of a vector function to a constant matrix, while the latter requires less effort than the former. To further reduce the design effort, the constant matrix $\boldsymbol{L}$ is selected as:

$$\boldsymbol{L} \triangleq \mathrm{diag}\,\{l_1, l_2, \cdots, l_n\}\,. \tag{7.13}$$

The next step is to determine the adaptive estimation algorithm for $s(x, t)$. Motivated by the structure in (7.10), we can construct an estimator in a similar manner:

$$\hat{s}_t(x, t) = \boldsymbol{k}(x)(\boldsymbol{y}(t) - \hat{\boldsymbol{y}}(t)), \tag{7.14}$$

where $\boldsymbol{k}(x) \in \mathbb{R}^{1\times n}$ denotes the output injection weights to be determined, and $\hat{s}_t(x, t) \triangleq \partial \hat{s}(x, t)/\partial t$ denotes the updating law of the estimation $\hat{s}(x, t)$ in the time domain. Unfortunately, this algorithm is not capable of completing this task. The reasons are twofold:

- First, the spatial distribution characteristic of $s(x, t)$ is unknown to us. Hence the algorithm (7.14) updates $s(x, t)$ according to a predefined $\boldsymbol{k}(x)$ without "feedback" signal of the spatial distribution error;
- Second, unlike the state variable $u(x, t)$, the abnormal source term's dynamic model is unknown and we cannot construct an observer for it. All we can use for the estimation is limited measurement, which is not enough for mapping $\boldsymbol{y}(t) - \hat{\boldsymbol{y}}(t)$ to $\hat{s}_t(x, t)$ for each specific location x.

Remark 7.2 Note that the PDE observer (7.5) uses only the estimation error at certain locations to achieve an exponential stable error system (7.6) while the adaptive estimation algorithm (7.14) fails to do so. The reason is that the diffusion term $\bar{u}_{xx}$ transmits the estimation error from limited sensor locations to the whole space.

Combining (7.11), (7.12), and (7.13), an adaptive Luenberger-type PDE observer is obtained as follows:

$$\begin{aligned} \hat{u}_t(x, t) &= \hat{u}_{xx}(x, t) + \lambda \hat{u}(x, t) + \hat{s}(x, t) + \boldsymbol{c}^{\mathrm{T}}(x)\boldsymbol{L}(\boldsymbol{y}(t) - \hat{\boldsymbol{y}}(t)), \\ \hat{u}(0, t) &= \hat{u}(L, t) = 0, \\ \hat{\boldsymbol{y}}(t) &= \int_0^L \boldsymbol{c}(x)\hat{u}(x, t)\mathrm{d}x. \end{aligned} \tag{7.15}$$

Define the state, abnormal source estimation, and output error as $e(x, t) \triangleq u(x, t) - \hat{u}(x, t)$, $\tilde{s}(x, t) \triangleq s(x, t) - \hat{s}(x, t)$, and $\tilde{\boldsymbol{y}}(t) \triangleq \boldsymbol{y}(t) - \hat{\boldsymbol{y}}(t)$, respectively. Then, an adaptive estimation algorithm which uses full-state measurement is designed in the following form:

$$\hat{s}_t(x, t) = ke(x, t) + \sigma(e_t(x, t) - e_{xx}(x, t) - \lambda e(x, t) + \boldsymbol{c}^{\mathrm{T}}(x)\boldsymbol{L}(\boldsymbol{y}(t) - \hat{\boldsymbol{y}}(t))) \tag{7.16}$$

where $k, \sigma > 0$ are constants to be determined.

Remark 7.3 Note that the adaptive estimation algorithm (7.16) requires full-state measurement, which restricts its potential of application. However, this is inevitable for accurate estimation of $s(x, t)$. This requirement holds for adaptive identification of spatially varying coefficients of DPSs, as discussed in [30]. Moreover, even for the sensor and actuator fault parameters estimation in [18], full-state measurement

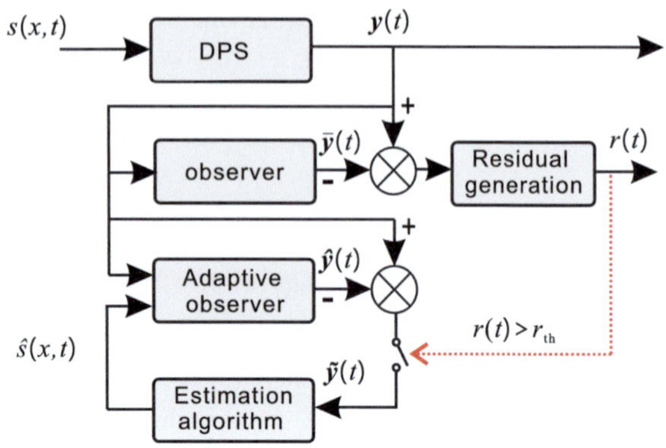

Fig. 7.1 Methodology framework

is still needed. However, from a control engineering perspective, one can use the method of interpolation alternatively, as mentioned in [31].

The following estimation error system is obtained by combining (7.1), (7.15) and (7.16):

$$e_t(x,t) = e_{xx}(x,t) + \lambda e(x,t) + \tilde{s}(x,t) - \mathbf{c}^{\mathrm{T}}(x)\boldsymbol{L}\tilde{\mathbf{y}}(t), \tag{7.17a}$$

$$e(0,t) = e(L,t) = 0, \tag{7.17b}$$

$$\tilde{\mathbf{y}}(t) = \int_0^L \mathbf{c}(x)e(x,t)\mathrm{d}x, \tag{7.17c}$$

$$\tilde{s}_t(x,t) = s_t(x,t) - ke(x,t) - \sigma(e_t(x,t) - e_{xx}(x,t) - \lambda e(x,t) + \mathbf{c}^{\mathrm{T}}(x)\boldsymbol{L}(\mathbf{y}(t) - \hat{\mathbf{y}}(t))). \tag{7.17d}$$

Combining (7.17a) and (7.17d), it can be obtained that

$$\tilde{s}_t(x,t) = s_t(x,t) - ke(x,t) - \sigma\tilde{s}(x,t),$$

which further leads to

$$\hat{s}_t(x,t) = -ke(x,t) - \sigma\tilde{s}(x,t).$$

Hence when $\hat{s}_t(x,t) = 0$, we have that

$$\tilde{s}(x,t) = -\frac{ke(x,t)}{\sigma},$$

which indicates that $\hat{s}(x,t) \to s(x,t)$ when $e(x,t) = 0$. Or in other words, $\hat{s}(x,t)$ keeps updating as long as $e(x,t) \neq 0$. Hence the adaptive estimation algorithm (7.16) inherits the merit of algorithm (7.10) for ODEs. A methodology framework which summarizes the main design procedures is presented in Fig. 7.1.

Remark 7.4 It is worth noticing that if $k = 0$ and $\sigma = 0$ are chosen for the estimation algorithm (7.16), we have that $s_t(x, t) = 0$. Selecting $\hat{s}_0(x) = 0$ leads to $\hat{s}(x, t) = 0$. Under this setting, the adaptive PDE observer (7.15) degrades to the state observer in (7.6) which only needs limited point-wise measurement. Furthermore, it is a trivial exercise to obtain that if the following LMIs can be satisfied:

$$\Upsilon_i \triangleq \begin{bmatrix} p\lambda - \frac{p\pi^2}{4\phi_i} & \star \\ \frac{p\pi^2}{4\phi_i} & -\frac{\bar{l}_i}{x_{i+1}-x_i} - \frac{p\pi^2}{4\phi_i} \end{bmatrix} < 0, i \in \mathcal{N}$$

then the estimation error system (7.17) is exponential stable in the sense of $||\cdot||_2$ when $s(x, t) = 0$. The observer gains l_i are given as:

$$l_i = p^{-1}\bar{l}_i, i \in \mathcal{N}.$$

In the same manner, one can design a residual signal $\bar{r}(t) = |\mathbf{y}(t) - \hat{\mathbf{y}}(t)|_2^2$ for abnormality detection in the time domain.

7.3.4 *Spatial Domain Decomposition-Based Convergence Analysis*

Now we are in the position to analyze the convergence property of the proposed adaptive PDE observer (7.15). Some necessary assumption and useful lemmas are introduced as follows.

Assumption 7.1 The time derivative of abnormal source $s_t(x, t)$ is bounded in the sense of spatial $\mathcal{L}^2$ norm, i.e.,

$$||s_t||_2 \le \bar{s}_1 < \infty,$$

where constant $\bar{s}_1 > 0$.

Lemma 7.2 *(Wirtinger's inequality [32, 33]) Let $u \in \mathcal{W}^{1,2}(0, L)$ be a scalar function with $u(0) = 0$ or $u(L) = 0$. Then*

$$\int_0^L u^2(s)\mathrm{d}s \le 4L^2\pi^{-2}\int_0^L (\mathrm{d}u(s)/\mathrm{d}s)^2\mathrm{d}s.$$

Lemma 7.3 *[34] Let $V(t)$ and $h(t)$ be real functions. Then*

$$\dot{V}(t) \le -\alpha V(t) + h(t), \forall t \ge 0$$

implies that

$$V(t) \le e^{-\alpha t}V(0) + \int_0^t e^{-\alpha(t-\tau)}h(\tau)\mathrm{d}\tau, \quad \forall t \ge 0$$

for any finite constant α.

Theorem 7.1 *Given a constant* $\sigma > 1/2$, *if there exist scalars* $p > 0, \bar{l}_i, i \in \mathcal{N}$, *and* $\bar{k}_i, i \in \mathcal{N}$ *satisfying the following LMIs:*

$$\boldsymbol{\Psi}_i \triangleq \begin{bmatrix} p\lambda - \frac{p\pi^2}{4\phi_i} & \star & \star \\ \frac{p\pi^2}{4\phi_i} & -\frac{\bar{l}_i}{x_{i+1}-x_i} - \frac{p\pi^2}{4\phi_i} & \star \\ \frac{p-\bar{k}}{2} & 0 & -\frac{p(2\sigma-1)}{2} \end{bmatrix} < 0, i \in \mathcal{N} \tag{7.18}$$

where $\phi_i \triangleq \max\{(\bar{x}_i - x_i)^2, (x_{i+1} - \bar{x}_i)^2\}$, *then the estimation errors* $e(x,t)$ *and* $\tilde{s}(x,t)$ *are UUB in the sense of* $\|\cdot\|_2$. *The observer gains* l_i *and* k *are given as:*

$$\begin{aligned} l_i &= p^{-1}\bar{l}_i, i \in \mathcal{N}, \\ k &= p^{-1}\bar{k}. \end{aligned} \tag{7.19}$$

Proof Consider the following Lyapunov function:

$$V(t) = \frac{1}{2}\int_0^L pe^2(x,t)\mathrm{d}x + \frac{1}{2}\int_0^L p\tilde{s}^2(x,t)\mathrm{d}x. \tag{7.20}$$

Set $\bar{l}_i = pl_i, i \in \mathcal{N}, \bar{k} = pk$, recalling the error dynamics in (7.17), the time derivative of $V(t)$ can be calculated as:

$$\begin{aligned} \dot{V}(t) &= \int_0^L pe(x,t)e_t(x,t)\mathrm{d}x + \int_0^L p\tilde{s}(x,t)\tilde{s}_t(x,t)\mathrm{d}x, \\ &= \underbrace{\int_0^L pe(x,t)e_{xx}(x,t)\mathrm{d}x}_{\text{Integration by parts+boundary conditions in (7.17b)}} + \lambda\int_0^L pe^2(x,t)\mathrm{d}x \\ &\quad - \int_0^L pe(x,t)\mathbf{c}^{\mathrm{T}}(x)\mathbf{L}\mathrm{d}x\int_0^L \mathbf{c}(x)e(x,t)\mathrm{d}x + \int_0^L pe(x,t)\tilde{s}(x,t)\mathrm{d}x + \int_0^L p\tilde{s}(x,t)s_t(x,t)\mathrm{d}x \\ &\quad - \int_0^L \bar{k}e(x,t)\tilde{s}(x,t)\mathrm{d}x - \int_0^L p\sigma\tilde{s}^2(x,t)\mathrm{d}x \\ &= -\int_0^L pe_x^2(x,t)\mathrm{d}x + \int_0^L p\lambda e^2(x,t)\mathrm{d}x - \sum_{i=1}^n \bar{l}_i e^2(\bar{x}_i,t) + \int_0^L (p-\bar{k})e(x,t)\tilde{s}(x,t)\mathrm{d}x \\ &\quad + \int_0^L p\tilde{s}(x,t)s_t(x,t)\mathrm{d}x - \int_0^L p\sigma\tilde{s}^2(x,t)\mathrm{d}x. \end{aligned} \tag{7.21}$$

Selecting $0 = x_1 < x_2 < \cdots < x_{n+1} = L$, then the integration interval $[0, L]$ is decomposed into n small intervals $[x_i, x_{i+1}], i \in \mathcal{N}$ while each measurement location $\bar{x}_i, i \in \mathcal{N}$ is contained in this interval. This is the main idea of the spatial domain decomposition approach [22], as illustrated in Fig. 7.2.

With the spatial domain decomposition approach, (7.21) can be rewritten in the following decomposed form:

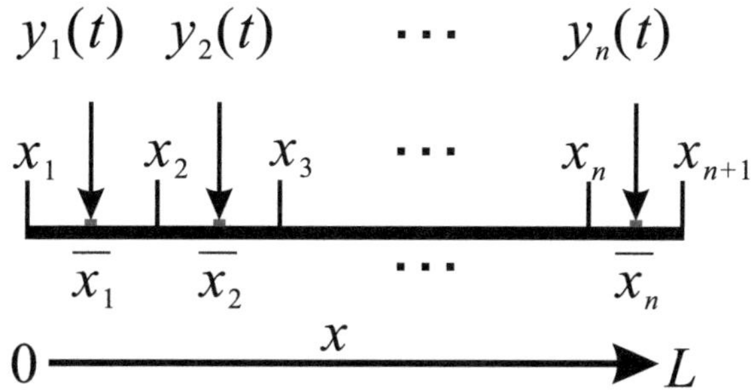

Fig. 7.2 Illustration of the spatial decomposition approach with pointwise measurement

$$\dot{V}(t) = -\sum_{i=1}^{n}\int_{x_i}^{x_{i+1}} pe_x^2(x,t)\mathrm{d}x + \sum_{i=1}^{n}\int_{x_i}^{x_{i+1}} p\lambda e^2(x,t)\mathrm{d}x - \sum_{i=1}^{n}\bar{l}_i e^2(\bar{x}_i,t)$$
$$+\sum_{i=1}^{n}\int_{x_i}^{x_{i+1}} (p-\bar{k})e(x,t)\tilde{s}(x,t)\mathrm{d}x + \sum_{i=1}^{n}\int_{x_i}^{x_{i+1}} p\tilde{s}(x,t)s_t(x,t)\mathrm{d}x - \sum_{i=1}^{n}\int_{x_i}^{x_{i+1}} p\sigma\tilde{s}^2(x,t)\mathrm{d}x. \tag{7.22}$$

Recalling Assumption 7.1, it can be obtained that the fifth term on the RHS of (7.22) satisfies that:

$$\begin{aligned}\sum_{i=1}^{n}\int_{x_i}^{x_{i+1}} p\tilde{s}(x,t)s_t(x,t)\mathrm{d}x &\le \sum_{i=1}^{n}\int_{x_i}^{x_{i+1}} \frac{p}{2}\tilde{s}^2(x,t)\,\mathrm{d}x + \int_0^L \frac{p}{2}s_t^2(x,t)\,\mathrm{d}x \\ &\le \sum_{i=1}^{n}\int_{x_i}^{x_{i+1}} \frac{p}{2}\tilde{s}^2(x,t)\,\mathrm{d}x + \frac{p\bar{s}_1^2}{2}\end{aligned} \tag{7.23}$$

by applying the Young's inequality.

Define a new variable $\tilde{e}_i(x,t) \triangleq e(x,t) - e(\bar{x}_i,t), i \in \mathcal{N}$, it can be obtained that $\tilde{e}_{i,x}(x,t) = e_x(x,t)$. Applying the spatial-domain decomposition technique once again, the first term on the RHS of (7.22) for any $i \in \mathcal{N}$ satisfies that:

$$\begin{aligned}&-\int_{x_i}^{x_{i+1}} pe_x^2(x,t)\mathrm{d}x \\ &= -\int_{x_i}^{\bar{x}_i} pe_x^2(x,t)\mathrm{d}x - \int_{\bar{x}_i}^{x_{i+1}} pe_x^2(x,t)\mathrm{d}x \\ &= -\int_{x_i}^{\bar{x}_i} p\tilde{e}_{i,x}^2(x,t)\mathrm{d}x - \int_{\bar{x}_i}^{x_{i+1}} p\tilde{e}_{i,x}^2(x,t)\mathrm{d}x \\ &\le -\frac{p\pi^2}{4(\bar{x}_i - x_i)^2}\int_{x_i}^{\bar{x}_i} \tilde{e}_i^2(x,t)\mathrm{d}x - \frac{p\pi^2}{4(x_{i+1}-\bar{x}_i)^2}\int_{\bar{x}_i}^{x_{i+1}} \tilde{e}_i^2(x,t)\mathrm{d}x \\ &\le -\frac{p\pi^2}{4\phi_i}\int_{x_i}^{x_{i+1}} \tilde{e}_i^2(x,t)\mathrm{d}x\end{aligned} \tag{7.24}$$

by applying Lemma 7.2 since $\tilde{e}_i(\bar{x}_i, t) = 0$ and $p > 0$.

Substituting (7.23) and (7.24) into (7.22) and recalling the definition of $\tilde{e}_i(x, t)$ leads to:

$$\begin{aligned}
\dot{V}(t) \le & -\sum_{i=1}^{n}\int_{x_i}^{x_{i+1}} \frac{p\pi^2}{4\phi_i}\tilde{e}_i^2(x,t)\mathrm{d}x + \sum_{i=1}^{n}\int_{x_i}^{x_{i+1}} p\lambda e^2(x,t)\mathrm{d}x - \sum_{i=1}^{n}\bar{l}_i e^2(\bar{x}_i, t) \\
& + \sum_{i=1}^{n}\int_{x_i}^{x_{i+1}} (p-\bar{k})e(x,t)\tilde{s}(x,t)\mathrm{d}x + \sum_{i=1}^{n}\int_{x_i}^{x_{i+1}} \frac{p-2p\sigma}{2}\tilde{s}^2(x,t)\,\mathrm{d}x + \frac{p\bar{s}_1^2}{2} \\
= & \sum_{i=1}^{n}\int_{x_i}^{x_{i+1}} (p\lambda - \frac{p\pi^2}{4\phi_i})e^2(x,t)\mathrm{d}x - \sum_{i=1}^{n}\int_{x_i}^{x_{i+1}} (\frac{\bar{l}_i}{x_{i+1}-x_i} + \frac{p\pi^2}{4\phi_i})e^2(\bar{x}_i,t)\mathrm{d}x \\
& + \sum_{i=1}^{n}\int_{x_i}^{x_{i+1}} (p-\bar{k})e(x,t)\tilde{s}(x,t)\mathrm{d}x + \sum_{i=1}^{n}\int_{x_i}^{x_{i+1}} \frac{p\pi^2}{2\phi_i}e(x,t)e(\bar{x}_i,t)\mathrm{d}x \\
& + \sum_{i=1}^{n}\int_{x_i}^{x_{i+1}} \frac{p-2p\sigma}{2}\tilde{s}^2(x,t)\,\mathrm{d}x + \frac{p\bar{s}_1^2}{2} \\
= & \sum_{i=1}^{n}\int_{x_i}^{x_{i+1}} \boldsymbol{\xi}_i^T(x,t)\boldsymbol{\Psi}_i\boldsymbol{\xi}_i(x,t)\mathrm{d}x + \frac{p\bar{s}_1^2}{2}
\end{aligned} \tag{7.25}$$

where $\boldsymbol{\xi}_i(x,t) \triangleq [e(x,t) \quad e(\bar{x}_i,t) \quad \tilde{s}(x,t)]^T$.

Since the LMIs in (7.18) hold, there exists an appropriate constant $\eta > 0$ such that

$$\boldsymbol{\Psi}_i + \frac{\eta p}{2}\boldsymbol{I} \le 0, i \in \mathcal{N}. \tag{7.26}$$

Substituting (7.26) into (7.25) further yields that

$$\begin{aligned}
\dot{V}(t) \le & -\frac{\eta p}{2}\sum_{i=1}^{n}\int_{x_i}^{x_{i+1}} \boldsymbol{\xi}_i^T(x,t)\boldsymbol{\xi}_i(x,t)\mathrm{d}x + \beta \le -\frac{\eta p}{2}\int_0^L (e^2(x,t) + \tilde{s}^2(x,t))\mathrm{d}x + \beta \\
= & -\eta V(t) + \beta,
\end{aligned} \tag{7.27}$$

where $\beta \triangleq \frac{p\bar{s}_1^2}{2}$.

Applying Lemma 7.3 to (7.27) yields that

$$V(t) \le V(0)\exp(-\eta t) + \frac{\beta}{\eta}, t \ge 0, \tag{7.28}$$

which further leads to:

$$||e||_2^2 + ||\tilde{s}||_2^2 \le \frac{2V(0)}{p}\exp(-\eta t) + \frac{2\beta}{p\eta}. \tag{7.29}$$

Hence it can be concluded that:

$$||e||_2 \leq \sqrt{\frac{2V(0)}{p}\exp(-\eta t) + \rho},$$
$$||\tilde{s}||_2 \leq \sqrt{\frac{2V(0)}{p}\exp(-\eta t) + \rho}, \tag{7.30}$$

where $\rho \triangleq \frac{\bar{s}_1}{\sqrt{\eta}}$. Recalling Assumption 7.1, it can be obtained that $0 < \rho < \infty$. Hence (7.30) implies that there exists a $T \to \infty$ such that $||e||_2 \leq \rho$, $||\tilde{s}||_2 \leq \rho$ holds for all $t > T$. That is to say, the estimation errors $e(x, t)$ and $\tilde{s}(x, t)$ are UUB in the sense of $|| \cdot ||_2$ with ultimate bound ρ.

This completes the proof.

Remark 7.5 The main theoretical contribution of this chapter lies in the adaptive PDE observer and the adaptive estimation algorithm design and analysis for ensuring the convergence of both the state and abnormality estimation errors. As for the detailed analytic method such as the Lyapunov stability analysis, it was also commonly used in other works.

Remark 7.6 The design parameters in the adaptive PDE observer design (7.15) and (7.16) include the n number of sensor placement locations $\bar{x}_i, i \in \mathcal{N}$, the spatial domain decomposition intervals $[x_i, x_{i+1}], i \in \mathcal{N}$, the design parameter of adaptive estimation algorithm-σ. The other parameters are calculated by solving the LMIs (7.7) and (7.18). For the parameters selection, we adopt a "trial and error manner" to obtain a satisfied performance. The optimal parameters selection is indeed an interesting yet challenging problem itself, it can be formulated as an optimal detection and estimation problem.

7.4 Illustrative Example

For simple illustration purpose, consider the linear PDE system (7.1)–(7.3) with $L = 1$ and $\lambda = 10 > L^{-2}\pi^2$, hence the system is unstable referring to Remark 7.1. The initial condition is selected as:

$$u_0(x) = 0.5\sin(\pi x/L), x \in [0, L],$$

which is **unknown** for both PDE observers.

Furthermore, the following abnormality is injected:

$$s(x, t) = b_s(x)s(t),$$

where

$$b_s(x) \triangleq H(x - 0.5L) - H(x - 0.75L)$$

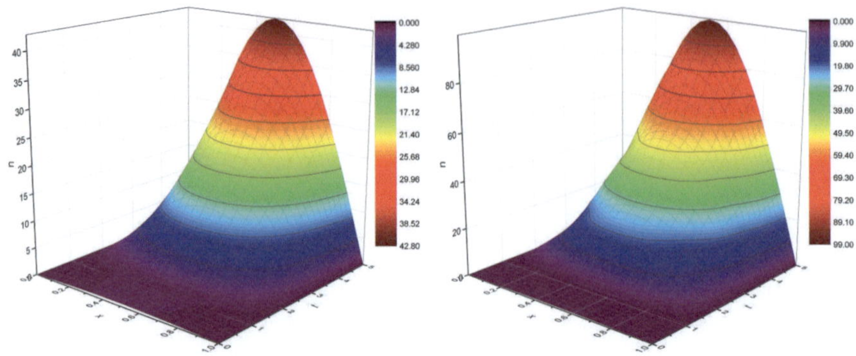

(a) Profile of evolution of $u(x,t)$ with incipient abnormality. (b) Profile of evolution of $u(x,t)$ with periodic abnormality.

Fig. 7.3 Simulation results of system (7.1)–(7.3) with abnormalities

is the shape function of the abnormality with $H(\cdot)$ denoting the standard Heaviside function. Two kinds of abnormal source are considered:

$$\text{Incipient abnormality}: s(t) = \begin{cases} 0, \ t \in [0, 1) \\ 20(1 - \exp(-2(t-5))), \ t \in [1, \infty) \end{cases}$$

$$\text{Periodic abnormality}: s(t) = \begin{cases} 0, \ t \in [0, 1) \\ 20(2 + \sin(5t)), \ t \in [1, \infty) \end{cases}$$

The profile of evolution of $u(x,t)$ with incipient and periodic abnormality are shown in Fig. 7.3a, respectively. The number of measurement sensors is chosen as $n = 2$, and the positions of two pointwise measurement sensors are selected as $\bar{x}_1 = 0.2L$ and $\bar{x}_2 = 0.7L$, respectively. To initiate the spatial domain decomposition-based observer design, the spatial domain $[0, L]$ is split into two subdomains $[x_1, x_2]$ and $[x_2, x_3]$ with $x_1 = 0$, $x_2 = 0.4L$, and $x_3 = L$. According to the definition of ϕ_i, it can be calculated that $\phi_1 = 0.04L^2$ and $\phi_2 = 0.09L^2$. Since the initial condition $u_0(x)$ is unknown for both observers, we select $\bar{u}_0(x) = \hat{u}_0(x) = 0$. Moreover, $\hat{s}_0(x) = 0$ is also set.

First, by solving LMIs (7.7), one can obtain an optimized attenuation level $\gamma^\star = 1.9573$, $q = 0.1523$, $\bar{h}_1 = 9.4493$, and $\bar{h}_2 = 7.6620$ via the *mincx* solver in the MATLAB LMI toolbox. Furthermore, it can be calculated that $h_1 = 62.0266$ and $h_2 = 50.2945$ by (7.8). Trajectories of $r(t)$ with incipient and periodic abnormality are shown in Fig. 7.4a, respectively. It can be observed that in the absence of abnormality, the residual signal $r(t) \to 0$ quickly, which indicates that the estimation error system (7.6) is stable. This result corresponds with the conclusion of Lemma 7.1. The abnormality detection time t_d for incipient and periodic abnormality is 1.004 and 1.002, respectively. Hence the adaptive observer (7.15) begins to work from $t = 1.004$ and $t = 1.002$ for incipient and periodic abnormality estimation, respectively.

According to Theorem 7.1, choosing $\sigma = 5$ and solving LMIs (7.18) via the *feasp* solver in the MATLAB LMI toolbox, one can obtain that $p = 0.0375$, $\bar{l}_1 = 0.9127$,

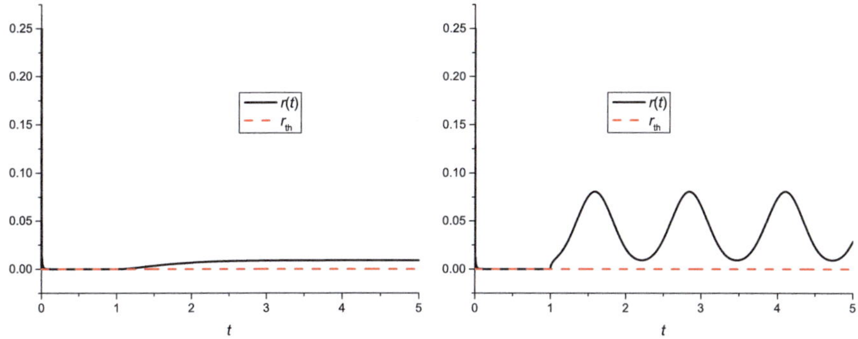

(a) Trajectory of $r(t)$ with incipient abnormality.

(b) Trajectory of $r(t)$ with periodic abnormality.

Fig. 7.4 Simulation results of the estimation error system (7.6) with abnormalities

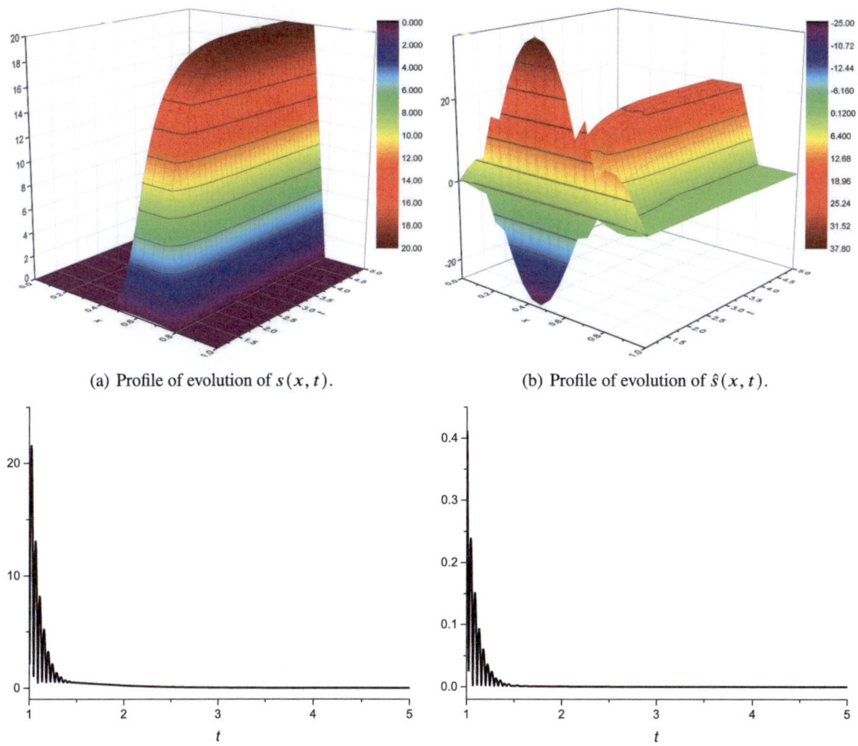

(a) Profile of evolution of $s(x, t)$.

(b) Profile of evolution of $\hat{s}(x, t)$.

(c) Trajectory of $||\tilde{s}||_2$.

(d) Trajectory of $||e||_2$.

Fig. 7.5 Simulation results of the adaptive estimation algorithm (7.16) and the estimation error system (7.17) with incipient abnormality from detection time ($t = 1.004$) and afterwards

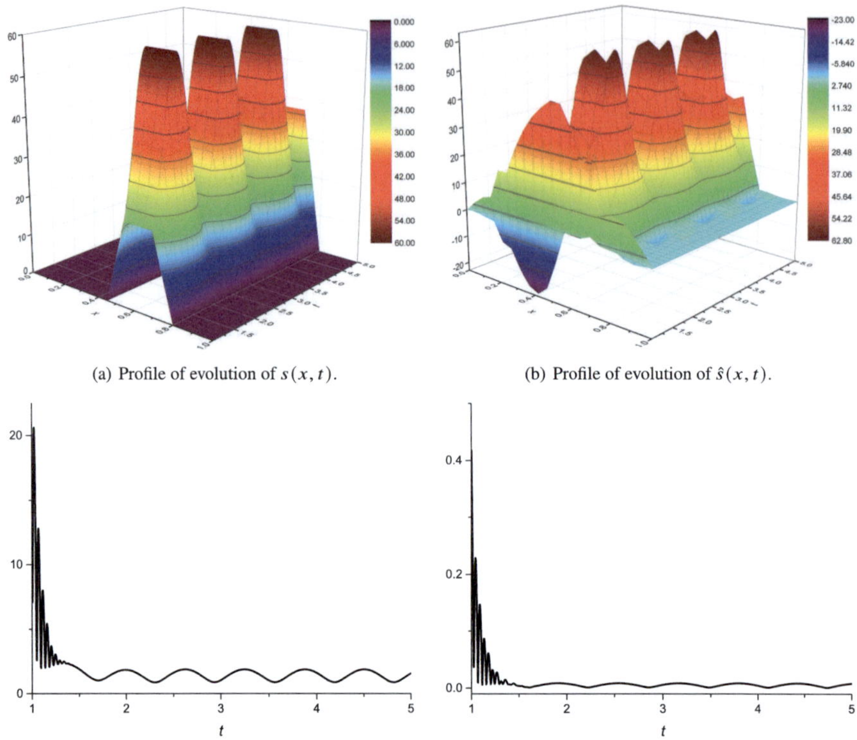

(a) Profile of evolution of $s(x,t)$.

(b) Profile of evolution of $\hat{s}(x,t)$.

(c) Trajectory of $||\tilde{s}||_2$.

(d) Trajectory of $||e||_2$.

Fig. 7.6 Simulation results of the adaptive estimation algorithm (7.16) and the estimation error system (7.17) with periodic abnormality from detection time ($t = 1.002$) and afterwards

$\bar{l}_2 = 1.4403$, and $\bar{k} = 0.2704$. Furthermore, it can be calculated that $l_1 = 24.3621$, $l_2 = 38.4468$ and $k = 7.2168$ by (7.19). The simulation results for estimation of incipient and periodic abnormality are shown in Figs. 7.5 and 7.6, respectively. It can be observed that for both abnormalities, $||\tilde{s}||_2$ and $||e||_2$ are UUB. To be noticed, there exists some error between $s(x,t)$ and $\hat{s}(x,t)$ in Figs. 7.5 and 7.6 initially and then vanishes to zero for incipient abnormality and nearly zero for periodic abnormality. Some minor and acceptable error exists for the periodic abnormality estimation due to the fact that for the considered periodic abnormality, the bound of $||s_t||_2$ (i.e., $\bar{s}_1$) is relatively high and it increases the bound of $||\tilde{s}||_2$ according to Theorem 7.1.

7.5 Conclusion

In this chapter, the detection and estimation problem of abnormal S-T source is studied. On the one hand, for the abnormality detection in the time domain, a residual signal is constructed using pointwise output estimation error; On the other hand, after an abnormality has been detected in the time domain, an adaptive Luenberger observer is triggered to estimate the abnormal source. Both rigorous theoretic analysis and numerical simulations are presented to validate the performance of the proposed method.

Moreover, the synchronization problem [35–37] and the delay phenomena [38–42] are also worth investigating.

References

1. Hu H, Yi T, Zou X (2020) On spatial-temporal dynamics of a Fisher-KPP equation with a shifting environment. Proc Am Math Soc 148(1):213–221
2. Feng Y, Li H-X (2018) Detection and spatial identification of fault for parabolic distributed parameter systems. IEEE Trans Ind Electron 66(9):7300–7309
3. Feng Y, Li H-X (2019) Dynamic spatial-independent-component-analysis-based abnormality localization for distributed parameter systems. IEEE Trans Ind Inform 16(5):2929–2936
4. George M, Jose BR, Mathew J (2020) Abnormal activity detection using shear transformed spatio-temporal regions at the surveillance network edge. Multimed Tools Appl 79(37):27511–27532
5. Zerkouk M, Chikhaoui B (2020) Spatio-temporal abnormal behavior prediction in elderly persons using deep learning models. Sensors 20(8):2359
6. Bouindour S, Snoussi H, Hittawe MM et al (2019) An on-line and adaptive method for detecting abnormal events in videos using spatio-temporal ConvNet. Appl Sci 9(4):757
7. George M, Jose BR, Mathew J et al (2018) Autoencoder-based abnormal activity detection using parallelepiped spatio-temporal region. IET Comput Vis 13(1):23–30
8. Geng Y, Du J, Liang M (2019) Abnormal event detection in tourism video based on salient spatio-temporal features and sparse combination learning. World Wide Web 22(2):689–715
9. Dey S, Perez HE, Moura SJ (2017) Model-based battery thermal fault diagnostics: algorithms, analysis, and experiments. IEEE Trans Control Syst Technol 27(2):576–587
10. Bandhauer TM, Garimella S, Fuller TF (2011) A critical review of thermal issues in lithium-ion batteries. J Electrochem Soc 158(3):R1–R25
11. Doughty DH, Roth EP (2012) A general discussion of Li ion battery safety. Electrochem Soc Interface 21(2):37–44
12. Wang B-C, Li H-X, Yang H-D (2019) Spatial correlation-based incremental learning for spatiotemporal modeling of battery thermal process. IEEE Trans Ind Electron 67(4):2885–2893
13. Wang B-C, Li H-X (2018) A sliding window based dynamic spatiotemporal modeling for distributed parameter systems with time-dependent boundary conditions. IEEE Trans Ind Inform 15(4):2044–2053
14. Ray WH (1981) Advanced process control. McGraw-Hill, Columbus
15. El-Farra NH, Ghantasala S (2007) Actuator fault isolation and reconfiguration in transport-reaction processes. AIChE J 53(6):1518–1537
16. Ghantasala S, El-Farra NH (2009) Robust actuator fault isolation and management in constrained uncertain parabolic PDE systems. Automatica 45(10):2368–2373
17. Krstic M, Smyshlyaev A (2008) Boundary control of PDEs: A course on backstepping designs. SIAM, Philadelphia
18. Cai J, Ferdowsi H, Sarangapani J (2016) Model-based fault detection, estimation, and prediction for a class of linear distributed parameter systems. Automatica 66:122–131
19. Dey S, Perez HE, Moura SJ (2019) Robust fault detection of a class of uncertain linear parabolic PDEs. Automatica 107:502–510

20. Wang H, Daley S (1996) Actuator fault diagnosis: an adaptive observer-based technique. IEEE Trans Autom Control 41(7):1073–1078
21. Feng Y, Li H-X, Yang H-D (2019) Abnormal source identification for parabolic distributed parameter systems. IEEE Trans Syst Man Cybern Syst 51(9):5698–5707
22. Wang J-W, Liu Y-Q, Hu Y-Y et al (2017) A spatial domain decomposition approach to distributed $\mathcal{H}_\infty$ observer design of a linear unstable parabolic distributed parameter system with spatially discrete sensors. Int J Control 90(12):2772–2785
23. Wang J-W, Liu Y-Q, Sun C-Y (2018) Pointwise exponential stabilization of a linear parabolic PDE system using non-collocated pointwise observation. Automatica 93:197–210
24. Wang J-W (2019) Observer-based boundary control of semi-linear parabolic PDEs with non-collocated distributed event-triggered observation. J Franklin Inst 356(17):10405–10420
25. Zhang Q (2002) Adaptive observer for multiple-input-multiple-output (MIMO) linear time-varying systems. IEEE Trans Autom Control 47(3):525–529
26. Jiang B, Wang JL, Soh YC (2002) An adaptive technique for robust diagnosis of faults with independent effects on system outputs. Int J Control 75(11):792–802
27. Jiang B, Staroswiecki M, Cocquempot V (2006) Fault accommodation for nonlinear dynamic systems. IEEE Trans Autom Control 51(9):1578–1583
28. Zhang K, Jiang B, Cocquempot V (2008) Adaptive observer-based fast fault estimation. Int J Control Autom Syst 6(3):320–326
29. Zhu J-W, Yang G-H, Wang H et al (2015) Fault estimation for a class of nonlinear systems based on intermediate estimator. IEEE Trans Autom Control 61(9):2518–2524
30. Orlov Y, Bentsman J (2000) Adaptive distributed parameter systems identification with enforceable identifiability conditions and reduced-order spatial differentiation. IEEE Trans Autom Control 45(2):203–216
31. Xiao T, Li H-X (2016) Eigenspectrum-based iterative learning control for a class of distributed parameter system. IEEE Trans Autom Control 62(2):824–836
32. Hardy GH, Littlewood JE, Pólya G (1952) Inequalities, 2nd edn. Cambridge University Press, Cambridge
33. Wang J-W, Wu H-N, Sun C-Y (2017) Local exponential stabilization via boundary feedback controllers for a class of unstable semi-linear parabolic distributed parameter processes. J Franklin Inst 354(13):5221–5244
34. Ioannou PA, Sun J (2012) Robust adaptive control. Courier Corporation, North Chelmsford
35. Pratap A, Raja R, Alzabut J et al (2020) Mittag-Leffler stability and adaptive impulsive synchronization of fractional order neural networks in quaternion field. Math Methods Appl Sci 43(10):6223–6253
36. Wang W, Huang C, Huang C et al (2020) Bipartite formation problem of second-order nonlinear multi-agent systems with hybrid impulses. Appl Math Comput 370:124926
37. Li L, Wang W, Huang L et al (2019) Some weak flocking models and its application to target tracking. J Math Anal Appl 480(2):123404
38. Huang C, Yang Z, Yi T et al (2014) On the basins of attraction for a class of delay differential equations with non-monotone bistable nonlinearities. J Differ Equ 256(7):2101–2114
39. Li W, Huang L, Ji J (2019) Periodic solution and its stability of a delayed Beddington-DeAngelis type predator-prey system with discontinuous control strategy. Math Methods Appl Sci 42(13):4498–4515
40. Long Z (2020) Tan Y (2020) Global attractivity for Lasota-Wazewska-type system with patch structure and multiple time-varying delays. Complexity 1:1947809
41. Wang F, Yao Z (2016) Approximate controllability of fractional neutral differential systems with bounded delay. Fixed Point Theory 17(2):495–508
42. Shi M, Guo J, Fang X et al (2020) Global exponential stability of delayed inertial competitive neural networks. Adv Difference Equ 1:1–12

Part II
Combined Model-Based and Data-Driven Abnormality Detection and Localization for Partially Known Industrial DPSs

Chapter 8
Computation-Efficient Fault Detection Framework for Partially-Known Nonlinear DPSs

Abstract In the field of distributed parameter systems, fault detection methods have typically relied on complete model knowledge. However, in many industrial scenarios, constructing accurate first-principle physical models is highly challenging, which limits the practical use of traditional model-based techniques. To address this issue, an adaptive neural network is developed to jointly estimate both the system states and the unknown nonlinear dynamics for a class of nonlinear distributed parameter systems with partial model information. Since full-state measurements are often impractical in real applications, the designed adaptive neural observer is built upon a reduced-order model, which also enhances computational efficiency. Fault detection is then realized through residual generation and evaluation based on the output estimation error of the proposed observer. Furthermore, to account for the neglected fast dynamics, a data-driven strategy for threshold generation is introduced. Comprehensive experimental studies are carried out to demonstrate and verify the effectiveness of the proposed approach.

8.1 Introduction

DPSs [1–3] are characterized by S-T dynamics, which cannot be adequately captured by ODEs. Typically, their first-principles models are governed by PDEs. Practical systems that can be categorized as DPSs include high-rise buildings [4], flexible mechanical structures [5–7], battery thermal processes [8], and industrial thermal processes [9, 10]. With the growing demand for safety and reliability in these applications, FD of DPSs has drawn increasing interest from both academic research [11, 12] and industrial practice [13]. As an illustration, actuator faults often arise in flexible manipulators, which may cause undesirable responses and system shutdowns, potentially leading to production accidents [14].

In contrast to the extensive studies on FD of LPSs represented by ODEs, which have been explored from both model-based [15] and data-driven [16] viewpoints, research on FD for DPSs remains relatively limited. The main challenge lies in the fact that the modeling and diagnosis techniques developed for LPSs cannot be directly extended to DPSs without substantial modifications.

© The Author(s), under exclusive license to Springer Nature Singapore Pte Ltd. 2026
Y. Feng et al., *Spatio-Temporal Abnormality Diagnosis for Industrial Distributed Parameter Systems*, https://doi.org/10.1007/978-981-95-3750-1_8

Unlike FD approaches for LPSs, the majority of existing methods for DPSs are model-based. In recent years, several data-driven statistical FD methods for DPSs have also been proposed [17, 18]. Nevertheless, similar to conventional SPC techniques, these methods exhibit poor adaptability to highly dynamic systems, which limits their applicability. Model-based approaches for DPSs can generally be divided into two categories: *early-lumping* and *late-lumping*, depending on whether modal approximation techniques are employed. The principle of *early-lumping* lies in converting the original PDE system into a set of ODEs, after which existing well-established LPSs methods can be applied [19, 20]. In contrast, most *late-lumping* approaches construct PDE observers directly through PDE backstepping techniques [21], with representative contributions found in [22–24].

Although significant progress has been made with model-based FD methods, a critical limitation of both early-lumping and late-lumping schemes is the requirement for explicit system models. However, this assumption is often impractical in industrial applications, where first-principles models are difficult to obtain due to inherent model uncertainties, unknown nonlinearities, and process disturbances or noises. In this chapter, we focus on a representative scenario: fault detection of nonlinear DPSs where only a linear nominal model is available while the nonlinear dynamics remain completely unknown. Since the unknown nonlinear terms depend on the infinite-dimensional system state, which is unobservable in practice, the natural idea is to first derive a reduced-order approximation model and then design a FDF.

Building upon the above motivations, this chapter introduces a novel and computation-efficient fault detection framework for partially known nonlinear DPSs. The framework begins with constructing a reduced-order model to facilitate online implementation. Based on this model, an adaptive neural observer is designed to simultaneously estimate both the system states and the unknown nonlinear dynamics. Fault detection is then achieved by generating and evaluating residuals derived from the observer's estimation errors. Furthermore, to address the influence of neglected fast dynamics—which are often ignored in existing studies—a data-driven approach is proposed for threshold generation.

The main contributions of this chapter can be summarized as follows:

- An adaptive neural network is developed to simultaneously estimate both the system states and the unknown nonlinear dynamics, thereby relaxing the requirement of complete model information in model-based fault detection approaches.
- A computationally efficient fault detection framework is established, which eliminates the need for full-state measurements.
- A data-driven scheme for fault detection threshold generation is proposed, where the influence of neglected fast dynamics is explicitly taken into account.

The remainder of this chapter is organized as follows. Section 8.2 presents the preliminaries and formulates the problem. Section 8.4 provides the detailed design of the proposed adaptive neural observer, together with the procedures for residual construction and evaluation. Experimental studies are reported in Sect. 8.5 to validate the effectiveness of the proposed approach. Finally, Sect. 8.6 concludes the chapter with a comprehensive summary.

8.2 Preliminaries and Problem Statement

Notations: $\mathbb{R}^n$ represents the n-dimensional Euclidean space, and $\|\cdot\|$ stands for the Euclidean norm. Define $\mathcal{H} \triangleq \mathcal{L}^2([0, L]; \mathbb{R})$ as the real Hilbert space consisting of square-integrable functions $\omega(z) : [0, L] \to \mathbb{R}$. The spatial $\mathcal{L}^2$-norm is given by $\|\omega(\cdot)\|_2 \triangleq \sqrt{\int_0^L \omega^2(z)\,\mathrm{d}z}$, and the corresponding inner product is defined as $\langle \omega_1(\cdot), \omega_2(\cdot)\rangle \triangleq \int_0^L \omega_1(z)\omega_2(z)\,\mathrm{d}z$. The notations for partial derivatives are $x_t \triangleq \partial x/\partial t$, $x_z \triangleq \partial x/\partial z$, and $x_{zz} \triangleq \partial^2 x/\partial z^2$. For a square matrix $\boldsymbol{A}$, $\lambda_{\min}(\boldsymbol{A})$ and $\lambda_{\max}(\boldsymbol{A})$ denote its smallest and largest eigenvalues, respectively. Similarly, $\sigma_{\min}(\cdot)$ and $\sigma_{\max}(\cdot)$ refer to the minimum and maximum singular values of a matrix. The operator $\mathrm{tr}(\cdot)$ denotes the trace of a matrix, while $\|\cdot\|_F$ stands for the Frobenius norm.

We focus on partially known nonlinear parabolic DPSs, which are governed by the following PDE:

$$x_t(z,t) = a_1 x_z(z,t) + a_2 x_{zz}(z,t) + g(x, \boldsymbol{u}(t)) + f_p(z,t), \tag{8.1a}$$

$$\boldsymbol{y}(t) = \int_0^1 \boldsymbol{c}(z) x(z,t)\mathrm{d}z + \boldsymbol{f}_s(t), \tag{8.1b}$$

subject to the following boundary conditions

$$b_1 x(0,t) + c_1 \frac{\partial x}{\partial z}(0,t) = d_1, \tag{8.2a}$$

$$b_2 x(L,t) + c_2 \frac{\partial x}{\partial z}(L,t) = d_2, \tag{8.2b}$$

with the initial condition

$$x(z,0) = x_0(z), \tag{8.3}$$

here, $x(\cdot, t) \in \mathcal{H}$ represents the system state; $z \in [0, L]$ is the spatial variable, and $t \in [0, \infty)$ denotes the (infinite-dimensional) time variable. The term $g(x, \boldsymbol{u}(t))$ is an *unknown* nonlinear function of the state $x(z,t)$ and the input $\boldsymbol{u}(t) \in \mathbb{R}^{n_u}$. The measured output is denoted by $\boldsymbol{y}(t) \in \mathbb{R}^{n_y}$, with $\boldsymbol{c}(z) \in \mathbb{R}^{n_y}$ characterizing the measurement mechanism. The system is subject to *distributed process faults* $f_p(z,t)$ and *sensor faults* $\boldsymbol{f}_s(t) \in \mathbb{R}^{n_y}$. These fault types correspond to actuator and sensor faults in ODE-based dynamical systems [15]. While sensor faults are straightforward, the process fault $f_p(z,t)$ encompasses both actuator faults (with identical spatial distribution) and additional process disturbances. The constants $a_1, a_2, b_1, b_2, c_1, c_2, d_1$, and d_2 are system parameters.

Remark 8.1 The systems described by (8.1)–(8.3) commonly arise in the modeling of transport-reaction processes where diffusive and dispersive effects are significant [25]. Typical application examples include rapid thermal processing systems [26], battery thermal processes [27], and tubular reactors [25], among others.

Remark 8.2 It is worth noting that the two most commonly employed measurement types are point-wise and piece-wise measurements. Detailed definitions and the associated measurement mechanisms can be found in [28].

The problem addressed in this chapter can be stated as follows:

Design a FDF for partially known nonlinear S-T systems, described by the state-space model in (8.1)–(8.3), using only a limited set of sensor measurements, *without requiring full-state observation.*

8.3 Methodology Framework

The major challenge of FDF design lies in the missing knowledge of the nonlinear dynamics $g(x, \boldsymbol{u}(t))$, which largely restricts the use of traditional model-based fault detection methods from both early-lumping [20] and late-lumping [23] perspectives.

To overcome this issue, one intuitive idea is to compensate for the unknown nonlinear dynamics in the FDF design. More specifically, a neural network (NN) is adopted to estimate the unknown nonlinear term in an on-line realization manner. One accompanying problem is that the full-state measurement $x(z, t)$ is *unavailable* in most industrial applications, hence late-lumping methods [35] which uses the state-variable $x(z, t)$ as the input for learning methods is not applicable. The physical meaning for full-state measurement is that the output equation in (8.1b) measures every single continuous point in the space $z \in [0, L]$ (point-wise measurements) and the dimension of $\boldsymbol{y}(t)$ is ∞. Obviously, this is impossible in applications, one can only use a large number of measurements along with the interpolation method. Instead, in the proposed framework, the early lumping methods are adopted to obtain a reduced-order approximation model in the first place. Then, an adaptive neural network (AdNN) is established for the observer design and it is on-line updating according to the *finite-dimensional output estimation error*. In this manner, the full-state measurement is not used in the whole design phase.

After that, the residual signal is constructed using the *finite-dimensional output estimation error* of the proposed AdNN. One accompanying problem of using the reduced-order model is the *observation spillover* phenomenon induced by neglecting the effects caused by the higher-order dynamics. Bearing this in mind, in the threshold generation phase, a kernel density estimation (KDE)-based data-driven approach is used to obtain the detection threshold rather than traditional model-based approaches which require information of the neglected fast dynamics to be known and restricts the usability of these approaches at large. It is worth noting that the threshold calculation is done in the normal condition (fault-free) and the threshold value keeps unchanged for on-line implementation.

In summary, the key idea of the proposed methodology is presented in Fig. 8.1. To help interesting readers to follow, the equations in the following sections are also put in the corresponding positions.

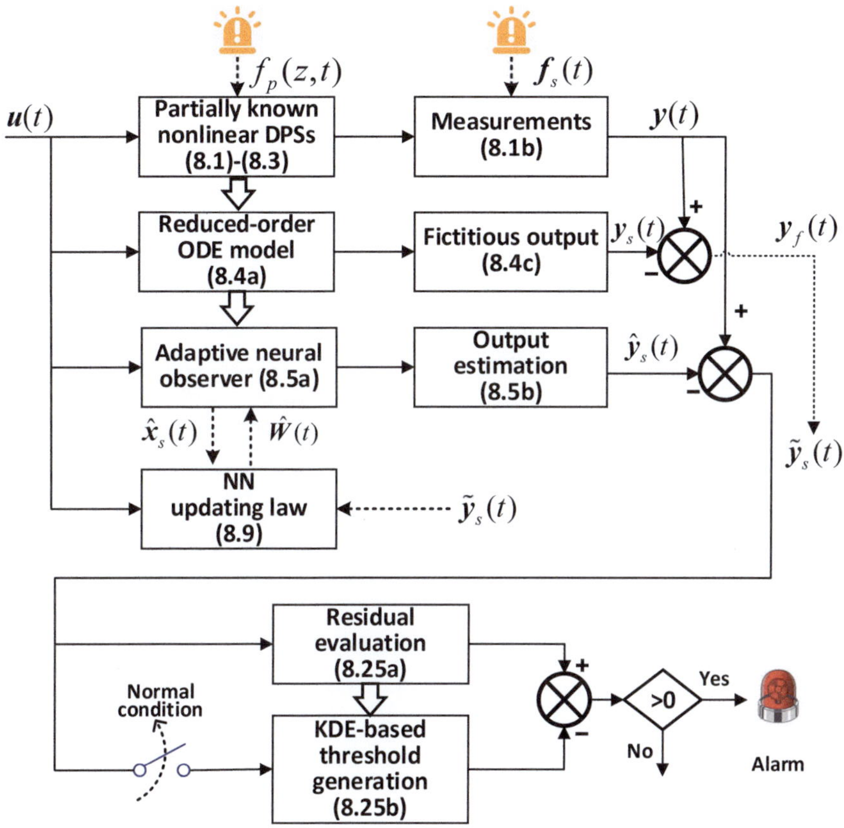

Fig. 8.1 Methodology framework

8.4 Neural Observer Design for Reduced-Order Model

8.4.1 Reduced-Order Model Acquisition

First, define the operator $\mathcal{A}$ on the Hilbert space $\mathcal{H}$ as:

$$\mathcal{A}x = a_1 \frac{\partial x}{\partial z} + a_2 \frac{\partial^2 x}{\partial z^2}, \; x \in \mathcal{S}(\mathcal{A}) \triangleq \{x \in \mathcal{H}, \text{ and the boundary conditions in (8.2a)-(8.2b) hold}\}.$$

For the spatial operator $\mathcal{A}$ defined above, the corresponding eigenvalue problem is given by:

$$\mathcal{A}\phi_j(z) = \lambda_j \phi_j(z), \;\; j = 1, \cdots, \infty,$$

here, ϕ_j denotes the eigenfunction associated with the eigenvalue λ_j. Let $\sigma(\mathcal{A})$ represent the set of all eigenvalues of $\mathcal{A}$, i.e., $\sigma(\mathcal{A}) = \{\lambda_1, \lambda_2, \cdots\}$. The following assumptions indicate that $\sigma(\mathcal{A})$ can be decomposed into a finite-dimensional subset containing n *slow* eigenvalues and an infinite-dimensional stable complement comprising the remaining *fast* eigenvalues, with a sufficiently large spectral gap separating the slow and fast modes.

Assumption 8.1 [25]
1. $Re\{\lambda_1\} \geqslant Re\{\lambda_2\} \geqslant \cdots \geqslant Re\{\lambda_j\} \geqslant \cdots$, where $Re\{\lambda_j\}$ denotes the real part of λ_j.
2. $\sigma(\mathcal{A})$ can be partitioned as $\sigma(\mathcal{A}) = \sigma_1(\mathcal{A}) + \sigma_2(\mathcal{A})$, where $\sigma_1(\mathcal{A})$ consists of the first n (with n finite) eigenvalues, that is $\sigma_1(\mathcal{A}) = \{\lambda_1, \cdots, \lambda_n\}$, and $|Re\{\lambda_1\}|/|Re\{\lambda_n\}| = O(1)$.
3. $Re\{\lambda_{n+1}\} < 0$ and $|Re\{\lambda_n\}|/|Re\{\lambda_{n+1}\}| = O(\varepsilon)$, where $\varepsilon < 1$ is a small positive number.

By applying the separation of variables theorem [29] together with Galerkin's method [25], the infinite-dimensional system can be approximated by the following set of ODEs:

$$\dot{\boldsymbol{x}}_s(t) = \boldsymbol{A}_s\boldsymbol{x}_s(t) + g_s\left(\boldsymbol{x}_s(t), \boldsymbol{x}_f(t), \boldsymbol{u}(t)\right) + \boldsymbol{f}_{ps}(t), \tag{8.4a}$$
$$\dot{\boldsymbol{x}}_f(t) = \boldsymbol{A}_f\boldsymbol{x}_f(t) + g_f\left(\boldsymbol{x}_s(t), \boldsymbol{x}_f(t), \boldsymbol{u}(t)\right) + \boldsymbol{f}_{pf}(t), \tag{8.4b}$$
$$\boldsymbol{y}_s(t) = \boldsymbol{C}_s\boldsymbol{x}_s(t), \tag{8.4c}$$
$$\boldsymbol{y}_f(t) = \boldsymbol{C}_f\boldsymbol{x}_f(t), \tag{8.4d}$$
$$\boldsymbol{y}(t) = \boldsymbol{y}_s(t) + \boldsymbol{y}_f(t) + \boldsymbol{f}_s(t), \tag{8.4e}$$

with the initial conditions:

$$\boldsymbol{x}_s(0) = \mathcal{P}_s\boldsymbol{x}_0(z) \triangleq < \boldsymbol{\phi}_s(\cdot), \boldsymbol{x}_0(\cdot) >,$$
$$\boldsymbol{x}_f(0) = \mathcal{P}_f\boldsymbol{x}_0(z) \triangleq < \boldsymbol{\phi}_f(\cdot), \boldsymbol{x}_0(\cdot) >,$$

where

$$\begin{aligned}
&\boldsymbol{x}_s(t) \triangleq \mathcal{P}_s x(z,t),\ \boldsymbol{x}_f(t) \triangleq \mathcal{P}_f x(z,t),\\
&\boldsymbol{A}_s \triangleq \mathcal{P}_s\mathcal{A}\boldsymbol{\phi}^{\mathrm{T}}(z) = diag\{\lambda_1, \cdots, \lambda_n\},\\
&\boldsymbol{A}_f \triangleq \mathcal{P}_f\mathcal{A}\boldsymbol{\phi}^{\mathrm{T}}(z) = diag\{\lambda_{n+1}, \cdots, \lambda_\infty\},\\
&\boldsymbol{\phi}_s(z) \triangleq [\phi_1(z), \cdots, \phi_n(z)],\ \boldsymbol{\phi}_f(z) \triangleq \left[\phi_{n+1}(z), \cdots, \phi_\infty(z)\right],\\
&g_s(\cdot) \triangleq \mathcal{P}_s g(x, \boldsymbol{u}(t)),\ g_f(\cdot) \triangleq \mathcal{P}_f g(x, \boldsymbol{u}(t)),\\
&\boldsymbol{f}_{ps}(t) \triangleq \mathcal{P}_s f_p(z,t),\ \boldsymbol{f}_{pf}(t) \triangleq \mathcal{P}_f f_p(z,t),\\
&\boldsymbol{C}_s \triangleq \int_0^L \boldsymbol{c}(z)\boldsymbol{\phi}_s^{\mathrm{T}}(z)\mathrm{d}z,\ \boldsymbol{C}_f \triangleq \int_0^L \boldsymbol{c}(z)\boldsymbol{\phi}_f^{\mathrm{T}}(z)\mathrm{d}z.
\end{aligned}$$

Note that the subscripts s and f correspond to the slow and fast subsystems, respectively. The derivation of this infinite-dimensional ODE system is omitted here for brevity, as it follows standard procedures; interested readers are referred to [25, 30] for further details.

8.4.2 Neural Observer Design and Analysis

The following adaptive observer is designed for the slow subsystem described by (8.4a)–(8.4e):

$$\dot{\hat{\boldsymbol{x}}}_s(t) = \boldsymbol{A}_s\hat{\boldsymbol{x}}_s(t) + \hat{g}_s\left(\hat{\boldsymbol{x}}_s(t), 0, \boldsymbol{u}(t)\right) + \boldsymbol{L}\left(\boldsymbol{y}(t) - \hat{\boldsymbol{y}}_s(t)\right), \tag{8.5a}$$

$$\hat{\boldsymbol{y}}_s(t) = \boldsymbol{C}_s\hat{\boldsymbol{x}}_s(t), \tag{8.5b}$$

here, $\hat{\boldsymbol{x}}_s(t)$ and $\hat{\boldsymbol{y}}_s(t)$ represent the estimates of the low-order state $\boldsymbol{x}_s(t)$ and the corresponding fictitious output $\boldsymbol{y}_s(t)$, respectively. The term $\hat{g}_s(\cdot) \in \mathbb{R}^n$ denotes the estimator of the unknown nonlinear function $g_s(\cdot)$, and $\boldsymbol{L} \in \mathbb{R}^{n\times n_y}$ is the observer gain matrix to be designed. The state and output estimation errors are defined as $\tilde{\boldsymbol{x}}_s(t) \triangleq \boldsymbol{x}_s(t) - \hat{\boldsymbol{x}}_s(t)$ and $\tilde{\boldsymbol{y}}_s(t) \triangleq \boldsymbol{y}(t) - \hat{\boldsymbol{y}}_s(t)$, respectively. Since the fast dynamics $\boldsymbol{x}_f(t)$ are *unknown* and typically neglected, they are set to zero in the adaptive observer.

Remark 8.3 It is worth noting that, as the true output $\boldsymbol{y}_s(t)$ of the reduced-order model (also referred to as the "fictitious output") is not available, the measured output $\boldsymbol{y}(t)$ of the PDE system (8.1a)–(8.3) is employed in the design of the adaptive neural observer. Consequently, it follows that

$$\tilde{\boldsymbol{y}}_s(t) = \boldsymbol{C}_s\tilde{\boldsymbol{x}}_s(t) + \boldsymbol{C}_f\boldsymbol{x}_f(t) + \boldsymbol{f}_s(t). \tag{8.6}$$

In some early early-lumping-based observer designs for DPSs [31, 32], the fast dynamics state $\boldsymbol{x}_f(t)$ was neglected. Specifically, these works typically assumed $\boldsymbol{x}_f(t) = 0$ and omitted this term in the observer design. As a result, the influence of $\boldsymbol{x}_f(t)$ on the observer error is manually removed, which gives rise to the phenomenon commonly referred to as "observation spillover" [33].

An AdNN is proposed for the construction of $\hat{g}_s(\cdot)$ as

$$\hat{g}_s\left(\hat{\boldsymbol{x}}_s(t), 0, \boldsymbol{u}(t)\right) = \hat{\boldsymbol{W}}^{\mathrm{T}}\boldsymbol{\Phi}(\hat{\boldsymbol{x}}_s(t), \boldsymbol{u}(t)), \tag{8.7}$$

here, $\hat{\boldsymbol{W}} \in \mathbb{R}^{m\times n}$ represents the weighting matrix of the neural network, while $\boldsymbol{\Phi}(\hat{\boldsymbol{x}}_s(t), \boldsymbol{u}(t)) : \mathbb{R}^n \times \mathbb{R}^{n_u} \to \mathbb{R}^m$ denotes a vector of Gaussian radial basis functions (RBFs), with the i-th component given by:

$$\boldsymbol{\Phi}_i(\boldsymbol{q}) = \exp\left[\frac{-\left(\boldsymbol{q}-\boldsymbol{\mu}_i\right)^{\mathrm{T}}\left(\boldsymbol{q}-\boldsymbol{\mu}_i\right)}{\sigma_i^2}\right], i = 1, \cdots, m \tag{8.8}$$

here, $\boldsymbol{q} \triangleq \begin{bmatrix} \hat{\boldsymbol{x}}_s(t) \\ \boldsymbol{u}(t) \end{bmatrix} \in \mathbb{R}^{n+n_u}$ represents the input vector. For the i-th element of $\hat{g}_s\big(\hat{\boldsymbol{x}}_s(t), 0, \boldsymbol{u}(t)\big)$, $\boldsymbol{\mu}_i \in \mathbb{R}^{n+n_u}$ and $\sigma_i > 0$ denote the center and width of the corresponding Gaussian RBF, respectively. The number of nodes in the neural network is m.

Remark 8.4 It should be emphasized that the Gaussian RBFs introduced above satisfy the condition $\|\boldsymbol{\Phi}_i(\boldsymbol{q})\|_{\text{peak}} \leq \Phi_b$, where $\Phi_b > 0$ is a constant bound, and $\|\cdot\|_{\text{peak}}$ denotes the peak norm [15].

According to the universal approximation theorem for RBF neural networks [34], for any continuous function $g_s\left(\hat{\boldsymbol{x}}_s(t), 0, \boldsymbol{u}(t)\right) \in \mathbb{R}^n$, one can always find a finite node number m and an ideal constant weight matrix $\boldsymbol{W}^\star \in \mathbb{R}^{m\times n}$ such that

$$g_s\left(\hat{\boldsymbol{x}}_s(t), 0, \boldsymbol{u}(t)\right) = \boldsymbol{W}^{\star^{\mathrm{T}}}\boldsymbol{\Phi}(\hat{\boldsymbol{x}}_s(t), \boldsymbol{u}(t)) + \boldsymbol{\delta}(\hat{\boldsymbol{x}}_s(t), \boldsymbol{u}(t)),$$

where $\boldsymbol{\delta}(\hat{\boldsymbol{x}}_s(t), \boldsymbol{u}(t)) \in \mathbb{R}^n$ is the approximation error for the NN satisfying that $\left\|\boldsymbol{\delta}(\hat{\boldsymbol{x}}_s(t), \boldsymbol{u}(t))\right\|_{\text{peak}} \leq \delta_b$.

Concerning the NN learning, the weight adaptation law is designed by utilizing the output estimation error $\tilde{\boldsymbol{y}}_s(t)$ as follows:

$$\dot{\hat{\boldsymbol{W}}}(t) = -\mu\boldsymbol{\Gamma}\hat{\boldsymbol{W}}(t) + \boldsymbol{\Gamma}\boldsymbol{\Phi}(\hat{\boldsymbol{x}}_s(t), \boldsymbol{u}(t))\tilde{\boldsymbol{y}}_s^{\mathrm{T}}(t)\boldsymbol{Q}, \tag{8.9}$$

where $\mu > 0$ is a constant scalar, $\boldsymbol{\Gamma} \in \mathbb{R}^{m\times m}$ is a positive definite matrix representing the learning rate, and $\boldsymbol{Q} \in \mathbb{R}^{n_y\times n}$ is an auxiliary constant matrix to be determined. It should be highlighted that, unlike full-state based schemes [35], the proposed approach relies only on the reduced-order state estimate $\hat{\boldsymbol{x}}_s(t)$, which substantially enhances its practicality. For notational convenience, we denote $\boldsymbol{\Phi}(t) \triangleq \boldsymbol{\Phi}(\hat{\boldsymbol{x}}_s(t), \boldsymbol{u}(t))$ in the sequel.

Recalling the definitions of the estimation errors, the resulting error dynamics can be expressed as:

$$\dot{\tilde{\boldsymbol{x}}}_s(t) = (\boldsymbol{A}_s - \boldsymbol{L}\boldsymbol{C}_s)\tilde{\boldsymbol{x}}_s(t) + \tilde{g}_s(t) - \boldsymbol{L}\boldsymbol{y}_f(t) + \tilde{\boldsymbol{W}}^{\mathrm{T}}(t)\boldsymbol{\Phi}(t) + \boldsymbol{f}_{ps}(t) - \boldsymbol{L}\boldsymbol{f}_s(t), \tag{8.10a}$$

$$\tilde{\boldsymbol{y}}_s(t) = \boldsymbol{C}_s\tilde{\boldsymbol{x}}_s(t) + \boldsymbol{y}_f(t) + \boldsymbol{f}_s(t), \tag{8.10b}$$

$$\dot{\tilde{\boldsymbol{W}}}(t) = \mu\boldsymbol{\Gamma}\hat{\boldsymbol{W}}(t) - \boldsymbol{\Gamma}\boldsymbol{\Phi}(t)\tilde{\boldsymbol{y}}_s^{\mathrm{T}}(t)\boldsymbol{Q}, \tag{8.10c}$$

by integrating (8.4) and (8.5) with (8.9), the error system can be expressed, where $\tilde{\boldsymbol{W}}(t) \triangleq \boldsymbol{W}^\star - \hat{\boldsymbol{W}}(t)$ represents the neural network weight estimation error, and $\tilde{g}_s(t) \triangleq g_s\left(\boldsymbol{x}_s(t), \boldsymbol{x}_f(t), \boldsymbol{u}(t)\right) - \boldsymbol{W}^{\star^{\mathrm{T}}}\boldsymbol{\Phi}(t)$ denotes the functional approximation error associated with the NN under ideal weights.

The subsequent assumptions and lemmas serve as fundamental prerequisites for the development of the proposed adaptive neural observer.

Assumption 8.2 [20] When no fault exists, the output and state of the $\boldsymbol{x}_f$-subsystem $\boldsymbol{y}_f(t)$ and $\boldsymbol{x}_f(t)$ satisfy that

$$\left\|\boldsymbol{y}_f(t)\right\|_{\text{peak}} \triangleq \sup_t ||\boldsymbol{y}_f(t)|| < \infty,$$
$$\left\|\boldsymbol{x}_f(t)\right\|_{\text{peak}} \triangleq \sup_t ||\boldsymbol{x}_f(t)|| < \infty. \ \forall t \geqslant 0$$

Assumption 8.3 [30] The unknown nonlinear function $g_s\left(\boldsymbol{x}_s(t), \boldsymbol{x}_f(t), \boldsymbol{u}(t)\right)$ satisfies that $\left\|g_s\left(\boldsymbol{x}_s(t), \boldsymbol{x}_f(t), \boldsymbol{u}(t)\right)\right\|_{\text{peak}} \leq g_b$ where $g_b > 0$ is a constant.

Assumption 8.4 The ideal network weight satisfies that $\left\|\boldsymbol{W}^\star\right\|_F \leq W_b$ where $W_b > 0$ is a constant.

Remark 8.5 It is not hard to verify that

$$\left\|\tilde{g}_s(t)\right\|_{\text{peak}} \leq \left\|g_s\left(\boldsymbol{x}_s(t), \boldsymbol{x}_f(t), \boldsymbol{u}(t)\right)\right\|_{\text{peak}} + \left\|\boldsymbol{W}^{\star^{\mathrm{T}}}\boldsymbol{\Phi}(t)\right\|_{\text{peak}} \leq g_b + W_b\Phi_b.$$

Lemma 8.1 *[36] Given regular real functions $V(t)$ and $h(t)$,*

$$\dot{V}(t) \leq -\beta V(t) + h(t), \forall t \geq 0,$$

implies that

$$V(t) \leq e^{-\alpha t}V(0) + \int_0^t e^{-\beta(t-\upsilon)}h(\upsilon)\mathrm{d}\upsilon, \quad \forall t \geq 0$$

for any finite positive constant α.

Theorem 8.1 *(Stability of the adaptive neural observer) Under Assumptions 8.1–8.4, given constants $\beta_0, \beta_1, \beta_2$, if there exist a positive scalar μ, symmetric positive definite matrices $\boldsymbol{P}_s \in \mathbb{R}^{n\times n}$ and $\boldsymbol{\Gamma} \in \mathbb{R}^{m\times m}$, as well as matrices $\boldsymbol{X} \in \mathbb{R}^{n\times n_y}$ and $\boldsymbol{Q} \in \mathbb{R}^{n_y\times n}$, such that the following conditions are satisfied:*

$$\boldsymbol{\Xi} \triangleq \begin{bmatrix} \boldsymbol{\Xi}_{11} & \star & \star \\ \boldsymbol{P} & -\beta_1\boldsymbol{I} & \star \\ -\boldsymbol{X}^{\mathrm{T}} & 0 & -\beta_2\boldsymbol{I} \end{bmatrix} < 0, \tag{8.11}$$

$$\boldsymbol{P} = \boldsymbol{C}_s^{\mathrm{T}}\boldsymbol{Q} \tag{8.12}$$

and

$$\mu - \Phi_b\sigma_{\max}(\boldsymbol{Q}) - \beta_0\lambda_{\max}(\boldsymbol{\Gamma}^{-1}) > 0, \tag{8.13}$$

where

$$\boldsymbol{\Xi}_{11} \triangleq \boldsymbol{P}\boldsymbol{A}_s + \boldsymbol{A}_s^{\mathrm{T}}\boldsymbol{P} - \boldsymbol{X}\boldsymbol{C}_s - \boldsymbol{C}_s^{\mathrm{T}}\boldsymbol{X}^{\mathrm{T}} + \beta_0\boldsymbol{P}, \tag{8.14}$$

then there exists an adaptive neural observer in the form of (8.5), where the observer gain is given by $\boldsymbol{L} = \boldsymbol{P}_s^{-1}\boldsymbol{X}$ and the weight update law follows (8.9), such that the reduced-order state estimation error $\tilde{\boldsymbol{x}}_s$ is UUB with the ultimate bound

$$\rho \triangleq \sqrt{\frac{\mu W_b^2 + \beta_1 \left\|\tilde{g}_s(t)\right\|_{\text{peak}}^2 + (\beta_2 + \Phi_b \sigma_{\max}(\boldsymbol{Q})) \left\|\boldsymbol{y}_f(t)\right\|_{\text{peak}}^2}{\beta_0 \lambda_{\min}(\boldsymbol{P})}}, \tag{8.15}$$

when no fault exists.

Proof Consider the following Lyapunov candidate:

$$V(t) \triangleq \tilde{\boldsymbol{x}}_s^{\mathrm{T}}(t)\boldsymbol{P}\tilde{\boldsymbol{x}}_s(t) + \operatorname{tr}(\tilde{\boldsymbol{W}}^{\mathrm{T}}(t)\boldsymbol{\Gamma}^{-1}\tilde{\boldsymbol{W}}(t)). \tag{8.16}$$

Recalling (8.10), the time derivative of $V(t)$ can be obtained as

$$\begin{aligned}\dot{V}(t) = {} & 2\tilde{\boldsymbol{x}}_s^{\mathrm{T}}(t)\boldsymbol{P}(\boldsymbol{A}_s - \boldsymbol{L}\boldsymbol{C}_s)\tilde{\boldsymbol{x}}_s(t) + 2\tilde{\boldsymbol{x}}_s^{\mathrm{T}}(t)\boldsymbol{P}\tilde{g}_s(t) - 2\tilde{\boldsymbol{x}}_s^{\mathrm{T}}(t)\boldsymbol{X}\boldsymbol{y}_f(t) + 2\mu \operatorname{tr}(\tilde{\boldsymbol{W}}^{\mathrm{T}}(t)\hat{\boldsymbol{W}}(t)) \\ & + 2\tilde{\boldsymbol{x}}_s^{\mathrm{T}}(t)\boldsymbol{P}\tilde{\boldsymbol{W}}^{\mathrm{T}}(t) - 2\operatorname{tr}(\tilde{\boldsymbol{W}}^{\mathrm{T}}(t)\boldsymbol{\Phi}(t)\tilde{\boldsymbol{y}}_s^{\mathrm{T}}(t)\boldsymbol{Q}).\end{aligned} \tag{8.17}$$

The forth term on the RHS of (8.17) is rewritten as

$$2\mu \operatorname{tr}(\tilde{\boldsymbol{W}}^{\mathrm{T}}(t)\hat{\boldsymbol{W}}(t)) = 2\mu \operatorname{tr}(\tilde{\boldsymbol{W}}^{\mathrm{T}}(t)\boldsymbol{W}^{\star}) - 2\mu \left\|\tilde{\boldsymbol{W}}(t)\right\|_F^2 \leq -\mu \left\|\tilde{\boldsymbol{W}}(t)\right\|_F^2 + \mu W_b^2, \tag{8.18}$$

by invoking Assumption 8.3 and applying Young's inequality [21].

Similarly, by integrating (8.10b) with (8.12), the last two terms of (8.17) can be expressed as

$$\begin{aligned}2\tilde{\boldsymbol{x}}_s^{\mathrm{T}}(t)\boldsymbol{P}\tilde{\boldsymbol{W}}^{\mathrm{T}}(t) - 2\operatorname{tr}(\tilde{\boldsymbol{W}}^{\mathrm{T}}(t)\boldsymbol{\Phi}(t)\tilde{\boldsymbol{y}}_s^{\mathrm{T}}(t)\boldsymbol{Q}) &= -2\boldsymbol{y}_f^{\mathrm{T}}(t)\boldsymbol{Q}\tilde{\boldsymbol{W}}^{\mathrm{T}}(t)\boldsymbol{\Phi}(t) \\ &\leq \Phi_b \sigma_{\max}(\boldsymbol{Q})(\left\|\boldsymbol{y}_f(t)\right\|_{\text{peak}}^2 + \left\|\tilde{\boldsymbol{W}}(t)\right\|_F^2).\end{aligned} \tag{8.19}$$

By substituting (8.18) and (8.19) into (8.17), it can be obtained that

$$\begin{aligned}\dot{V}(t) \leq {} & \boldsymbol{\xi}^{\mathrm{T}}(t)\Xi\boldsymbol{\xi}(t) - \beta_0\tilde{\boldsymbol{x}}_s^{\mathrm{T}}(t)\boldsymbol{P}\tilde{\boldsymbol{x}}_s(t) + \beta_1 \left\|\tilde{g}_s(t)\right\|^2 + \beta_2 \left\|\tilde{\boldsymbol{y}}_f(t)\right\|^2 + \mu W_b^2 + \Phi_b\sigma_{\max}(\boldsymbol{Q}) \left\|\boldsymbol{y}_f(t)\right\|_{\text{peak}}^2 \\ & - (\mu - \Phi_b\sigma_{\max}(\boldsymbol{Q})) \left\|\tilde{\boldsymbol{W}}(t)\right\|_F^2 \leq \boldsymbol{\xi}^{\mathrm{T}}(t)\Xi\boldsymbol{\xi}(t) - \beta_0\tilde{\boldsymbol{x}}_s^{\mathrm{T}}(t)\boldsymbol{P}\tilde{\boldsymbol{x}}_s(t) + \alpha - (\mu - \Phi_b\sigma_{\max}(\boldsymbol{Q})) \left\|\tilde{\boldsymbol{W}}(t)\right\|_F^2,\end{aligned} \tag{8.20}$$

where

$$\boldsymbol{\xi} \triangleq \begin{bmatrix} \tilde{\boldsymbol{x}}_s(t) \\ \tilde{g}_s(t) \\ \boldsymbol{y}_f(t) \end{bmatrix},$$

$$\alpha \triangleq \mu W_b^2 + \beta_1 \left\|\tilde{g}_s(t)\right\|_{\text{peak}}^2 + (\beta_2 + \Phi_b\sigma_{\max}(\boldsymbol{Q})) \left\|\boldsymbol{y}_f(t)\right\|_{\text{peak}}^2.$$

Recalling the Lyapunov function $V(t)$ defined in (8.16), it follows that

$$-\beta_0\tilde{\boldsymbol{x}}_s^{\mathrm{T}}(t)\boldsymbol{P}\tilde{\boldsymbol{x}}_s(t) = -\beta_0 V(t) + \beta_0 \operatorname{tr}(\tilde{\boldsymbol{W}}^{\mathrm{T}}(t)\boldsymbol{\Gamma}^{-1}\tilde{\boldsymbol{W}}(t)) \leq -\beta_0 V(t) + \beta_0\lambda_{\max}(\boldsymbol{\Gamma}^{-1}) \left\|\tilde{\boldsymbol{W}}(t)\right\|_F^2. \tag{8.21}$$

Substituting (8.21) into (8.20) leads to

$$\dot{V}(t) \le \boldsymbol{\xi}^{\mathrm{T}}(t)\Xi\boldsymbol{\xi}(t) - \beta_0 V(t) + \alpha - (\mu - \Phi_b\sigma_{\max}(\boldsymbol{Q}) - \beta_0\lambda_{\max}(\boldsymbol{\Gamma}^{-1}))\left\|\tilde{\boldsymbol{W}}(t)\right\|_F^2. \tag{8.22}$$

Once $\Xi < 0$ and $\mu - \Phi_b\sigma_{\max}(\boldsymbol{Q}) - \beta_0\lambda_{\max}(\boldsymbol{\Gamma}^{-1}) > 0$ are both satisfied, one has

$$\dot{V}(t) \le -\beta_0 V(t) + \alpha.$$

By Lemma 8.1, it can be derived that

$$V(t) \le \exp(-\beta_0 t)V(0) + \frac{\alpha}{\beta_0}.$$

Furthermore, by similar analysis in [37], one can obtain that

$$\|\tilde{\boldsymbol{x}}_s(t)\|^2 \le \frac{1}{\lambda_{\min}(\boldsymbol{P})}(\exp(-\beta_0 t)V(0) + \frac{\alpha}{\beta_0}), \tag{8.23}$$

which implies that $\tilde{\boldsymbol{x}}_s(t)$ is UUB with ultimate bound ρ in the absence of faults. This concludes the proof. □

Corollary 8.1 *Recalling the mode decomposition technique introduced in the previous subsection, the original state estimate can be expressed as* $\hat{x}(z,t) \triangleq \boldsymbol{\phi}_s^{\mathrm{T}}(z)\hat{\boldsymbol{x}}_s(t)$, *and the corresponding estimation error is defined by* $\tilde{x}(z,t) \triangleq x(z,t) - \hat{x}(z,t)$. *Furthermore, it follows that*

$$\begin{aligned}\|\tilde{x}(\cdot,t)\|_2^2 &= \left\|\boldsymbol{\phi}_s^{\mathrm{T}}(z)\tilde{\boldsymbol{x}}_s(t) + \boldsymbol{\phi}_f^{\mathrm{T}}(z)\boldsymbol{x}_f(t)\right\|_2^2 = \|\tilde{\boldsymbol{x}}_s(t)\|^2 + \left\|\boldsymbol{x}_f(t)\right\|^2 \\ &\le \frac{1}{\lambda_{\min}(\boldsymbol{P})}(\exp(-\beta_0 t)V(0) + \frac{\alpha}{\beta_0}) + \left\|\boldsymbol{x}_f(t)\right\|_{\text{peak}}^2,\end{aligned}$$

by invoking Assumption 8.1 and the result in (8.23), it can be concluded that the original state estimation error $\tilde{x}(z,t)$ *is bounded in the spatial* $\mathcal{L}^2$ *norm.*

8.4.3 Residual Evaluation and Threshold Computation

With the adaptive neural observer developed in the previous section, fault detection can be integrated with the system modeling framework. It is well known that the core of fault detection lies in the construction and evaluation of residual signals. In the following, the design and analysis of the FDF are presented in detail.

Remark 8.6 Since full-state measurements are not available, the residual signal can be constructed using the output measurement error $\tilde{\boldsymbol{y}}_s(t)$. Note that the reduced-order state estimation error $\tilde{\boldsymbol{x}}_s(t)$ is not employed in the FDF design because the actual reduced-order state $\boldsymbol{x}_s(t)$ is unavailable, even though its estimate $\hat{\boldsymbol{x}}_s(t)$ can be obtained from the adaptive neural observer.

For the purpose of fault detection, the residual signal is intuitively defined as

$$\boldsymbol{r}(t) \triangleq \tilde{\boldsymbol{y}}_s(t). \tag{8.24}$$

Subsequently, the following *lumped* residual evaluation scheme is presented:

$$J(t) = \int_t^{t+\Delta} \|\boldsymbol{r}(\tau)\|^2 \,\mathrm{d}\tau, \tag{8.25a}$$

$$J_{\text{th}} = \sup_{f_p(z,t)=0,\, \boldsymbol{f}_s(t)=0} J(t), \tag{8.25b}$$

here, $\Delta > 0$ denotes the length of the evaluation time window, while $J(t)$ and J_{th} represent the residual evaluation function and its corresponding threshold, respectively. Several residual evaluation functions are commonly employed in fault detection, including the peak value, average value, RMS value, and the $\mathcal{L}_2$ norm [15, 38]. The RMS and $\mathcal{L}_2$ norm are closely related, typically designed to reduce the FAR at the expense of fault detectability. The residual evaluation function presented here is a general form of the $\mathcal{L}_2$ norm applied to the signal $\boldsymbol{r}(t)$, capturing the signal energy over the interval $[t, t+\Delta]$. Such evaluation functions have been widely adopted in the literature on FD of dynamic systems; see, for example, [39, 40].

The fault detection threshold J_{th} is determined using the KDE method [18, 41] under a *fault-free scenario*. The procedure first approximates the probability density function (PDF) of the residual evaluation function $J(t)$, and then selects the threshold based on a chosen significance level α_0 (commonly 0.05 or 0.01).

Moreover, the decision logic for fault detection is as follows:

$$\begin{cases} J(t) > J_{\text{th}} \Longrightarrow \text{ faulty at time } t \\ \text{Otherwise} \Longrightarrow \text{ fault-free at time } t. \end{cases} \tag{8.26}$$

Remark 8.7 This chapter addresses the detection of both distributed process faults $f_p(z, t)$ and sensor faults $\boldsymbol{f}_s(t)$. The proposed fault detection framework can, however, be extended to tackle distributed fault localization by employing the spatial mapping filter approach presented in [20, 23].

Corollary 8.2 *With the above residual evaluation function in (8.25a), if the conditions in Theorem 8.1 are satisfied, the residual evaluation function $J(t)$ is bounded by*

$$J(t) \le \int_t^{t+\Delta} (\sigma_{\max}(\boldsymbol{C}_s)\upsilon(t) + \left\|\boldsymbol{y}_f(t)\right\|_{\text{peak}}^2)\mathrm{d}\tau, \tag{8.27}$$

where

$$\upsilon(t) \triangleq \frac{1}{\lambda_{\min}(\boldsymbol{P})}(\exp(-\beta_0 t)V(0) + \frac{\alpha}{\beta_0}),$$

and α is defined in Theorem 8.1.

Proof A sketch of the proof is as follows:

$$
\begin{aligned}
J(t) &= \int_t^{t+\Delta} \left\| \tilde{\boldsymbol{y}}_s(\tau) \right\|^2 \mathrm{d}\tau = \int_t^{t+\Delta} \left\| \boldsymbol{C}_s \tilde{\boldsymbol{x}}_s(\tau) + \boldsymbol{y}_f(\tau) \right\|^2 \mathrm{d}\tau \\
&\leq \int_t^{t+\Delta} (\sigma_{\max}(\boldsymbol{C}_s) \left\| \tilde{\boldsymbol{x}}_s(\tau) \right\|^2 + \left\| \boldsymbol{y}_f(\tau) \right\|_{\text{peak}}^2) \mathrm{d}\tau,
\end{aligned}
$$

by invoking (8.23) under the fault-free scenario and applying the triangle inequality. This completes the proof. □

Remark 8.8 It is worth noting that in the traditional model-based FD framework [15], the fault detection threshold J_{th} can typically be computed offline before the observer is implemented online. However, since the nonlinear function $g(x, \boldsymbol{u}(t))$ is completely unknown, the ideal network weight bound W_b and the functional estimation error bound $\left\| \tilde{g}_s(t) \right\|_{\text{peak}}$ cannot be determined offline. Consequently, the results in Corollary 8.2 cannot be used to compute J_{th}, as the value of α depends on these bounds, as indicated in (8.20). Therefore, the KDE technique, implemented online under a fault-free scenario, is employed to obtain the threshold J_{th}.

8.4.4 On-Line Implementation of FDF

In this subsection, we summarize the key procedures, including the selection of design parameters, to facilitate implementation of the proposed method for interested readers.

In practice, the implementation generally follows an iterative trial-and-error approach. For a given application, an initial set of design parameters is selected based on experience and by solving LMIs to satisfy the conditions in Theorem 8.1. The performance of the proposed adaptive neural observer is then evaluated. Based on these results, the design parameters are adjusted to improve performance. Once satisfactory performance is achieved, the selected set of parameters is used for online implementation.

The design parameters are required to satisfy the matrix inequalities (8.11), (8.13) and the matrix equality (8.12). Typically, (8.11) is a standard LMI and can be readily solved using the MATLAB LMI toolbox. Regarding (8.13), once the parameters μ, β_0, and the learning rate matrix $\boldsymbol{\Gamma}$ are chosen, it can be reformulated as the following equivalent LMI:

$$
\begin{bmatrix} \eta_0 \boldsymbol{I} & \star \\ \boldsymbol{Q}^{\mathrm{T}} & \eta_0 \boldsymbol{I} \end{bmatrix} > 0, \tag{8.28}
$$

where

$$\eta_0 \triangleq \frac{\mu - \beta_0 \lambda_{\max}(\boldsymbol{\Gamma}^{-1})}{\Phi_b}. \tag{8.29}$$

As the condition in (8.12) is a matrix equality, it cannot be directly handled by the LMI toolbox. Nevertheless, it can be reformulated as the following optimization problem for practical implementation:

$$\begin{aligned} \min \quad & \eta_1 \\ \text{s.t.} \quad & \begin{bmatrix} \eta_1 \boldsymbol{I} & \star \\ (\boldsymbol{P} - \boldsymbol{C}_s^{\mathrm{T}} \boldsymbol{Q})^{\mathrm{T}} & \eta_1 \boldsymbol{I} \end{bmatrix} > 0, \end{aligned} \tag{8.30}$$

which can be solved via the *mincx* solver in the MATLAB LMI toolbox.

A summary of the proposed fault detection framework is presented in the form of pseudo-code in Algorithm 8.1.

Algorithm 8.1: Computation-efficient fault detection framework

---

Input: Measured input-output (I/O) data $(\boldsymbol{u}(t), \boldsymbol{y}(t))$
Output: Faulty time set t_f
/****Initialization*****/
  Step 1. Select parameters μ, β_0, β_1, β_2 and learning rate matrix $\boldsymbol{\Gamma}$;
  Step 2. Determine the number of neural nodes m in the hidden layer, the center $\boldsymbol{\mu}_i$ and width σ_i of the NN for $i = 1, \cdots, m$;
  Step 3. Calculate the observer gain matrix $\boldsymbol{L}$ and auxiliary matrix $\boldsymbol{Q}$ by solving (8.11)-(8.14) ;
  Step 4. Set both the initial value of $\hat{\boldsymbol{x}}_s(t)$ and $\hat{\boldsymbol{W}}(t)$ as 0 and activate the AdNN according to (8.5), and (8.7)-(8.9) ;
/****Threshold calculation under fault-free scenario*****/
  Step 1. Choose the evaluation time window Δ and the significance level α_0;
  Step 2. **for** $t \in [0, t_0]$ **do**
    Residual evaluation: $J(t) \leftarrow$ (8.25a);
  Step 3. Approximate the PDF of $J(t), t \in [0, t_0]$ and determine the threshold J_{th} ;
/****Fault detection*****/
for $t \in [t_{\min}, t_{\max}]$ **do**
  Residual evaluation: $J(t) \leftarrow$ (8.25a);
  if $J(t) > J_{th}$ **then**
    $t \in t_f$
  else
    $t \notin t_f$

8.5 Illustrative Example

The temperature profile along a catalytic rod in a reactor [25] is used to validate the proposed method. As illustrated in Fig. 8.2, the reactor is supplied with pure species A, and a zero-order exothermic catalytic reaction, A $\rightarrow$ B, takes place along the rod. Additionally, a cooling medium interacts with the rod to remove heat. Typically, several assumptions are made regarding the physical parameters of this process, including an excess of species A in the furnace, constant rod conductivity, constant rod density and heat capacity, and fixed temperatures at both ends of the rod. Under these assumptions, the temperature distribution along the catalytic rod can be described by the following parabolic PDE:

$$\frac{\partial x(z,t)}{\partial t} = \frac{\partial^2 x(z,t)}{\partial z^2} + \beta_T \left(e^{-(\gamma/(1+x(z,t)))} - e^{-\gamma}\right) + \beta_U (b_u(z)u(t) - x(z,t)) + f_p(z,t), \tag{8.31}$$

$$\mathbf{y}(t) = \begin{bmatrix} \int_0^{\pi} \delta(z - \frac{\pi}{4})x(z,t)\mathrm{d}z + f_{s1}(t) \\ \int_0^{\pi} \delta(z - \frac{3\pi}{4})x(z,t)\mathrm{d}z + f_{s2}(t) \end{bmatrix} = \begin{bmatrix} x(\frac{\pi}{4}, t) + f_{s1}(t) \\ x(\frac{3\pi}{4}, t) + f_{s2}(t) \end{bmatrix}, \tag{8.32}$$

subject to the Dirichlet boundary conditions

$$x(0,t) = 0, \;\; x(\pi,t) = 0, \tag{8.33}$$

here, $x(z,t)$ represents the dimensionless temperature of the rod; β_T is the dimensionless heat of reaction; γ denotes the dimensionless activation energy; β_U is the dimensionless heat transfer coefficient; $u(t)$ represents the manipulated input (temperature of the cooling medium) with spatial distribution function $b_u(z)$; $f_p(z,t)$ denotes the *distributed process fault* affecting the system; $\delta(\cdot)$ is the Dirac delta function; and $\mathbf{y}(t)$ corresponds to thermocouple measurements at locations $z = \pi/4$

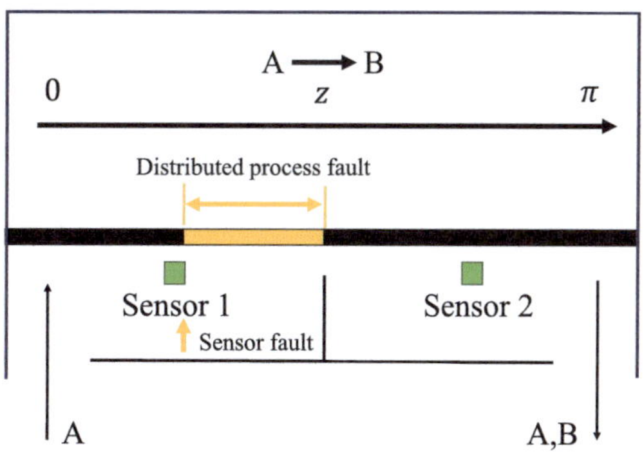

Fig. 8.2 Schematic plot of the catalytic rod

and $z = 3\pi/4$. The vector $\boldsymbol{f}_s(t) = [f_{s1}(t), f_{s2}(t)]^{\mathrm{T}}$ represents potential *sensor faults* in the system.

For practical implementation, the distributed process fault $f_p(z, t)$ is chosen in the following separable form:

$$f_p(z, t) = b_p(z) f_p(t),$$

here, $f_p(t)$ represents the temporal component, while $b_p(z)$ denotes its corresponding spatial distribution.

The typical values of the process parameters are

$$\beta_T = 50.0, \quad \beta_U = 2.0, \quad \gamma = 4.0.$$

The actuator distribution function is set as

$$b_u(z) = \sqrt{\frac{2}{\pi}} \sin(z),$$

while the input $u(t)$ is chosen as

$$u(t) = \begin{cases} 0.5(1 + \sin(t)) & \text{for} \quad 0 \le t \le 27(\text{sec}) \\ 4 + \cos(2t) & \text{for} \quad 27 < t \le 80(\text{sec}) \end{cases} \tag{8.34}$$

the total simulation time is set to 80 sec. After 27 sec, both the magnitude and frequency of the input are varied to emulate a *change in operating conditions*. In the subsequent analysis, the impact of this variation will be examined in detail.

As described in Sect. 8.4.3, the threshold J_{th} is determined using the KDE approach. The cumulative distribution function (CDF) of the residual evaluation function $J(t)$ is shown in Fig. 8.3, and J_{th} is chosen for two different significance levels, namely $\alpha_0 = 0.05$ and $\alpha_0 = 0.01$.

For validation purposes, the faults considered are summarized in Table 8.1, where Fault Scenario 0 corresponds to the fault-free case. Notably, under Fault Scenario 1, the distributed process fault $f_p(z, t) = 8_{[\frac{\pi}{4}, \frac{\pi}{2}]}$ indicates that the fault is uniformly distributed over the spatial domain $[\frac{\pi}{4}, \frac{\pi}{2}]$, i.e., its spatial distribution function is given by

$$b_p(z) = H(z - \pi/4) - H(z - \pi/2),$$

where $H(\cdot)$ is the standard Heaviside function and the temporal variable $f_p(t)$ is

$$f_p(t) = \begin{cases} 0 & \text{for} \quad 0 \le t \le 50(\text{sec}) \\ 8 & \text{for} \quad 50 < t \le 80(\text{sec}). \end{cases}$$

In Fault Scenario 2, the constant-bias sensor fault is expressed as

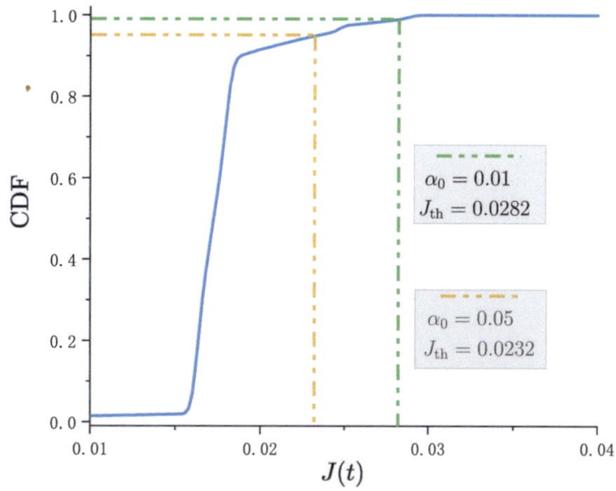

Fig. 8.3 Threshold J_{th} calculation under fault-free scenario

Table 8.1 Settings of fault scenarios

Fault scenarios	Description	Duration time (sec)
0	$f_p(z,t) = 0$, $\boldsymbol{f}_s(t) = 0$	$0 \to 80$
1	$f_p(z,t) = 8_{[\frac{\pi}{4},\frac{\pi}{2}]}$, $\boldsymbol{f}_s(t) = 0$	$50 \to 80$
2	$f_p(z,t) = 0$, $\boldsymbol{f}_s(t) = [1.4, 0]^{\mathrm{T}}$	$30 \to 80$
3	$f_p(z,t) = 0$, $\boldsymbol{f}_s(t) = [3, 0]^{\mathrm{T}}$	$40 \to 80$
4	$f_p(z,t) = 10_{[\frac{\pi}{4},\frac{\pi}{2}]}$, $\boldsymbol{f}_s(t) = [1.5, 0]^{\mathrm{T}}$	$30 \to 50$, $40 \to 80$
5	$f_p(z,t) = 10(1 - \exp(-0.2(t - 40)))_{[\frac{\pi}{4},\frac{\pi}{2}]}$, $\boldsymbol{f}_s(t) = [0, 0]$	$40 \to 80$

$$f_{s1}(t) = 0, \;\; f_{s2}(t) = \begin{cases} 0 & \text{for} \;\; 0 \le t \le 30(\text{sec}) \\ 1.4 & \text{for} \;\; 30 < t \le 80(\text{sec}). \end{cases}$$

For comparison, Fault Scenarios 2 and 3 involve two abrupt sensor faults with different magnitudes. In Fault Scenario 4, both abrupt process and sensor faults are introduced with varying durations. Additionally, Fault Scenario 5 considers a time-varying incipient process fault.

The eigenvalue problem associated with the spatial differential operator

$$\begin{aligned} \mathcal{A}x &= \frac{\partial^2 x}{\partial z^2}, x \in \mathcal{S}(\mathcal{A}) \\ &= \{x \in \mathcal{L}_2([0,\pi];\mathbb{R}); x(0,t) = 0, x(\pi,t) = 0\}, \end{aligned}$$

can be directly solved as

$$\lambda_j = -j^2, \;\; \phi_j(z) = \sqrt{\frac{2}{\pi}} \sin(jz), \;\; j = 1, \cdots, \infty.$$

For detailed procedures on solving this type of eigenvalue problem, see Chap. 4 of [42]. The first two eigenvalues are treated as the dominant modes, yielding $\epsilon = |\lambda_1| / |\lambda_3| \approx 0.111 \ll 1$.

The reduced-order approximation model is then derived as follows:

$$\dot{\boldsymbol{x}}_s(t) = \boldsymbol{A}_s \boldsymbol{x}_s(t) + g_s\left(\boldsymbol{x}_s(t), \boldsymbol{x}_f(t), \boldsymbol{u}(t)\right) + \boldsymbol{f}_{ps}(t), \tag{8.35a}$$

$$\boldsymbol{y}(t) = \boldsymbol{C}_s \boldsymbol{x}_s(t) + \boldsymbol{C}_f \boldsymbol{x}_f(t) + \boldsymbol{f}_s(t), \tag{8.35b}$$

where

$$\boldsymbol{x}_s(t) = \begin{bmatrix} x_1(t) \\ x_2(t) \end{bmatrix}, \boldsymbol{A}_s = \begin{bmatrix} -1 & 0 \\ 0 & -4 \end{bmatrix},$$

$$\boldsymbol{C}_s = \int_0^{\pi} \begin{bmatrix} \delta\left(z - \frac{\pi}{4}\right) \\ \delta\left(z - \frac{3\pi}{4}\right) \end{bmatrix} [\phi_1(z)\phi_2(z)]\,\mathrm{d}z = \begin{bmatrix} \sqrt{\frac{1}{\pi}} & \sqrt{\frac{2}{\pi}} \\ \sqrt{\frac{1}{\pi}} & -\sqrt{\frac{2}{\pi}} \end{bmatrix},$$

$$g_s(\boldsymbol{x}_s(t), \boldsymbol{x}_f(t), \boldsymbol{u}(t)) =$$
$$\begin{bmatrix} -\beta_U x_1(t) + \beta_U u(t) + \beta_T \int_0^{\pi} \phi_1(z)(e^{-(\gamma/(1+x_1(t)\phi_1(z)+x_2(t)\phi_2(z)+\boldsymbol{\phi}_f^{\mathrm{T}}(z)\boldsymbol{x}_f(t)))} - e^{-\gamma})\mathrm{d}z \\ -\beta_U x_2(t) + \beta_T \int_0^{\pi} \phi_2(z)(e^{-(\gamma/(1+x_1(t)\phi_1(z)+x_2(t)\phi_2(z)+\boldsymbol{\phi}_f^{\mathrm{T}}(z)\boldsymbol{x}_f(t)))} - e^{-\gamma})\mathrm{d}z \end{bmatrix}. \tag{8.36}$$

It should be noted that only $\boldsymbol{A}_s$ and $\boldsymbol{C}_s$ are known, whereas $g_s(\boldsymbol{x}_s(t), \boldsymbol{x}_f(t), \boldsymbol{u}(t))$ remains completely unknown for the FDF design. Following Theorem 8.1, the design parameters are chosen as $\mu = 20$, $\beta_0 = 8$, $\beta_1 = 0.0001$, and $\beta_2 = 0.45$. Additionally, the hidden layer of the neural network contains $m = 20$ nodes, with centers $\boldsymbol{\mu}_i$ uniformly distributed over a regular lattice $[-9, 9] \times [-0.1, 0.1] \times [0, 2]$ and widths set as $\sigma_i = 10$ for $i = 1, 2, \ldots, 20$. The learning rate matrix $\boldsymbol{\Gamma}$ is chosen as a 20×20 diagonal matrix with each diagonal entry set to 10. As outlined in Algorithm 8.1, by solving (8.11)–(8.14) using the MATLAB LMI toolbox, the observer gain matrix $\boldsymbol{L}$ and the auxiliary matrix $\boldsymbol{Q}$ are determined as

$$\boldsymbol{L} = \begin{bmatrix} 87.6525 & 87.6521 \\ 81.9105 & -81.9103 \end{bmatrix}, \boldsymbol{Q} = \begin{bmatrix} 0.0026 & 0.0027 \\ 0.0026 & -0.0027 \end{bmatrix}.$$

The original PDE system in (8.31)–(8.33) is numerically solved using the FDM [43] with a sampling interval of $\Delta t = 0.01$ sec and 19 spatial grid points. The threshold J_{th} is computed numerically over an evaluation window of $\Delta = 0.8$ sec. A summary of the system and design parameters is provided in Table 8.2.

Remark 8.9 The design parameters listed in Table 8.2 primarily involve the selection of sensors and the configuration of the NN. Typically, the sensor type (e.g., point-wise, piece-wise) is determined first, followed by the choice of the number of sensors, their spatial locations, and the order of the reduced-order model to satisfy the compatibility condition in Assumption 8.2. Subsequently, the NN-related

Table 8.2 System parameters and design parameters for catalytic rod

Parameters	Type	Whether known	Value (Unit)
Heat of reaction (β_T)	System parameter	**Unknown**	50.0
Activation energy (γ)	System parameter	**Unknown**	4.0
Heat transfer coefficient (β_U)	System parameter	**Unknown**	2.0
Number of sensors (n_y)	System parameter	Known	2
Sensor location	System parameter	Known	$z = \pi/4$ and $z = 3\pi/4$
Number of dominant mode (m)	Design parameter	Known	2
Coefficient of AdNN (μ)	Design parameter	Known	20
Coefficient of AdNN (β_0)	Design parameter	Known	8
Coefficient of AdNN (β_1)	Design parameter	Known	0.0001
Coefficient of AdNN (β_2)	Design parameter	Known	0.45
Number of neural nodes in the hidden layer (m)	Design parameter	Known	20
Center of the NN ($\boldsymbol{\mu}_i$)	Design parameter	Known	$[-9, 9] \times [-0.1, 0.1] \times [0, 2]$
Width of the NN (σ_i)	Design parameter	Known	10
Learning rate matrix ($\boldsymbol{\Gamma}$)	Design parameter	Known	$diag(10, \cdots, 10) \in \mathbb{R}^{20\times 20}$
Observer gain matrix ($\boldsymbol{L}$)	Design parameter	Known	$\begin{bmatrix} 87.6525 & 87.6521 \\ 81.9105 & -81.9103 \end{bmatrix}$
Auxiliary matrix $\boldsymbol{Q}$	Design parameter	Known	$\begin{bmatrix} 0.0026 & 0.0027 \\ 0.0026 & -0.0027 \end{bmatrix}$
Evaluation time window length (Δ)	Design parameter	Known	0.16 (sec)

parameters, including the number of nodes, centers, and widths, are chosen through a trial-and-error procedure. Specifically, these parameters are adjusted based on the tracking performance of the measured output $\boldsymbol{y}(t)$ for the adaptive neural network, as neither the full-state variable $x(z, t)$ nor the reduced-order state $\boldsymbol{x}_s(t)$ is directly accessible.

Table 8.3 Fault detection results on catalytic rod

Index/fault scenarios		0	1	2	3	4	5
Fault detection rate (FDR %)	$\alpha_0 = 0.05$	NA	**100**	**100**	**100**	**100**	97.93
False alarm rate (FAR %)	-	5.00	26.34	12.17	20.63	11.30	19.95
Fault detection rate (FDR %)	$\alpha_0 = 0.01$	NA	33.52	30.55	**100**	49.71	38.99
False alarm rate (FAR %)	-	1.00	1.54	3.57	2.68	1.43	0.70

To evaluate the effectiveness of the proposed approach, key performance indices under the fault scenarios listed in Table 8.1 are presented and analyzed. In fault detection studies, two critical metrics are typically considered: the FDR and the FAR [15], which are defined as follows:

$$\text{FDR} = \text{prob}\left(J > J_{\text{th}} \mid f \neq 0\right),$$
$$\text{FAR} = \text{prob}\left(J > J_{\text{th}} \mid f = 0\right).$$

The FDR represents the likelihood that a fault is correctly identified by the detection method when it occurs, whereas the FAR quantifies the probability of an incorrect alarm when the system is actually fault-free.

Remark 8.10 Ideally, a fault detection method should achieve a high FDR while maintaining a low FAR. In practice, however, these two metrics often conflict, making their trade-off both important and challenging. Chapter 12 of [15] discusses the dual problems of maximizing FDR for a given FAR and minimizing FAR for a given FDR.

Using the computationally efficient fault detection framework described in Algorithm 8.1, the detection results corresponding to the fault scenarios listed in Table 8.1 are summarized in Table 8.3. For comparison, results with significance levels $\alpha_0 = 0.05$ and $\alpha_0 = 0.01$ are both presented. In the fault-free case (scenario 0), the observed FAR values are 5.00% and 1.00%, respectively, which align with the chosen significance levels α_0.

In general, for the same fault scenarios, the FDR obtained with $\alpha_0 = 0.05$ is higher or at least comparable to that with $\alpha_0 = 0.01$, whereas the FAR exhibits the opposite trend. This is expected, as a larger significance level α_0 allows the detection of more

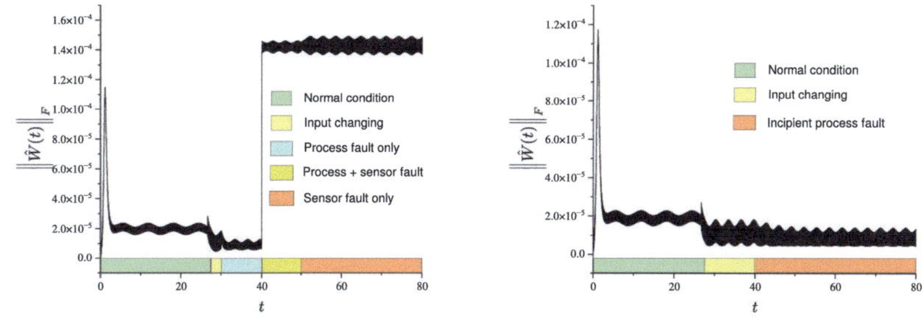

(a) Trajectory of NN weights norm $\|\hat{W}(t)\|_F$ under fault scenario 4.

(b) Trajectory of NN weights norm $\|\hat{W}(t)\|_F$ under fault scenario 5.

Fig. 8.4 Trajectory of state estimation error $\tilde{x}_s(t)$ for reduced-order model under different faulty conditions

faults but at the expense of increased false alarms. Specifically, with $\alpha_0 = 0.05$, the FDRs for fault scenarios 1–5 are generally high, though the corresponding FARs are also elevated due to the detection of changes in working conditions as spurious faults. Conversely, adopting $\alpha_0 = 0.01$ results in more acceptable FARs, while the FDRs are somewhat lower in comparison.

Remark 8.11 Although variations in operating conditions or setpoints are common in industrial applications, studies addressing the differentiation between such changes and actual faults are relatively limited. Effectively distinguishing them requires identifying indices that exhibit significant differences between these two phenomena, which in turn demands additional information regarding both operational changes and fault characteristics. In this chapter, we focus on illustrating how changes in operating conditions affect fault detection results. It should be noted that the residual evaluation function $J(t)$ following the change in operating conditions at $t = 27$ sec is not incorporated into the threshold calculation, leading to a relatively high FAR for $\alpha_0 = 0.05$. Nevertheless, the FAR could be reduced if the post-change $J(t)$ information were available and used in determining the threshold. Here, we consider this as a worst-case scenario to assess the robustness of the proposed method.

Furthermore, to provide a more comprehensive demonstration of the performance of the proposed reduced-order adaptive neural observer and fault detection framework, additional experimental results under fault scenario 4 (simultaneous process and sensor faults) and fault scenario 5 (incipient process fault) are presented in Fig. 8.5a, b.

The evolution of the neural network weights is depicted in Fig. 8.4a, b, showing that the weight estimation error remains bounded in the fault-free scenario. The tracking performance of the proposed reduced-order adaptive neural observer is presented in Fig. 8.5a, b, it can be observed that the reduced-order state estimation error $\tilde{x}_s(t)$ under fault-free scenario satisfies the UUB condition, which coincides with the results in Theorem 8.1. Additionally, variations in the input and the occurrence of faults have noticeable effects on the state estimation performance.

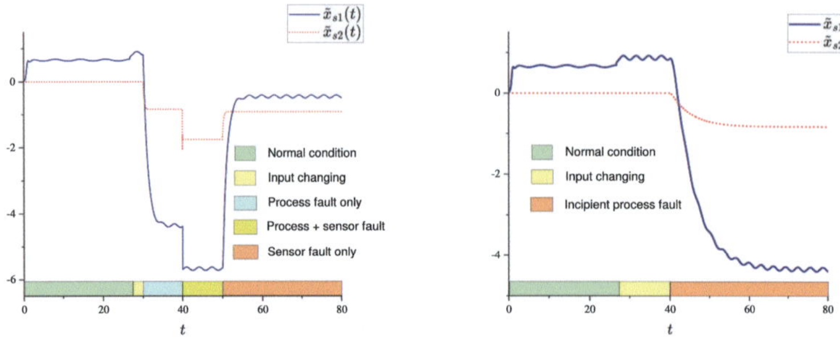

(a) Trajectory of reduced-order state estimation error $\tilde{x}_s(t)$ under fault scenario 4.

(b) Trajectory of state estimation error $\tilde{x}_s(t)$ under fault scenario 5.

Fig. 8.5 Trajectory of state estimation error $\tilde{x}_s(t)$ for reduced-order model under different faulty conditions

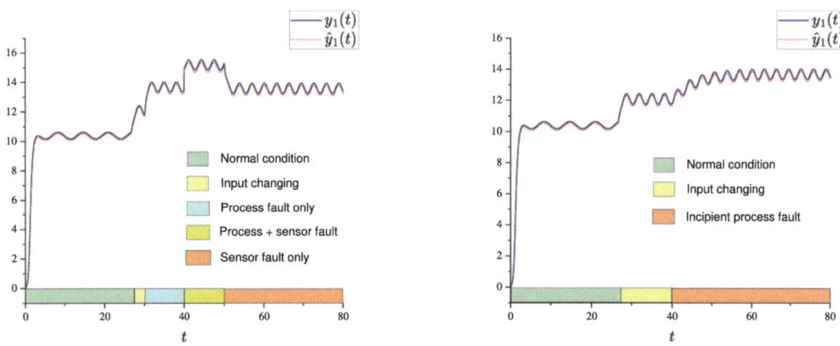

(a) Trajectory of the first sensor measurement $y_1(t)$ and its estimation $\hat{y}_1(t)$ under fault scenario 4.

(b) Trajectory of the first sensor measurement $y_1(t)$ and its estimation $\hat{y}_1(t)$ under fault scenario 5.

Fig. 8.6 Output estimation performance under different faulty conditions

The output tracking performance of the proposed adaptive neural observer under fault scenarios 4 (simultaneous process and sensor faults) and 5 (incipient process fault) is illustrated in Fig. 8.6a, b, respectively. The results indicate that the observer achieves satisfactory output tracking under fault-free conditions, during input variations, and in the presence of single or multiple faults, demonstrating the adaptive capability of the proposed neural observer.

Furthermore, the original state $x(z, t)$ and its estimation using the proposed adaptive neural observer are depicted in Figs. 8.7a, b and 8.8a, b, respectively. It is evident that distinguishing between faults and changes in working conditions is not possible by examining the state variable $x(z, t)$ or the output measurement $\boldsymbol{y}(t)$ alone. As illustrated in Figs. 8.8a, 8.4b, the estimation error $\tilde{x}(z, t)$ remains bounded in the spatial $\mathcal{L}^2$ norm under fault-free conditions, which corroborates the findings in Corollary 8.1.

The temporal trajectories of the residual evaluation function $J(t)$ under fault scenarios 4 and 5 are shown in Fig. 8.9a, b, respectively. Consistent with the results

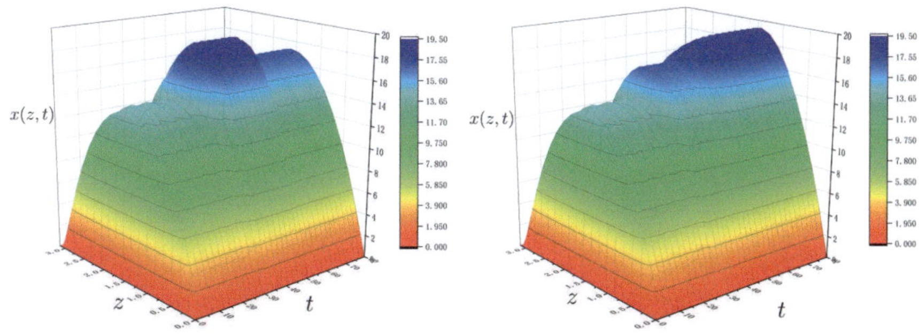

(a) Profile of evolution of original state $x(z, t)$ under fault scenario 4.

(b) Profile of evolution of original state $x(z, t)$ under fault scenario 5.

Fig. 8.7 Profile of evolution of original state $x(z, t)$ under different faulty conditions

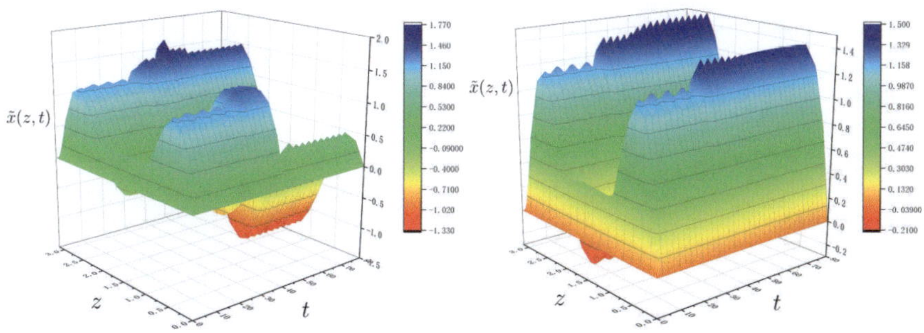

(a) Profile of evolution of original state estiamtion error $\tilde{x}(z, t)$ under fault scenario 4.

(b) Profile of evolution of original state estiamtion error $\tilde{x}(z, t)$ under fault scenario 5.

Fig. 8.8 Original state estimation performance under different faulty conditions

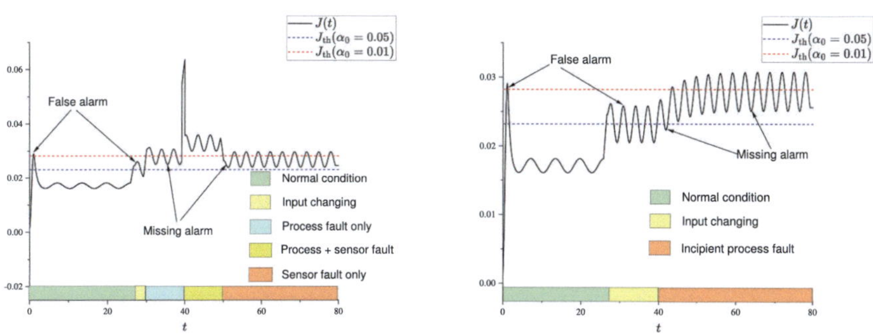

(a) Fault detection results in fault scenario 4.

(b) Fault detection results in fault scenario 5.

Fig. 8.9 Fault detection results of the proposed method under different faulty conditions in Table 8.3, adopting the threshold with significance level $\alpha_0 = 0.05$ leads to a higher FDR, but also a higher FAR, compared to using $\alpha_0 = 0.01$. Additionally, the observed false alarms are primarily induced by variations in the input.

8.6 Conclusion

This chapter addresses the fault detection problem for partially known nonlinear DPSs. To handle the unknown nonlinearities associated with both the infinite-dimensional state and the finite-dimensional input, a reduced-order ODE-based adaptive neural observer is developed. The residuals are generated and evaluated using the estimation provided by the proposed observer. To account for the influence of neglected fast dynamics, a data-driven approach for threshold determination is also introduced. Extensive simulation studies are conducted to validate the method, and the impact of changes in operating conditions on fault detection performance is analyzed.

References

1. Luo B, Wu H-N, Li H-X (2014) Adaptive optimal control of highly dissipative nonlinear spatially distributed processes with neuro-dynamic programming. IEEE Trans Neural Netw Learn Syst 26(4):684–696
2. Luo B, Huang T, Wu H-N et al (2015) Data-driven $\mathcal{H}_\infty$ control for nonlinear distributed parameter systems. IEEE Trans Neural Netw Learn Syst 26(11):2949–2961
3. Wang Z-P, Li Q-Q, Wu H-N et al (2023) Pinning spatiotemporal sampled-data synchronization of coupled reaction-diffusion neural networks under deception attacks. IEEE Trans Neural Netw Learn Syst 34(10):7967–7977
4. Feng J, Liu Z, He X et al (2023) Vibration suppression of a high-rise building with adaptive iterative learning control. IEEE Trans Neural Netw Learn Syst 34(8):4261–4272
5. Liu Z, Liu J, He W (2017) Modeling and vibration control of a flexible aerial refueling hose with variable lengths and input constraint. Automatica 77:302–310
6. Liu Z, He X, Zhao Z et al (2021) Vibration control for spatial aerial refueling hoses with bounded actuators. IEEE Trans Ind Electron 68(5):4209–4217
7. Zhao Z, Liu Y, Zou T et al (2023) Robust adaptive fault-tolerant control for a riser-vessel system with input hysteresis and time-varying output constraints. IEEE Trans Cybern 53(6):3939–3950
8. Wang B-C, Li H-X, Yang H-D (2020) Spatial correlation-based incremental learning for spatiotemporal modeling of battery thermal process. IEEE Trans Ind Electron 67(4):2885–2893
9. Chen N, Li B, Luo B et al (2023) Event-triggered optimal control for temperature field of roller kiln based on adaptive dynamic programming. IEEE Trans Cybern 53(5):2805–2817
10. Wang Z, Li H-X, Chen C (2020) Reinforcement learning-based optimal sensor placement for spatiotemporal modeling. IEEE Trans Cybern 50(6):2861–2871
11. Fischer F, Deutscher J (2020) Flatness-based algebraic fault diagnosis for distributed-parameter systems. Automatica 117:108987
12. Fischer F, Deutscher J (2022) Fault diagnosis for linear heterodirectional hyperbolic ODE-PDE systems using backstepping-based trajectory planning. Automatica 135:109952

13. Hu X, Zhang K, Liu K et al (2020) Advanced fault diagnosis for lithium-ion battery systems, a review of fault mechanisms, fault features, and diagnosis procedures. IEEE Ind Electron Mag 14(3):65–91
14. Zhao Z, Liu Z, He W et al (2021) Boundary adaptive fault-tolerant control for a flexible Timoshenko arm with backlash-like hysteresis. Automatica 130:109690
15. Ding SX (2013) Model-based fault diagnosis techniques: design schemes, algorithms, and tools, 2nd edn. Springer-Verlag, London
16. Ding SX (2014) Data-driven design of fault diagnosis and fault-tolerant control systems. Springer, London
17. Feng Y, Li H-X (2018) Detection and spatial identification of fault for parabolic distributed parameter systems. IEEE Trans Ind Electron 66(9):7300–7309
18. Feng Y, Li H-X (2020) Dynamic spatial independent component analysis based abnormality localization for distributed parameter systems. IEEE Trans Ind Inf 16(5):2929–2936
19. Ghantasala S, El-Farra NH (2009) Robust actuator fault isolation and management in constrained uncertain parabolic PDE systems. Automatica 45(10):2368–2373
20. Feng Y, Wang Y, Wang J-W et al (2021) Spatio-temporal fault localization for nonlinear spatially distributed processes, A spatial mapping filter-based framework. Int J Robust Nonlinear Control 31(14):6953–6971
21. Krstic M, Smyshlyaev A (2008) Boundary Control of PDEs: A Course on Backstepping Designs. SIAM, Philadelphia
22. Cai J, Ferdowsi H, Sarangapani J (2016) Model-based fault detection, estimation, and prediction for a class of linear distributed parameter systems. Automatica 66:122–131
23. Feng Y, Wang Y, Wang J-W et al (2022) Backstepping-based distributed abnormality localization for linear parabolic distributed parameter systems. Automatica 135:109930
24. Xu X, Yuan Y, Wang M et al (2019) Fault detection and estimation for a class of PIDE systems based on boundary observers. Int J Robust Nonlinear Control 29(17):5867–5885
25. Christofides PD (2012) Nonlinear and robust control of PDE systems: Methods and applications to transport-reaction processes. Springer Science+Business Media, Berlin
26. Xiao T, Li H-X (2016) Eigenspectrum-based iterative learning control for a class of distributed parameter system. IEEE Trans Autom Control 62(2):824–836
27. Wang B-C, Li H-X, Yang H-D (2019) Spatial correlation-based incremental learning for spatiotemporal modeling of battery thermal process. IEEE Trans Ind Electron 67(4):2885–2893
28. Wang J-W (2019) A unified Lyapunov-based design for a dynamic compensator of linear parabolic MIMO PDEs. Int J Control 94(7):1804–1811
29. MacCluer CR (1994) Boundary value problems and orthogonal expansions: physical problems from a Sobolev viewpoint. IEEE Press, New York
30. Wu H-N, Li H-X (2011) Robust adaptive neural observer design for a class of nonlinear parabolic PDE systems. J Process Control 21(8):1172–1182
31. El-Farra NH (2006) Integrated fault detection and fault-tolerant control architectures for distributed processes. Ind Eng Chem Res 45(25):8338–8351
32. El-Farra NH, Ghantasala S (2007) Actuator fault isolation and reconfiguration in transport-reaction processes. AIChE J 53(6):1518–1537
33. Harkort C (2014) Early-lumping based controller synthesis for linear infinite-dimensional systems. FAU Forschungen, Reihe B, Medizin, Naturwissenschaft, Technik 1. FAU University Press, Erlangen
34. Farrell JA, Polycarpou MM (2006) Adaptive Approximation Based Control: Unifying Neural. Fuzzy and Traditional Adaptive Approximation Approaches. John Wiley & Sons, Hoboken
35. Wang J-W (2020) Indirect adaptive distributed fuzzy control of semi-linear parabolic PDE systems. In: Proc. 2020 7th Int Conf Inf Cybern Comput Soc Syst (ICCSS), pp. 286–291
36. Ioannou PA, Sun J (2012) Robust Adaptive Control. Courier Corporation, Chelmsford
37. Feng Y, Li H-X, Yang H-D (2021) Abnormal source identification for parabolic distributed parameter systems. IEEE Trans Syst Man Cybern Syst 51(9):5698–5707
38. Li L, Ding SX, Peng K et al (2019) Fuzzy fault detection filter design for nonlinear distributed parameter systems. IEEE Access 7:11105–11113

39. Li L, Ding SX, Qiu J et al (2016) Fuzzy observer-based fault detection design approach for nonlinear processes. IEEE Trans Syst Man Cybern Syst 47(8):1941–1952
40. Li L, Ding SX (2020) Gap metric techniques and their application to fault detection performance analysis and fault isolation schemes. Automatica 118:109029
41. Silverman BW (2018) Density Estimation for Statistics and Data Analysis. Routledge, London
42. Ray WH (1981) Advanced Process Control. McGraw-Hill, New York
43. Strikwerda JC (2004) Finite Difference Schemes and Partial Differential Equations. SIAM, Philadelphia

Chapter 9
PDE Model-Based On-Line Cell-Level Thermal Fault Localization Framework for Batteries

Abstract Early-stage distributed thermal fault detection and localization are essential for battery safety but have been relatively underexplored compared to state-of-charge and state-of-health estimation. To address this, a backstepping-based fault localization filter is proposed. Traditional fault localization requires full-state temperature measurements, which are impractical in real applications. Using an interpolation-based approximation, the number of required sensors is reduced to a few, enhancing the fault localization filter's practical applicability. A structured framework, including fault localization filter design, distributed residual evaluation, and threshold determination, ensures robust and reliable real-time performance. The effectiveness of the approach is demonstrated through theoretical analysis and experimental validation.

9.1 Introduction

Li-ion batteries are widely utilized in EVs and HEVs due to their high energy capacity and long-term durability [1]. Ensuring their safe operation has become a major concern in both research and industrial practice. Thermal effects can significantly constrain battery performance [2], and under abusive conditions, cells frequently generate heat and gas [3]. Therefore, identifying thermal faults in Li-ion batteries is essential for maintaining safe operation.

Over the past few decades, numerous studies have addressed battery temperature estimation, as summarized in [4–9]. For instance, [2] introduced an incremental learning approach for distributed modeling of battery temperature. Additionally, [10] was the first to investigate robust temperature estimation of battery systems using sparse sensors.

Beyond these notable studies, research on battery FD remains limited. Most existing approaches rely on limit checking of measured surface temperature or terminal voltage, without fully accounting for the intrinsic thermal dynamics. As a result, their performance is constrained when inputs vary sharply or when faults are small, keeping the measured variables within normal operating ranges. Furthermore, few

© The Author(s), under exclusive license to Springer Nature Singapore Pte Ltd. 2026

Y. Feng et al., *Spatio-Temporal Abnormality Diagnosis for Industrial Distributed Parameter Systems*, https://doi.org/10.1007/978-981-95-3750-1_9

model-based FD methods address both issues [11, 12], and their applicability to distributed faults is limited due to the use of lumped-parameter thermal models.

Typically, the thermal behavior of Li-ion batteries is modeled as a DPS, also referred to as an infinite-dimensional system [13, 14], using PDEs. The S-T dynamics [15] make battery FD more challenging compared to lumped-parameter models based on ODEs. Instead of employing a lumped thermal model, [16] introduced a PDE-based scheme for detecting thermal faults in Li-ion batteries. Considering the S-T dynamics, diagnosing distributed faults [17–20] that vary across time and space is a natural research problem. From an engineering perspective, fault localization at the pack, module, and cell levels is crucial for fault mitigation and system recovery. Nevertheless, previous studies mainly focus on temporal fault detection, leaving spatial localization underexplored. Recently, [21] addressed fault localization for large-format batteries by dividing the two-dimensional space into multiple zones and designing a set of diagnostic filters for each zone. However, the localization was performed as a fault isolation scheme, and its accuracy depends on the number of isolator filters.

Therefore, this chapter addresses the problem of localizing unknown distributed faults in Li-ion batteries. To tackle this, a fault localization filter (FLF) is introduced, leveraging the advantages of backstepping. Specifically, a boundary observer is designed as a core component of the FLF, using only surface temperature measurements based on backstepping techniques. In addition, a distributed residual is constructed employing only a limited number of thermocouples, eliminating the need for full-state measurements.

The key contributions of this study are threefold:

- A comprehensive methodology is presented, encompassing the design of the FLF, distributed residual evaluation, and threshold determination, to ensure reliable and robust real-time fault localization for Li-ion batteries.
- The proposed FLF produces residual signals in a distributed manner using only a few temperature measurements, eliminating the need for full-state temperature information across the entire spatial domain.
- Unlike approximated lumped-parameter models, the design avoids any observation spillover issues.

The chapter is structured as follows. Section 9.2 presents the problem statement, while Sect. 9.3 introduces the methodological framework along with rigorous theoretical analysis. Experimental validation and discussions on a Li-ion battery are provided in Sect. 9.4. Finally, Sect. 9.5 synthesizes the key findings and presents the chapter's conclusions.

9.2 Preliminaries and Problem Statement

This chapter focuses on the thermal modeling of a cylindrical battery cell [16, 22] in 1-D:

$$\rho C_p T_\tau(\xi, \tau) = \lambda T_{\xi\xi}(\xi, \tau) + q(\xi, \tau),$$

the system is subject to these boundary conditions:

$$T_\xi(0, \tau) = 0,$$
$$T_\xi(R, \tau) = \frac{h}{\lambda}(T_\infty - T(R, t)),$$

where $T(\xi, \tau) \in \mathcal{L}^2([0, R])$ represents the temperature distribution; ρ denotes the cell's mass density; C_p is the specific heat; λ indicates thermal conductivity; $\xi \in [0, R]$ is the radial spatial coordinate; $\tau \in [0, \infty)$ denotes time; $q(\xi, \tau)$ corresponds to volumetric heat generation; and h and T_∞ denote the convective heat transfer coefficient and ambient temperature, respectively.

By applying the scaling transformation [23] $x = \frac{\xi}{R}$ and $t = \frac{\lambda}{\rho C_p R^2}\tau$, the model can be expressed in the following normalized form:

$$T_t(x, t) = T_{xx}(x, t) + \frac{R^2}{\lambda} q(x, t).$$

Furthermore, taking into account faults and disturbances in the system, the model can be represented as follows:

$$T_t(x, t) = T_{xx}(x, t) + \frac{R^2}{\lambda} q(x, t) + f(x, t) + d(x, t), \tag{9.1a}$$

$$\boldsymbol{T}_m(t) = \int_0^1 \boldsymbol{c}(x) T(x, t) \mathrm{d}x \in \mathbb{R}^{N+1}, \tag{9.1b}$$

$$\boldsymbol{c}(x) = [\delta(x - x_1), \cdots, \delta(x - x_N), \delta(x - 1)]^T \in \mathbb{R}^{N+1}, \tag{9.1c}$$
$$0 < x_1 < \cdots < x_N < 1,$$

with the following boundary conditions imposed:

$$T_x(0, t) = 0, \tag{9.2a}$$

$$T_x(1, t) = \frac{hR}{\lambda}(T_\infty - T(1, t)), \tag{9.2b}$$

and the initial condition is

$$T(x, 0) = T_0(x),$$

where $f(x, t)$ and $d(x, t)$ represent unknown spatially distributed fault and disturbance, respectively; $\boldsymbol{T}_m(t) = \mathrm{col}[\boldsymbol{T}_{\mathrm{in}}(t), T(1, t)]$ includes N in-domain point-wise measurements $\boldsymbol{T}_{\mathrm{in}}(t) \in \mathbb{R}^N$ at locations $x_1, \cdots, x_N$, together with a boundary measurement $T(1, t)$ at the cell surface, with $\delta(\cdot)$ denoting the Dirac delta function. It should be noted that the model in (9.1) is normalized, with the spatial variable x scaled to $[0, 1]$, facilitating the subsequent observer design and analysis. For convenience, the fault $f(x,t)$ and disturbance $d(x,t)$ are also defined within this normalized model [16].

Remark 9.1 As introduced in [16], the unknown distributed fault $f(x,t)$ models the effect of several physical failures such as abnormal heat generation owing to electrochemical side reactions or internal failure caused by thermal or mechanical-abuse [3, 25]. These failures can be caused by the following factors:

- Internal short circuit
- Overcharging or overdischarging
- External physical damage (puncture, crush, vibration, or shock)

Moreover, the unknown distributed disturbance $d(x,t)$ introduced in (9.1) captures the unmodeled dynamics, model uncertainties and noises exist in the thermal dynamics. Without loss of generality, both fault and disturbance exist in the system, which is a common setting in the literature [26, 27]. One major difference between the fault and disturbance is that the value of fault is zero in fault-free cases while the disturbance is generally unavoidable with a nonzero value.

Without loss of generality, the volumetric heat generation $q(x,t)$ considered is as follows [10, 28]:

$$q(x,t) = I(t)(V_{\mathrm{OC}}(\mathrm{SOC}) - V_{\mathrm{term}}(t) - T\frac{\partial V_{\mathrm{OC}}}{\partial T})/V_o.$$

where $I(t)$ denotes the current, V_{OC} and V_{term} denote the open-circuit voltage (OCV) and terminal voltage, respectively. V_o denotes the volume of battery. For simplicity, the effect of entropy change [29] is not considered in this chapter.

Remark 9.2 (**Measurements**) To facilitate the FLF design, the following measurements are essential:

- The current $I(t)$;
- The terminal voltage $V_{\mathrm{term}}(t)$;
- Temperature measurements $\boldsymbol{T}_m(t) = \mathrm{col}[\boldsymbol{T}_{in}(t), T(1,t)]$ including a measurement at the surface $T(1,t)$.

Remark 9.3 Note that both the surface measurement $T(1,t)$ and N in-domain measurements exist in system (9.1). The motivations are twofold.

- First, for the fault localization, in-domain measurements are necessary for constructing a distributed residual.
- Moreover, in speaking of the essential observer design, the surface measurement can serve this purpose well with the aid of backstepping techniques in a straightforward manner and the observer gains can be obtained explicitly. However, if in-domain measurements are used to achieve this goal, the observer gain matrix is obtained by solving N LMIs [30]. Therefore, the observer design efforts can be largely reduced for the backstepping-based design.

Assumption 9.1 The spatially distributed disturbance $d(x, t)$ satisfies that

$$||d||_2 \leq \bar{d}_2, \ ||d||_\infty \leq \bar{d}_\infty,$$

where $\bar{d}_2, \bar{d}_\infty > 0$ are **known** constants. This is a common setting in FDF design, as can be found in [20, 31].

The problem studied can be concluded as:

By utilizing **few sensor measurements** (including one surface sensor) to design a FLF for **distributed thermal fault localization** of batteries

whose thermal dynamics are described by the form in (9.1)–(9.4). Furthermore, to simplify the procedures related to the essential observer design, use only the surface sensor to achieve this purpose.

Remark 9.4 Compared with [11], the current chapter differs in both the system model and problem setting. First, the equivalent electrical circuit plus the lumped thermal dynamics model were adopted in [11], which is described by ODEs. However, such a model neglects the spatial distribution characteristic of temperature. Second, since the temperature distribution is not considered, only the sensor FD in the time domain was considered in [11].

Remark 9.5 To be mentioned, both fault detection and localization are considered within a battery cell scope. However, for the battery module and pack, the proposed methodology still applies since the temperature distribution physics are similar. One evident difference is that the spatial dimension for the battery module and pack is 2-D and 3-D, respectively. Moreover, since the key idea is the backstepping-based FLF design, the boundary observer design techniques for high-dimension PDEs [32] can be inherited. One possible difficulty maybe the distributed residual design with a finite number of measurements in high-dimensional space.

9.3 Intelligent Optimization-Based On-Line Localization

9.3.1 Parameter Identification Based on Intelligent Optimization

Accurate identification of thermal parameters plays a crucial role in predicting the battery's temperature distribution. As indicated in (9.1), the parameters to be determined consist of the specific heat capacity C_p, thermal conductivity λ, and convective heat transfer coefficient h.

The initial stage involves determining the thermal parameters required for the subsequent FLF development. Specifically, the surface temperature deviation $\widehat{T}(1, t)$ between the model prediction and experimental measurement is adopted as the criterion for parameter identification. By defining $\boldsymbol{\theta} \triangleq [C_p, \lambda, h]^{\mathrm{T}}$, the cost function is formulated as:

$$f(\boldsymbol{\theta}) = \sqrt{\sum_{l=1}^{N_t} \widehat{T}(1, l, \boldsymbol{\theta})^2}, \tag{9.3}$$

where N_t denotes the number of snapshots.

To obtain the parameter set that minimizes the cost function $f(\boldsymbol{\theta})$, the improved teaching–learning-based optimization (ITLBO) method [33] is employed, owing to its rapid convergence and ability to avoid local minima. In addition, ITLBO has demonstrated superior performance compared with other advanced constrained evolutionary optimization techniques [33]. Since the parameter identification task can be reformulated as a constrained optimization problem, ITLBO is selected to address it. The corresponding pseudo-code is provided in Algorithm 9.1.

9.3.2 Backstepping-Based Boundary Observer Design

To begin the FLF design, an infinite-dimensional boundary observer is initially established as described in [23]:

$$\hat{T}_t(x, t) = \hat{T}_{xx}(x, t) + \frac{R^2}{\lambda} q(x, t) + p_1(x)(T(1, t) - \hat{T}(1, t)), \tag{9.4a}$$

$$\hat{\boldsymbol{T}}_m(t) = \int_0^1 \boldsymbol{c}(x)\hat{T}(x, t)\mathrm{d}x, \tag{9.4b}$$

$$\hat{T}_x(0, t) = 0, \tag{9.4c}$$

$$\hat{T}_x(1, t) = \frac{hR}{\lambda}(T_\infty - T(1, t)) + p_{10}(T(1, t) - \hat{T}(1, t)), \tag{9.4d}$$

$p_1(x)$ and p_{10} represent the observer gains that need to be determined. By introducing the state estimation error $e(x, t) \triangleq T(x, t) - \hat{T}(x, t)$, the resulting error system can be formulated as follows:

$$e_t(x, t) = e_{xx}(x, t) - p_1(x)(T(1, t) - \hat{T}(1, t)) + f(x, t) + d(x, t), \tag{9.5a}$$

$$\tilde{\boldsymbol{T}}_m(t) = \int_0^1 \boldsymbol{c}(x)e(x, t)\mathrm{d}x, \tag{9.5b}$$

$$e_x(0, t) = 0, \tag{9.5c}$$

$$e_x(1, t) = -p_{10}e(1, t), \tag{9.5d}$$

by combining (9.1) with (9.4).

Lemma 9.1 *([23]). By applying the Volterra integral transformation $\mathcal{T}$:*

$$e(x, t) = \tilde{\omega}(x, t) - \int_x^1 p(x, \xi)\tilde{\omega}(\xi, t)\mathrm{d}\xi, \tag{9.6}$$

Algorithm 9.1: ITLBO

Input: upper bound of $\boldsymbol{\theta}$, lower bound of $\boldsymbol{\theta}$, population size (number of possible solutions) NP, maximum generation G, number of sunpopulations Ns

Output: the optimized $\boldsymbol{\theta}$

/****Initialization****/

Randomly generate $\{\boldsymbol{\theta}_1^0, \cdots, \boldsymbol{\theta}_{NP}^0\}$ between the upper and lower bounds: ;

Evaluate $\{\boldsymbol{\theta}_1^0, \cdots, \boldsymbol{\theta}_{NP}^0\}$ based on the objective function;

$g \leftarrow 0$;

while $g < G$ **do**

- Sort the population and divide it into Ns subpopulations uniformly;
- **for** $i = 1 : Ns$ **do**
  - Denote the best solution in the i-th subpopulation as $\boldsymbol{\theta}_{best,i}^g$;
  - Denote the centeriod of the i-th subpopulation as $\boldsymbol{\theta}_{mean,i}^g$;
  - **for** $j = 1 : \frac{NP}{Ns}$ **do**
    - $t \leftarrow (i-1) \cdot \frac{NP}{Ns} + j$;
    - **if** $rand < 0.5$ **then**
      - /****teacher phase****/
      - $TF \leftarrow [1 + rand]$;
      - $\hat{\boldsymbol{\theta}}_t^g \leftarrow \boldsymbol{\theta}_{best,i}^g - TF \cdot \frac{\boldsymbol{\theta}_{mean,i}^g + \boldsymbol{\theta}_t^g}{2}$;
    - **else**
      - /****learner phase****/
      - Randomly select $\boldsymbol{\theta}_{xr1}^g$, $\boldsymbol{\theta}_{xr2}^g$, and $\boldsymbol{\theta}_{xr3}^g$;
      - **if** $f(\boldsymbol{\theta}_{xr1}^g) < f(\boldsymbol{\theta}_t^g)$ **then**
        - $\hat{\boldsymbol{\theta}}_t^g \leftarrow rand \cdot (\boldsymbol{\theta}_{xr1}^g - \boldsymbol{\theta}_t^g) + rand \cdot (\boldsymbol{\theta}_{xr2}^g - \boldsymbol{\theta}_{xr3}^g)$;
      - **else**
        - **if** $f(\boldsymbol{\theta}_{xr2}^g) < f(\boldsymbol{\theta}_{xr3}^g)$ **then**
          - $\boldsymbol{v}_t^g \leftarrow \boldsymbol{\theta}_{xr2}^g - \boldsymbol{\theta}_{xr3}^g$;
        - **else**
          - $\boldsymbol{v}_t^g \leftarrow \boldsymbol{\theta}_{xr3}^g - \boldsymbol{\theta}_{xr2}^g$;
        - **for** $d = 1 : D$ **do**
          - **if** $rand < rand$ **then**
            - $\hat{\theta}_{t,d}^g \leftarrow \theta_{t,d}^g$;
          - **else**
            - $TF \leftarrow 2 * rand - 1$;
            - $\hat{\theta}_{t,d}^g \leftarrow \theta_{xr1,d}^g + TF \cdot v_{t,d}^g$;
- /****selection****/
- Evaluate $\{\hat{\boldsymbol{\theta}}_1^g, \cdots, \hat{\boldsymbol{\theta}}_{NP}^g\}$ according to the objective function;
- **for** $i = 1 : NP$ **do**
  - Replace $\boldsymbol{\theta}_i^g$ by $\boldsymbol{\theta}_i^g$, if the latter is better;
  - $\boldsymbol{\theta}_i^{g+1} \leftarrow \boldsymbol{\theta}_i^g$;
- $g \leftarrow g + 1$;

the error system (9.5) converts to the target $\tilde{\omega}$-system as follows:

$$\tilde{\omega}_t(x,t) = \tilde{\omega}_{xx}(x,t) - c\tilde{\omega}(x,t) + f_{\tilde{\omega}}(x,t) + d_{\tilde{\omega}}(x,t), \tag{9.7a}$$

$$\tilde{\omega}_x(0,t) = 0, \tag{9.7b}$$

$$\tilde{\omega}_x(1,t) = 0, \tag{9.7c}$$

where $c > 1$ *denotes a predefined parameter and* $f(x,t) = f_{\tilde{\omega}}(x,t) - \int_0^x p(x,\xi) f_{\tilde{\omega}}(\xi,t)\mathrm{d}\xi$, $d(x,t) = d_{\tilde{\omega}}(x,t) - \int_0^x p(x,\xi) d_{\tilde{\omega}}(\xi,t)\mathrm{d}\xi$ *where* $f_{\tilde{\omega}}(x,t)$ *and* $d_{\tilde{\omega}}(x,t)$ *denote the transformed fault and disturbance in the* $\tilde{\omega}$*-system.*

Lemma 9.2 *([23]).* $p(x,\xi)$ *denotes the gain kernel and it subjects to*

$$p_{xx}(x,\xi) - p_{\xi\xi}(x,\xi) = -cp(x,\xi), \tag{9.8a}$$

$$p_x(0,\xi) = 0, \tag{9.8b}$$

$$p(x,x) = -\frac{c}{2}x, \tag{9.8c}$$

and the observer gains can be calculated as

$$p_1(x) = -p_\xi(x,1), \tag{9.9a}$$

$$p_{10} = -p(1,1). \tag{9.9b}$$

Remark 9.6 As introduced in ([23]), a unique solution of the Klein-Gordon PDEs (9.8) can be obtained as

$$p(x,\xi) = -c\xi\frac{I_1(\sqrt{c(\xi^2 - x^2)})}{\sqrt{c(\xi^2 - x^2)}}, \tag{9.10}$$

by the method of successive approximations. Recalling (9.9), the observer gains can be further calculated as

$$p_1(x) = -p_\xi(x,1) = \frac{c}{1 - x^2} I_2(\sqrt{c(1 - x^2)}), \tag{9.11a}$$

$$p_{10} = -p(1,1) = \frac{c}{2}. \tag{9.11b}$$

Note that I_n denotes the modified Bessel Function with order n.

Lemma 9.3 *The inverse transformation* $\mathcal{T}^{-1}$ *of (9.6) is*

$$\tilde{\omega}(x,t) = e(x,t) + \int_x^1 k(x,\eta)e(\eta,t)\mathrm{d}\eta,$$

where $k(x,\eta)$ *can also be calculated as*

$$k(x,\eta) = -c\eta\frac{J_1(\sqrt{c(\eta^2 - x^2)})}{\sqrt{c(\eta^2 - x^2)}},$$

where J_n *denotes the Bessel Function with order* n*.*

Proof See Chap. 4 of reference [23].

Remark 9.7 It is worth mentioning that

$$d_{\tilde{\omega}}(x,t) = d(x,t) + \int_x^1 k(x,\eta)d(\eta,t)\mathrm{d}\eta,$$

then

$$||d_{\tilde{\omega}}(\cdot,t)||_2 \le (1+\bar{k})\bar{d}_2,$$

holds where

$$\bar{k} = \max_{0\le x\le 1} ||k(x,\cdot)||_2.$$

Lemma 9.4 *[34] For regular real functions $V(t)$ and $h(t)$.*

$$\dot{V}(t) \le -\alpha V(t) + h(t), \forall t \ge 0$$

implies that

$$V(t) \le e^{-\alpha t}V(0) + \int_0^t e^{-\alpha(t-\tau)}h(\tau)\mathrm{d}\tau, \quad \forall t \ge 0$$

for any finite constant α.

Theorem 9.1 *The target $\tilde{\omega}$-system is exponentially stable in $\mathcal{L}^\infty$ without faults or disturbances ($f_{\tilde{\omega}}(x,t) = 0, d_{\tilde{\omega}}(x,t) = 0$). Furthermore, under the circumstances with disturbances, $\tilde{\omega}(x,t)$ is bounded by the following expression:*

$$||\tilde{\omega}(\cdot,t)||_\infty \le \tilde{\omega}_b(t),$$

where

$$\begin{aligned} \tilde{\omega}_b(t) &\triangleq (2e^{-2(c-1)t}V(0) + \frac{1}{\gamma(c-1)}(1+\bar{k})^2\bar{d}_2^2 + \frac{1}{2\gamma(c-1)}\bar{d}_\infty^2)^{\frac{1}{2}}, \\ V(0) &\triangleq \frac{1}{2}\left(\tilde{\omega}^2(1,0) + \|\tilde{\omega}_0\|_2^2 + \|\tilde{\omega}_{x0}\|_2^2\right), \end{aligned} \tag{9.12}$$

with $0 < \gamma \le 2$.

Proof Consider the following Lyapunov candidate:

$$V(t) = \frac{1}{2}\tilde{\omega}^2(1,t) + \frac{1}{2}\int_0^1 (\tilde{\omega}^2(x,t) + \tilde{\omega}_x^2(x,t))\mathrm{d}x. \tag{9.13}$$

Then we have

$$\dot{V}(t) = \tilde{\omega}(1,t)\tilde{\omega}_t(1,t) + \int_0^1 \tilde{\omega}(x,t)\tilde{\omega}_t(x,t)\mathrm{d}x + \int_0^1 \tilde{\omega}_x(x,t)\tilde{\omega}_{xt}(x,t)\mathrm{d}x. \tag{9.14}$$

With integration by parts, the second term in (9.14) can be further decomposed as:

$$\begin{aligned}\int_0^1 \tilde{\omega}\tilde{\omega}_t \mathrm{d}x &= \int_0^1 \tilde{\omega}\tilde{\omega}_{xx}\mathrm{d}x - c\int_0^1 \tilde{\omega}^2\mathrm{d}x + \int_0^1 d_{\tilde{\omega}}\tilde{\omega}\mathrm{d}x \\ &= -\int_0^1 \tilde{\omega}_x^2\mathrm{d}x - c\int_0^1 \tilde{\omega}^2\mathrm{d}x + \int_0^1 d_{\tilde{\omega}}\tilde{\omega}\mathrm{d}x,\end{aligned} \tag{9.15}$$

by combining with (9.7).

Moreover, the third term in (9.14) can be rewritten as

$$\begin{aligned}\int_0^1 \tilde{\omega}_x\tilde{\omega}_{xt}\mathrm{d}x &= -\int_0^1 \tilde{\omega}_{xx}^2\mathrm{d}x + c\int_0^1 \tilde{\omega}\tilde{\omega}_{xx}\mathrm{d}x - \int_0^1 d_{\tilde{\omega}}\tilde{\omega}_{xx}\mathrm{d}x \\ &= -\int_0^1 \tilde{\omega}_{xx}^2\mathrm{d}x - c\int_0^1 \tilde{\omega}_x^2\mathrm{d}x - \int_0^1 d_{\tilde{\omega}}\tilde{\omega}_{xx}\mathrm{d}x,\end{aligned} \tag{9.16}$$

through using integration by parts and combining it with (9.7).

Combining (9.14), (9.15) and (9.16), we have

$$\begin{aligned}\dot{V}(t) &= -(1+c)||\tilde{\omega}_x(\cdot,t)||_2^2 - c||\tilde{\omega}(\cdot,t)||_2^2 - ||\tilde{\omega}_{xx}(\cdot,t)||_2^2 \\ &+ \int_0^1 d_{\tilde{\omega}}\tilde{\omega}\mathrm{d}x - \int_0^1 d_{\tilde{\omega}}\tilde{\omega}_{xx}\mathrm{d}x - c\tilde{\omega}^2(1) + \tilde{\omega}(1)d_{\tilde{\omega}}(1).\end{aligned} \tag{9.17}$$

Part I. First, we investigate the situations without disturbances, i.e., $d_{\tilde{\omega}}(x,t) = 0$. We have that

$$\begin{aligned}\dot{V}(t) &= -||\tilde{\omega}_x(\cdot,t)||_2^2 - c||\tilde{\omega}_x(\cdot,t)||_2^2 - c||\tilde{\omega}(\cdot,t)||_2^2 - ||\tilde{\omega}_{xx}(\cdot,t)||_2^2 - c\tilde{\omega}^2(1) \\ &\le -c(||\tilde{\omega}(\cdot,t)||_2^2 + ||\tilde{\omega}_x(\cdot,t)||_2^2 + \tilde{\omega}^2(1)) = -2cV(t),\end{aligned} \tag{9.18}$$

which further yields

$$V(t) \le e^{-2ct}V(0), \tag{9.19}$$

by referring to Lemma 9.4.

Applying Agmon's inequality and Young's inequality [23] and considering (9.19) yields

$$\begin{aligned}||\tilde{\omega}(\cdot,t)||_\infty^2 &\le 2||\tilde{\omega}(\cdot,t)||_2||\tilde{\omega}_x(\cdot,t)||_2 + \tilde{\omega}^2(1,t) \\ &\le ||\tilde{\omega}(\cdot,t)||_2^2 + ||\tilde{\omega}_x(\cdot,t)||_2^2 + \tilde{\omega}^2(1,t) \\ &\le e^{-2ct}(||\tilde{\omega}_0||_2^2 + ||\tilde{\omega}_{x0}||_2^2 + \tilde{\omega}^2(1,t)),\end{aligned} \tag{9.20}$$

which indicates the exponentially convergence of $\tilde{\omega}(x,t)$.

Part II. When there exists a disturbance, first apply the Cauchy-Schwarz inequality to (9.17) leads to

$$\begin{aligned}\dot{V}(t) \leq & -(1+c)||\tilde{\omega}_x(\cdot,t)||_2^2 - c||\tilde{\omega}(\cdot,t)||_2^2 - ||\tilde{\omega}_{xx}(\cdot,t)||_2^2 \\ & + ||d_{\tilde{\omega}}(\cdot,t)||_2||\tilde{\omega}(\cdot,t)||_2 + ||d_{\tilde{\omega}}(\cdot,t)||_2||\tilde{\omega}_{xx}(\cdot,t)||_2 - c\tilde{\omega}^2(1,t) + \tilde{\omega}(1,t)d_{\tilde{\omega}}(1,t).\end{aligned} \tag{9.21}$$

Moreover, by using Young's inequality, (9.21) can be further decomposed as

$$\begin{aligned}\dot{V}(t) \leq & -(1+c)||\tilde{\omega}_x(\cdot,t)||_2^2 - c||\tilde{\omega}(\cdot,t)||_2^2 - ||\tilde{\omega}_{xx}(\cdot,t)||_2^2 \\ & + \frac{1}{2\gamma}||d_{\tilde{\omega}}(\cdot,t)||_2^2 + \frac{\gamma}{2}||\tilde{\omega}(\cdot,t)||_2^2 + \frac{1}{2\gamma}||d_{\tilde{\omega}}(\cdot,t)||_2^2 \\ & + \frac{\gamma}{2}||\tilde{\omega}_{xx}(\cdot,t)||_2^2 - c\tilde{\omega}^2(1,t) + \frac{\gamma}{2}\tilde{\omega}^2(1,t) + \frac{1}{2\gamma}d_{\tilde{\omega}}^2(1,t),\end{aligned} \tag{9.22}$$

with $\gamma > 0$ denoting a positive constant.

For $0 < \gamma \leq 2$, we have that

$$\begin{aligned}\dot{V}(t) & \leq -(1+c)||\tilde{\omega}_x(\cdot,t)||_2^2 + (1-c)||\tilde{\omega}(\cdot,t)||_2^2 + (1-c)\tilde{\omega}^2(1) + \frac{1}{\gamma}||d_{\tilde{\omega}}(\cdot,t)||_2^2 + \frac{1}{2\gamma}d_{\tilde{\omega}}^2(1,t) \\ & \leq (1-c)(||\tilde{\omega}_x(\cdot,t)||^2 + ||\tilde{\omega}(\cdot,t)||^2 + \tilde{\omega}^2(1,t)) + \frac{1}{\gamma}||d_{\tilde{\omega}}(\cdot,t)||_2^2 + \frac{1}{2\gamma}d_{\tilde{\omega}}^2(1,t) \\ & = 2(1-c)V(t) + \frac{1}{\gamma}||d_{\tilde{\omega}}(\cdot,t)||_2^2 + \frac{1}{2\gamma}d_{\tilde{\omega}}^2(1,t).\end{aligned} \tag{9.23}$$

Recalling Remark 5, we can obtain that

$$\dot{V}(t) \leq 2(1-c)V(t) + \frac{1}{\gamma}(1+\bar{k})^2\bar{d}_2^2 + \frac{1}{2\gamma}\bar{d}_\infty^2. \tag{9.24}$$

Since $c > 1$, using Lemma 9.4, we have that

$$V(t) \leq e^{-2(c-1)t}V(0) + \frac{1}{2\gamma(c-1)}(1+\bar{k})^2\bar{d}_2^2 + \frac{1}{4\gamma(c-1)}\bar{d}_\infty^2. \tag{9.25}$$

Recalling the expression of $V(t)$, it can be observed that

$$||\tilde{\omega}(\cdot,t)||_\infty \leq \tilde{\omega}_b(t). \tag{9.26}$$

Remark 9.8 It is worth mentioning that the selected Lyapunov candidate in (9.13) indicates the "energy" in the space domain and is widely used in the analysis of PDE observers [23].

9.3.3 FLF Design and Analysis

A essential component of model-based FD [26] involves constructing a residual signal that reflects the discrepancy between measured data and analytical redundancy.

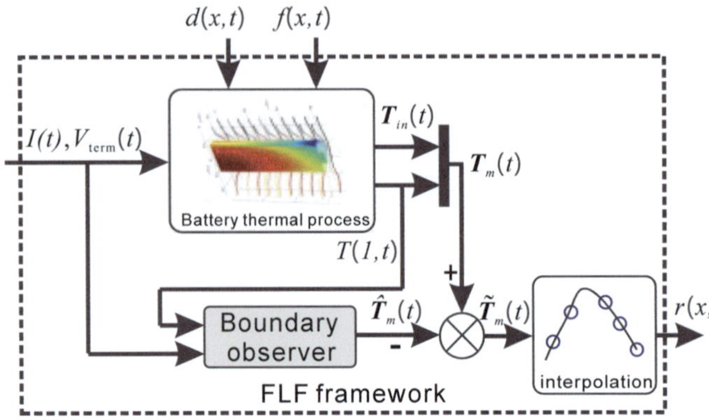

Fig. 9.1 Methodology framework

A straightforward choice for the residual signal in temporal fault detection is given as follows:

$$r(t) = e(1, t),$$

by using only the surface estimation error as it has been done in [16, 28, 31].

For the fault localization task addressed in this chapter, boundary measurements alone are insufficient; it is necessary to formulate the following distributed residual:

$$r(x, t) \triangleq e(x, t) = T(x, t) - \hat{T}(x, t).$$

Remark 9.9 It is worth noting that the FLF requires full-state measurement, which restricts the usability of the proposed method. However, one can adopt the interpolation method from a control engineering perspective [35].

After deriving the distributed residual from the output estimation error $\tilde{\boldsymbol{T}}_m(t)$, the subsequent step is to perform residual evaluation (Fig. 9.1).

The **distributed** residual evaluation index and threshold are designed as follows:

$$J(x, t) = \int_t^{t+T_w} r^2(x, \tau)\mathrm{d}\tau, \tag{9.27a}$$

$$J_{\mathrm{th}}(t) = \sup_{f(x,t)=0} J(x, t), \tag{9.27b}$$

where $T_w > 0$ denotes the evaluation time window length.

Moreover, the following fault localization logic is presented

$$\begin{cases} J(x, t) > J_{\mathrm{th}}(t) \Longrightarrow \text{faulty for location } x \\ J(x, t) \leq J_{\mathrm{th}}(t) \Longrightarrow \text{fault-free for location } x. \end{cases} \tag{9.28}$$

Theorem 9.2 *(Threshold computation) With the residual evaluation scheme in (9.27a), since the conditions in Theorem 9.1 can be fulfilled, then we have that*

$$J_{th}(t) = \int_t^{t+T_w} \left(1 + \bar{p}^2\right) \tilde{\omega}_b^2(\tau)\mathrm{d}\tau, \tag{9.29}$$

where

$$\bar{p} = \max_{0 \leq x, \xi \leq 1} ||p(x, \cdot)||_2. \tag{9.30}$$

Proof According to the definition of $J(x, t)$ and $J_{\mathrm{th}}(t)$, one has that

$$J(x, t) = \int_t^{t+T_w} e^2(x, \tau)\mathrm{d}\tau \leq \int_t^{t+T_w} ||e(\cdot, \tau)||_\infty^2 \mathrm{d}\tau.$$

Combining (9.6) and Theorem 9.1, we have that

$$||e(\cdot, t)||_\infty^2 \leq (1 + \bar{p}^2)||\tilde{\omega}(\cdot, t)||_\infty^2 \leq (1 + \bar{p}^2)\tilde{\omega}_b^2(t)$$

by triangle inequality.

Hence the localization threshold $J_{th}(t)$ can be calculated according to (9.29). This completes the proof.

A summary of the proposed FLF framework is presented in Figure 9.1 and the pseudo code can be found in Algorithm 9.2.

Algorithm 9.2: FLF

Input: current $I(t)$, terminal voltage $V_{\mathrm{term}}(t)$, temperature measurements $\boldsymbol{T}_m(t)$
Output: Faulty locations set x_f

/****Initialization****/
Select $c > 1$ (convergence rate), $0 < \gamma \leq 2$, $\hat{T}_0(x)$ (initial condition), and T_w (evaluation time window length);
/****Off-line threshold generation****/
Observer gains: $(p_1(x), p_{10}) \leftarrow (9.11)$;
for $t \in [0, t_{\max}]$ **do**
    Localization threshold: $J_{\mathrm{th}}(t) \leftarrow (9.27\mathrm{b})$;
/****On-line distributed residual evaluation****/
for $t \in [0, t_{\max}]$ **do**
    for $x \in [0, 1]$ **do**
        Residual evaluation: $J(x, t) \leftarrow (9.27\mathrm{a})$;
        if $J(x, t) > J_{th}(t)$ **then**
            $x \in x_f$
        else
            $x \notin x_f$

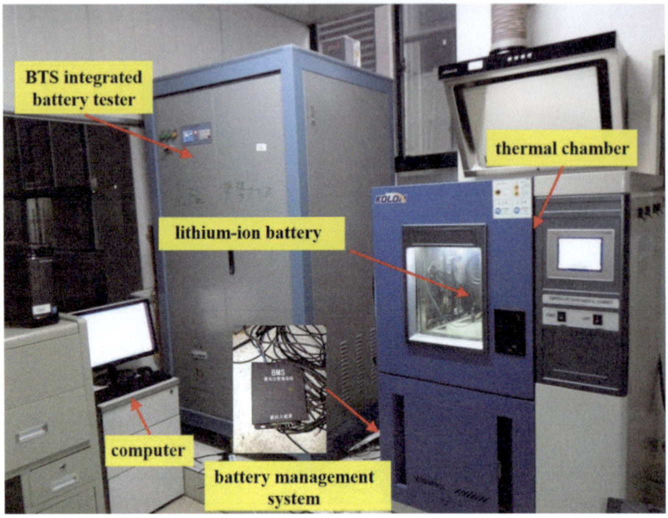

Fig. 9.2 Experiment platform

9.4 Experiment Validation

9.4.1 Experiment Setup

To demonstrate the effectiveness of the proposed FLF, experimental validation is carried out on a cylindrical Li-ion battery. The experimental setup is illustrated in Fig. 9.2. Input currents are generated using a Battery thermal system (BTS) equipped tester, while the cell is placed inside a thermal chamber. The BMS records the measurement data, and the FLF is implemented in MATLAB on the host computer.

Figure 9.3 displays the 38650 battery utilized in this study. Key specifications of the battery are listed in Table 9.1. Data were collected using three thermocouples placed evenly across the battery surface.

Table 9.1 Some nominal parameters of the battery

Parameter	Value
Radius	32 mm
Height	65 mm
Weight	150 g
Capacity	6.5 Ah
Nominal voltage	3.7 V
Charge cut-off voltage	3.0 V
Discharge cut-off voltage	4.2 V

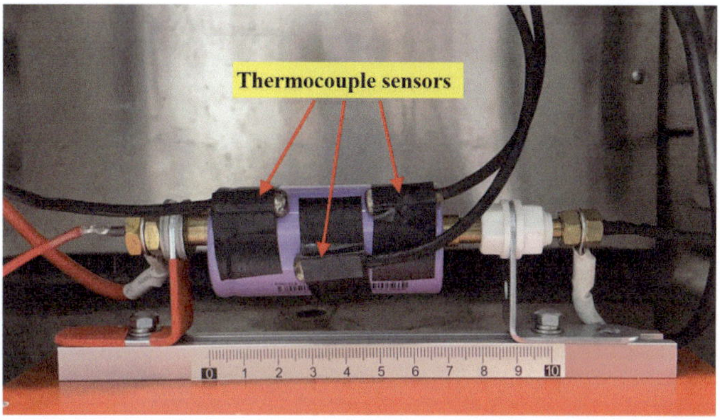

Fig. 9.3 32650 battery

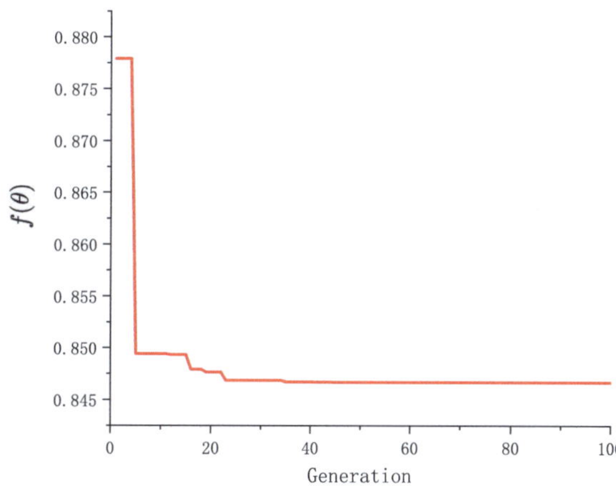

Fig. 9.4 Parameters identification results

9.4.2 Thermal Parameter Identification

The identification procedure is illustrated in Fig. 9.4, where $f(\boldsymbol{\theta})$ is defined in (9.3). Figure 9.5a shows the terminal voltage predicted by the equivalent circuit model [16], with an RMSE of 0.0277 V; here, $\tilde{V}_{\text{term}}(t)$ represents the discrepancy between measured and modeled voltages. Similarly, Fig. 9.5b presents the surface temperature predicted by the thermal model in (9.1), with an RMSE of 0.0448 K, where $\widehat{T}(1, t)$ indicates the difference between the average measured temperature and the model. These results demonstrate that the proposed parameter identification method keeps surface temperature errors within an acceptable range, making the estimated parameters from Table 9.2 suitable for the subsequent design of the FLF.

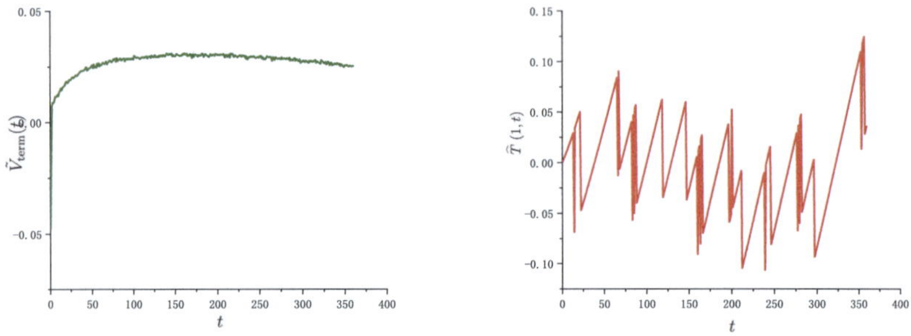

(a) Terminal voltage estimation error (V).

(b) Surface temperature error between experiment and identified thermal model (K).

Fig. 9.5 Thermal model validation

Table 9.2 Identified thermal parameters

	Definition (unit)	Value
λ	Thermal conductivity ([W/m K])	0.99
c	Specific heat ([J/Kg K])	1227.81
h	Convective coefficient ([W/m^2 K])	5.00

9.4.3 *Fault Localization Results*

Using the identified thermal parameters, the fault localization capability of the proposed method was evaluated through simulations. The following distributed fault is considered:

$$f(x,t) = b_f(x) f(t),$$

where

$$b_f(x) = H(x - 1/2) - H(x - 3/4),$$

$$f(t) = \begin{cases} 0, & 0 \leqslant t < 2.77(\text{sec}) \\ 3, & 2.77 \leqslant t \leqslant 5.54(\text{sec}) \end{cases}$$

with $H(\cdot)$ is the standard Heaviside function. Moreover, the distributed disturbance adopts the form

$$d(x,t) = 0.005 e^{(-0.5(x-0.5)^2)} \sin(t).$$

It should be noted that the fault is introduced in a simulation setup where the thermal model is built using the identified parameters (thermal conductivity λ, specific heat c, and convective coefficient h). Since performing such a fault injection in reality is hazardous and could lead to fire, numerical simulations are employed instead.

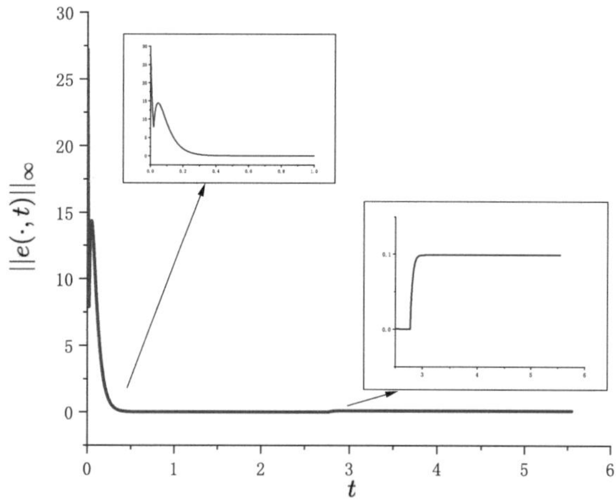

Fig. 9.6 Backstepping observer performance

The design parameters are set as $c = 20$, $\gamma = 1$, $T_w = 1.10$ sec. The evolution of $||e(\cdot, t)||\infty$ is illustrated in Fig. 9.6, confirming the result of Theorem 9.1; that is, the state estimation error is bounded in the spatial $\mathcal{L}\infty$ norm. After the fault occurs, the state estimation error $e(x, t)$ and the distributed residual signal $r(x, t)$ are shown in Fig. 9.7a, b, respectively. The distributed residual is constructed using spline interpolation with 10 in-domain sensors (21 grid points in total), and the discrepancy between $e(x, t)$ and $r(x, t)$ is within an acceptable range. To quantify the interpolation performance, the root of mean square error (RMSE) is adopted as:

$$\text{RMSE} = \sqrt{\frac{1}{ML}\sum_{i=1}^{M}\sum_{j=1}^{L}\left(e\left(x_i, t_j\right) - r\left(x_i, t_j\right)\right)^2},$$

where M and L represent the number of outputs and the duration, respectively. As shown in Fig. 9.7c, the computed RMSE is 0.0097, indicating that the interpolation error falls within an acceptable range.

The fault localization results obtained using interpolation are shown in Fig. 9.8b, where $\bar{J}(x, t)$ and $\bar{J}_{\text{th}}(t)$ represent the averages over the time interval (2.77sec–5.54sec). For comparison, Fig. 9.8a presents the fault localization results based on full-state measurements. It can be observed that both the proposed interpolation-based method and the full-state approach can identify the faulty region, though with some deviations. To further assess fault localization performance, the metrics "detection rate" and "misdetection rate" are introduced as follows:

$$\text{Detection rate} = \frac{\text{Detected fault area} - \text{Misdetected fault area}}{\text{Actual fault area}},$$

$$\text{Misdetection rate} = \frac{\text{Misdetected fault area}}{\text{Detected fault area}}.$$

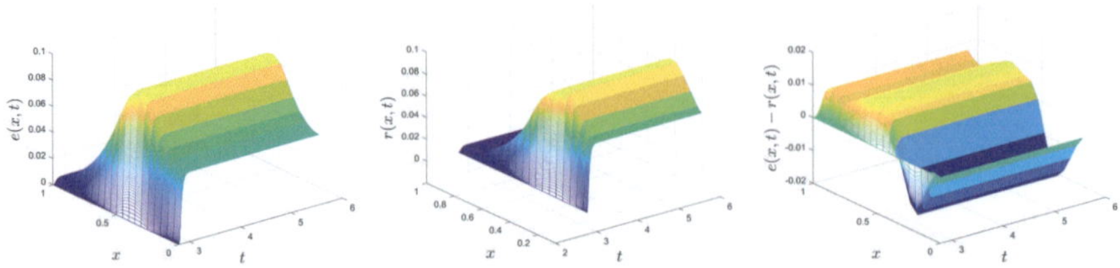

(a) $e(x,t)$. (b) $r(x,t)$. (c) $e(x,t) - r(x,t)$.

Fig. 9.7 State estimation error and the distributed residual signal

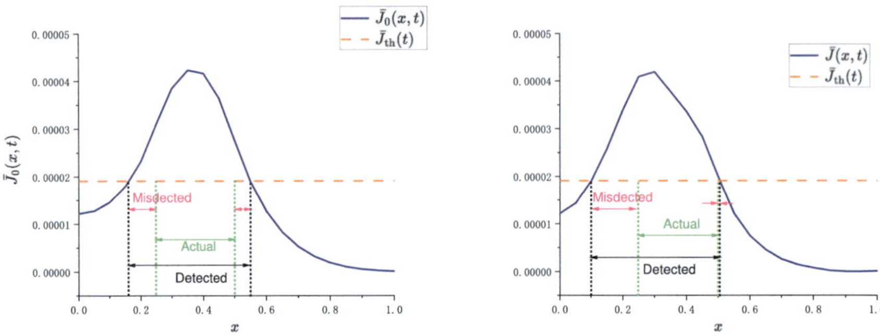

(a) Full-state measurement. (b) Proposed method with interpolation.

Fig. 9.8 Fault localization results

Table 9.3 Fault localization results using full-state measurement and interpolation

	Full-state measurement	Interpolation
Detection rate(%)	**100**	**100**
Misdetection rate(%)	**35.57**	38.12

Table 9.3 presents the fault localization performance. It can be seen that both methods are capable of identifying the faulty region, although the accuracy of the proposed approach is slightly lower than that of the full-state measurement method. This indicates that the performance of the proposed method is close to that achieved with full-state measurements.

9.5 Conclusion

This chapter introduces a FLF approach to tackle the problem of incipient thermal fault localization in batteries. Specifically, a boundary observer is designed using backstepping techniques with only surface measurements, avoiding observation spillover during the design phase. In addition, a FDF is proposed by generating a distributed residual signal without requiring full-state measurements. The effectiveness of the proposed framework is demonstrated through rigorous theoretical analysis and experimental validation.

References

1. Lu L, Han X, Li J et al (2013) A review on the key issues for lithium-ion battery management in electric vehicles. J Power Sources 226:272–288
2. Wang B-C, Li H-X, Yang H-D (2019) Spatial correlation-based incremental learning for spatiotemporal modeling of battery thermal process. IEEE Trans Ind Electron 67(4):2885–2893
3. Doughty DH, Roth EP (2012) A general discussion of Li ion battery safety. Electrochem Soc Interface 21(2):37–44
4. Zhou Y, Deng H, Li H-X et al (2021) Dual separation-based spatiotemporal modeling methodology for battery thermal process under nonhomogeneous boundary conditions. IEEE Trans Transp Electrification 7(4):2260–2268
5. Wang B-C, Li H-X (2019) A sliding window based dynamic spatiotemporal modeling for distributed parameter systems with time-dependent boundary conditions. IEEE Trans Ind Inform 15(4):2044–2053
6. Meng X-B, Li H-X, Yang H-D (2019) Evolutionary design of spatiotemporal learning model for thermal distribution in lithium-ion batteries. IEEE Trans Ind Inform 15(5):2838–2848
7. Xu K-K, Li H-X, Yang H-D (2018) Local-properties-embedding-based nonlinear spatiotemporal modeling for lithium-ion battery thermal process. IEEE Trans Ind Electron 65(12):9767–9776
8. Xu K-K, Li H-X, Liu Z (2017) ISOMAP-based spatiotemporal modeling for lithium-ion battery thermal process. IEEE Trans Ind Inform 14(2):569–577
9. Xu K, Fan B, Yang H et al (2022) Locally weighted principal component analysis-based multimode modeling for complex distributed parameter systems. IEEE Trans Cybern 52(10):10504–10514
10. Lin X, Perez HE, Siegel JB et al (2020) Robust estimation of battery system temperature distribution under sparse sensing and uncertainty. IEEE Trans Control Syst Technol 28(3):753–765
11. Dey S, Mohon S, Pisu P et al (2016) Sensor fault detection, isolation, and estimation in lithium-ion batteries. IEEE Trans Control Syst Technol 24(6):2141–2149
12. Sidhu A, Izadian A, Anwar S (2014) Adaptive nonlinear model-based fault diagnosis of Li-ion batteries. IEEE Trans Ind Electron 62(2):1002–1011
13. Wang J-W, Li H-X, Wu H-N (2016) A membership-function-dependent approach to design fuzzy pointwise state feedback controller for nonlinear parabolic distributed parameter systems with spatially discrete actuators. IEEE Trans Syst Man Cybern Syst 47(7):1486–1499
14. Wang Z-P, Wu H-N, Li H-X (2018) Estimator-based $\mathcal{H}_\infty$ sampled-data fuzzy control for nonlinear parabolic PDE systems. IEEE Trans Syst Man Cybern Syst 50(7):2491–2500
15. Wang Z-P, Zhang X, Wu H-N et al (2023) Fuzzy boundary control for nonlinear delayed DPSs under boundary measurements. IEEE Trans Cybern 53(3):1547–1556
16. Dey S, Perez HE, Moura SJ (2017) Model-based battery thermal fault diagnostics: Algorithms, analysis, and experiments. IEEE Trans Control Syst Technol 27(2):576–587
17. Feng Y, Li H-X (2018) Detection and spatial identification of fault for parabolic distributed parameter systems. IEEE Trans Ind Electron 66(9):7300–7309
18. Feng Y, Li H-X (2020) Dynamic spatial independent component analysis based abnormality localization for distributed parameter systems. IEEE Trans Ind Inform 16(5):2929–2936
19. Feng Y, Li H-X, Yang H-D (2021) Abnormal source identification for parabolic distributed parameter systems. IEEE Trans Syst Man Cybern Syst 51(9):5698–5707

20. Feng Y, Wang Y, Wang B-C et al (2022) Spatial decomposition-based fault detection framework for parabolic-distributed parameter processes. IEEE Trans Cybern 52(8):7319–7327
21. Sattarzadeh S, Roy T, Dey S (2021) Thermal fault detection and localization framework for large format batteries. J Power Sources 512:230400
22. Gomadam PM, White RE, Weidner JW (2003) Modeling heat conduction in spiral geometries. J Electrochem Soc 150(10):A1339–A1345
23. Krstic M, Smyshlyaev A (2008) Boundary control of PDEs: A course on backstepping designs. SIAM, Philadelphia
24. Wang J-W, Liu Y-Q, Sun C-Y (2018) Pointwise exponential stabilization of a linear parabolic PDE system using non-collocated pointwise observation. Automatica 93:197–210
25. Bandhauer TM, Garimella S, Fuller TF (2011) A critical review of thermal issues in lithium-ion batteries. J Electrochem Soc 158(3):R1–R25
26. Ding SX (2013) Model-based fault diagnosis techniques: design schemes, algorithms, and tools, 2nd edn. Springer-Verlag, London
27. Deutscher J (2016) Fault detection for linear distributed-parameter systems using finite-dimensional functional observers. Int J Contro 89(3):550–563
28. Dey S, Perez HE, Moura SJ (2019) Robust fault detection of a class of uncertain linear parabolic PDEs. Automatica 107:502–510
29. Kim Y, Mohan S, Siegel JB et al (2014) The estimation of temperature distribution in cylindrical battery cells under unknown cooling conditions. IEEE Trans Control Syst Technol 22(6):2277–2286
30. Wang J-W, Liu Y-Q, Hu Y-Y et al (2017) A spatial domain decomposition approach to distributed $\mathcal{H}_\infty$ observer design of a linear unstable parabolic distributed parameter system with spatially discrete sensors. Int J Control 90(12):2772–2785
31. Cai J, Ferdowsi H, Sarangapani J (2016) Model-based fault detection, estimation, and prediction for a class of linear distributed parameter systems. Automatica 66:122–131
32. Meurer T (2012) Control of higher-dimensional PDEs: Flatness and backstepping designs. Springer Science+Business Media, Berlin
33. Wang B-C, Li H-X, Feng Y (2018) An improved teaching-learning-based optimization for constrained evolutionary optimization. Inf Sci 456:131–144
34. Ioannou PA, Sun J (2012) Robust Adaptive Control. Courier Corporation, Chelmsford
35. Xiao T, Li H-X (2016) Eigenspectrum-based iterative learning control for a class of distributed parameter system. IEEE Trans Autom Control 62(2):824–836

Chapter 10
Independent Component Analysis-Based Fault Detection and Localization for Partially Known DPSs

Abstract In this chapter, a new approach is proposed for fault detection and spatial localization in parabolic distributed parameter systems with limited sensors. The baseline behavior of the distributed parameter systems is first characterized under restricted sensor availability. Subsequently, the spatio-temporal dynamics are decoupled through time–space separation. The temporal coefficients are then further processed using independent component analysis, from which the dominant temporal modes are extracted. Based on these, spatial residuals are constructed to derive two monitoring statistics. By employing kernel density estimation, the confidence intervals of these statistics in the fault-free case are obtained as reference thresholds. Distinct from traditional model-based techniques that rely on explicit mathematical formulations, the proposed strategy is data-driven and exploits only the separable property of parabolic distributed parameter systems. Its effectiveness is validated through studies on a representative parabolic process and a snap-curing oven.

10.1 Introduction

Many industrial processes, such as transport–reaction, thermal, and fluid systems, can be modeled as DPSs. In these systems, inputs, outputs, and parameters may vary across both space and time, resulting in S-T dynamics [1–3]. Among them, parabolic DPSs are particularly representative, where the spatial differential operator can be decomposed into a finite-dimensional slow subsystem and an infinite-dimensional fast subsystem. Over the past decades, research on DPSs has mainly concentrated on system modeling [4, 5] and controller design [6–9].

With the growing demand for high productivity, product quality, and cost-effective as well as environmentally friendly operation of complex industrial systems, increasing attention has been directed to system reliability, dependability, and safety. Consequently, process monitoring and FD [10–13] have become essential. The mainstream approaches can be broadly classified into model-based [14] and data-driven methods [13]. Model-based techniques rely on precise dynamic models of the system, whereas data-driven ones utilize large volumes of noisy process data in combination with statistical tools. With recent advances in sensor technology, data-driven mon-

© The Author(s), under exclusive license to Springer Nature Singapore Pte Ltd. 2026
Y. Feng et al., *Spatio-Temporal Abnormality Diagnosis for Industrial Distributed Parameter Systems*, https://doi.org/10.1007/978-981-95-3750-1_10

itoring has attracted significant interest and achieved wide industrial application. Among these, principal component analysis (PCA) [15] and partial least squares (PLS) [10, 16] are two representative methods that have proven highly effective in SPC. Independent component analysis (ICA) [17], initially developed in signal processing, is an unsupervised technique that extracts statistically independent sources from measurements without prior knowledge of the mixing process. Unlike PCA, which is limited to second-order statistics, ICA exploits higher-order statistics, making it suitable for handling non-Gaussian variables that are prevalent in industrial processes [18, 19]. In recent years, ICA has been increasingly applied to SPC for fault detection and diagnosis [20–22], demonstrating clear advantages over PCA-based monitoring approaches.

Recently, fault detection in DPSs has gained increasing attention [23–28]. In [25], finite-dimensional residual generators were directly designed for Riesz-spectral systems with bounded input and output operators to enable fault detection in linear DPSs. A model-based framework for fault detection, estimation, and prediction in a class of linear DPSs was further proposed in [28]. Although these studies provide valuable insights, they give limited consideration to spatial fault localization. Since DPSs inherently exhibit S-T dynamics, accurate spatial identification of faults is critical for effective fault elimination and process recovery. More recently, a modulating function-based approach for estimating space–time-dependent unknowns in one-dimensional PDEs has been introduced in [29].

The main challenges in fault detection and spatial identification of DPSs can be summarized as follows:

- Unlike LPS, which can be mathematically described by ODEs, DPSs possess spatially distributed characteristics that result in infinite-dimensional dynamics, making fault detection and spatial identification more challenging.
- In practical industrial applications, the number of available sensors is limited due to cost constraints, which makes accurate spatial fault localization with sparse measurements particularly challenging.
- The inherent strong nonlinearities in many DPSs further complicate the fault detection and identification problem.

Motivated by the above issues, this work develops a fault detection and spatial localization method for a class of parabolic DPSs under limited sensor availability. The proposed framework consists of two stages: reference signal modeling (offline) and fault estimation (real-time).

- Reference signal modeling: In this phase, the fault-free dynamics of the DPSs are examined via time–space separation to derive spatial basis functions and temporal coefficients. The temporal coefficients are further processed via ICA, from which the dominant components are retained. Using these, spatial residuals are constructed to generate two monitoring statistics. Their probability distributions are then estimated using kernel density functions, and the corresponding confidence intervals under normal (fault-free) conditions are obtained. These intervals serve as reference signals for subsequent fault detection.

- Fault estimation: In this phase, when a fault arises in the system, the real-time monitoring statistics deviate from the reference signals, thereby indicating the presence of a fault. Inspired by the variable contribution concept in SPC, a location contribution index is introduced for spatial fault identification. Specifically, the spatial position with the highest contribution is determined as the fault location.

The main advantages of this method can be summarized as :

- This approach is data-driven and only requires knowledge of the spatial operator, boundary conditions (to compute spatial basis functions), and sensor measurements at specific positions. It is assumed that the dominant dynamics of the parabolic DPS can be represented by a finite set of variables. Fault detection is then performed on this unknown finite-dimensional ODE system using ICA. Unlike the traditional early-lumping strategy, which relies on state or functional observers and explicit models for residual generation and threshold design, the proposed method avoids such requirements. When the system model is unavailable, reliable detection becomes challenging for early-lumping, whereas the proposed method further achieves spatial fault localization by exploiting space–time separation.
- It is applicable to industrial processes with state variables that follow non-Gaussian distributions.
- It requires only a limited number of sensors.

The remainder of this chapter is structured as follows. Section 10.2 presents the problem formulation along with the time–space separation and synthesis methods. The fault detection and spatial localization approaches are detailed in Sect. 10.3. Experimental results are reported in Sect. 10.4, and concluding remarks are provided in Sect. 10.5.

10.2 Basics of ICA

We focus on a class of faulty parabolic DPSs in one spatial dimension, which can be represented by the following mathematical formulation:

$$\begin{aligned}
&\frac{\partial y}{\partial t} = a_1\frac{\partial y}{\partial z} + a_2\frac{\partial^2 y}{\partial z^2} + f(y, \boldsymbol{u}(t)) + k_w \boldsymbol{b}_w^{\mathrm{T}}(z)\boldsymbol{w}(t) + k_u \boldsymbol{b}_u^{\mathrm{T}}(z)\boldsymbol{u}(t) + H(t - t_f)d(z, t),\\
&\boldsymbol{y}_m(t) = [y(z_1, t), y(z_2, t), \cdots, y(z_N, t)]^{\mathrm{T}},
\end{aligned} \tag{10.1}$$

subject to the following boundary conditions:

$$\begin{aligned}
c_1 y(\alpha_1, t) + d_1\frac{\partial y}{\partial z}(\alpha_1, t) = r_1,\\
c_2 y(\alpha_2, t) + d_2\frac{\partial y}{\partial z}(\alpha_2, t) = r_2,
\end{aligned} \tag{10.2}$$

and with the following initial condition:

$$y(z, 0) = y_0(z), \tag{10.3}$$

where $y = y(z, t) \in \mathbb{R}$ represents the system state, and $y_0(z)$ denotes the initial condition. The spatial domain is given by $[\alpha_1, \alpha_2] \subset \mathbb{R}$, with $z \in [\alpha_1, \alpha_2]$ as the spatial coordinate, and $t \in [0, \infty)$ representing time. The process disturbance is denoted by $\boldsymbol{w}(t) \in \mathbb{R}^{n_w \times 1}$, which in this chapter is assumed to be random noise. The manipulated input is represented by $\boldsymbol{u}(t) \in \mathbb{R}^{n_u \times 1}$, and $\boldsymbol{y}_m(t) \in \mathbb{R}^{N \times 1}$ corresponds to the measured outputs from sensors. $H(\cdot)$ denotes the standard Heaviside function; t_f indicates the fault occurrence time, and $d(z, t)$ represents the unknown fault affecting either actuators or the process. The function $f(y, \boldsymbol{u}(t))$ is an unknown continuous nonlinear mapping. Constants include $k_w, k_u, a_1, a_2, c_1, c_2, d_1, d_2, r_1$, and r_2. The functions $\boldsymbol{b}_w(z) \in \mathbb{R}^{n_w \times 1}$ and $\boldsymbol{b}_u(z) \in \mathbb{R}^{n_u \times 1}$ are smooth and known, where the i-th element specifies the spatial location of the i-th disturbance input $\boldsymbol{w}(t)$ and control input $\boldsymbol{u}(t)$ over the domain $[\alpha_1, \alpha_2]$.

For the simplicity of notation, here we define the following Hilbert Space:

$$\mathcal{H} \triangleq \mathcal{L}_2([\alpha_1, \alpha_2]; \mathbb{R}).$$

Also, inner product and norm under this Hilbert Space are defined as:

$$< y_1(\cdot), y_2(\cdot) > \triangleq \int_{\alpha_1}^{\alpha_2} y_1(z) y_2(z) \mathrm{d}z,$$
$$\|y_1(\cdot)\|_2 \triangleq < y_1(\cdot), y_1(\cdot) >^{1/2},$$

where $y_1(\cdot)$ and $y_2(\cdot)$ are two elements of $\mathcal{H}$.

Then, we can apply Galerkin's method [2, 30, 31] to obtain the finite-dimensional slow system described by the following ODE:

$$\dot{\boldsymbol{a}}_s(t) = \boldsymbol{A}_s \boldsymbol{a}_s(t) + \boldsymbol{f}_s(\boldsymbol{a}_s(t), 0, \boldsymbol{u}(t)) + \boldsymbol{B}_{w,s} \boldsymbol{w}(t) + \boldsymbol{B}_{u,s} \boldsymbol{u}(t) + H(t - t_f) \boldsymbol{d}_s(t), \tag{10.4}$$

$$\boldsymbol{y}_m(t) = \boldsymbol{\varphi} \boldsymbol{a}_s(t), \tag{10.5}$$

where

$$\boldsymbol{\varphi} = \begin{pmatrix} \phi_1(z_1) & \dots & \phi_m(z_1) \\ \vdots & \ddots & \vdots \\ \phi_1(z_N) & \cdots & \phi_m(z_N) \end{pmatrix} \in \mathbb{R}^{N \times m}. \tag{10.6}$$

Further details are provided in Sect. A.1 of the appendix. For brevity, $\boldsymbol{a}(t)$ and $\phi(z)$ are adopted to denote $\boldsymbol{a}_s(t)$ and $\phi_s(z)$, respectively, in the subsequent discussion of this chapter.

As illustrated in Fig. 10.1, the final step of this approximation involves constructing the DPS model based on time–space separation through time–space synthesis as follows:

$$\hat{y}(z, t) = \phi^{\mathrm{T}}(z) \boldsymbol{a}(t). \tag{10.7}$$

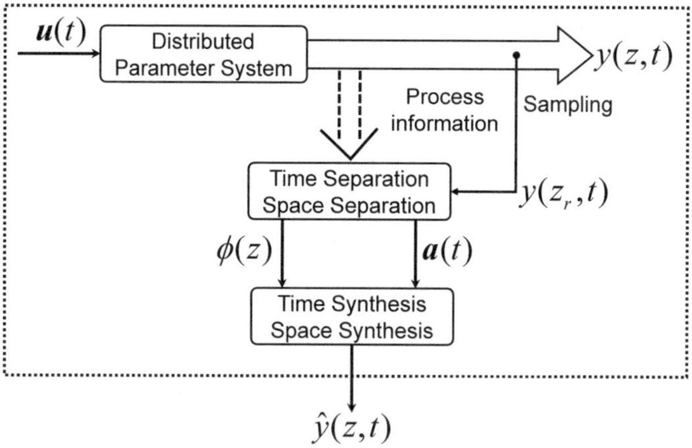

Fig. 10.1 Time/Space separation and synthesis

Compared with Eq. 10.5, here the continuous spatial basis function $\phi(z)$ is used rather than $\phi_s(z_r)$, which is a discrete vector of certain sensor locations. Therefore, through time/space synthesis, the full space state estimation is obtained.

Figure 10.2 illustrates a typical DPS structure. Here, $\boldsymbol{u}(t)$ represents the temporal input; $y(z, t)$ and $d(\boldsymbol{z}_f, \boldsymbol{t}_f)$ denote the S-T process output and the faults affecting the system, respectively, where $\boldsymbol{t}_f$ and $\boldsymbol{z}_f$ indicate the sets of fault occurrence times and spatial locations. We assume that the system has n_u actuators, each characterized by a temporal signal $\boldsymbol{u}(t)$ and a corresponding spatial distribution. Due to cost and resource limitations, only a limited number of sensors are deployed to monitor the system in practical applications.

The fault detection and spatial localization problem addressed in this chapter can be formulated as follows: For a class of systems governed by the parabolic DPSs in Eq. 10.1, how can faults be identified in both time and space using only the sampled system outputs $\boldsymbol{y}_m(t)$? Specifically, this involves two tasks: detecting faults in the time domain and localizing faults in the spatial domain, which correspond to identifying $\boldsymbol{t}_f$ and $\boldsymbol{z}_f$ in Fig. 10.2, respectively.

Remark 10.1 Without loss of generality, the parabolic DPSs in Eq. 10.1 are considered on a one-dimensional spatial domain, and this formulation can be directly extended to three-dimensional spatial domains. For such systems, once transformed into the finite-dimensional slow system as in Eq. 10.4, the resulting problem remains equivalent. In the experiments, a three-dimensional parabolic DPS—a snap curing oven—is employed, and the corresponding derivation is provided in Sect. A.3 of the appendix.

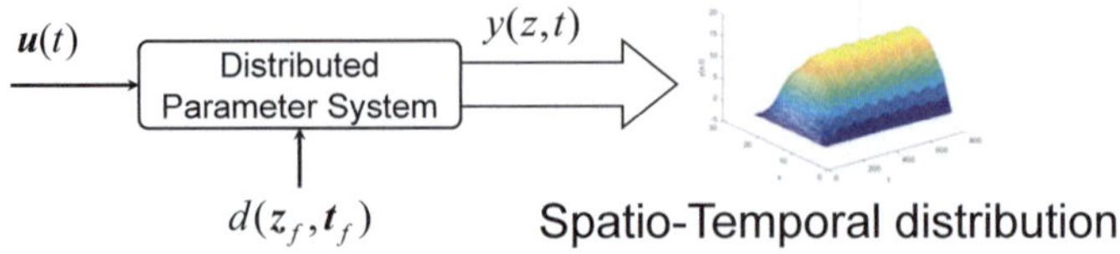

Fig. 10.2 Spatial/Temporal dynamics of DPS

10.3 ICA Modeling

The fault detection and spatial localization procedure consists of three components: construction of reference signals, fault detection based on these references, and spatial fault identification.

10.3.1 Reference Signal Construction

When the system operates under fault-free conditions, the collected data can be utilized for constructing reference signals. The initial step involves establishing the ICA-based monitoring model using historical S-T data that reflect only common-cause variations, denoted as $\boldsymbol{y}_m(t)$.

Step 1. ICA under Time/Space Separation of DPSs
For normal operation data, the first task is to apply the spectral method introduced in Sect. II.A to extract the dominant spatial basis functions $\phi(z) = [\phi_1(z), \cdots, \phi_m(z)]^{\mathrm{T}}$. The associated temporal coefficients $\boldsymbol{a}(t) = [a_1(t), \cdots, a_m(t)]^{\mathrm{T}}$ are then obtained by solving Eq. 10.5.

$$\boldsymbol{a}(t) = \boldsymbol{\varphi}^{\dagger}\boldsymbol{y}_m(t),$$

where $\boldsymbol{\varphi}^{\dagger}$ denotes the Moore-Penrose inverse of $\boldsymbol{\varphi}$. This requires $\boldsymbol{\varphi}$ to be full-row rank and it can be satisfied by appropriate selection of the number of the slow eigenvalues m. In practice, to ensure that the temporal coefficients $\boldsymbol{a}(t)$ can be solved through Eq. 10.5, the number of the slow eigenvalues m should be selected larger than or at least equal to the number of point-wise measurements N.

The next step is to view $\boldsymbol{a}(t)$ as the *mixed signal* and use ICA to extract the *independent components* $\tilde{s}(t)$:

$$\begin{aligned}\tilde{s}(t) &= \tilde{\boldsymbol{A}}^{\mathrm{T}}\boldsymbol{Q}\boldsymbol{a}(t)\\ &= \boldsymbol{W}\boldsymbol{a}(t),\end{aligned}$$

where $\tilde{\boldsymbol{A}}$ denotes the orthogonal mixing matrix, $\boldsymbol{Q}$ is the whitening matrix, and $\boldsymbol{W}$ represents the de-mixing matrix. To estimate $\tilde{\boldsymbol{A}}$, the FastICA algorithm [32] is employed, which utilizes a fixed-point iteration to maximize the non-Gaussianity of the independent components. Dimensionality reduction can be achieved by selecting the rows of $\boldsymbol{W}$ with the largest sum of squares [33], denoted as $\boldsymbol{W}_d$, while the remaining rows form a reduced matrix $\boldsymbol{W}_e$. Correspondingly, a matrix $\tilde{\boldsymbol{A}}_d$ can be constructed by choosing the columns of $\tilde{\boldsymbol{A}}$ associated with the selected rows of $\boldsymbol{W}$. Consequently, the independent components are given by:

$$\begin{aligned}\tilde{\boldsymbol{s}}_d(t) &= \boldsymbol{W}_d\boldsymbol{a}(t),\\ \tilde{\boldsymbol{s}}_e(t) &= \boldsymbol{W}_e\boldsymbol{a}(t),\end{aligned}$$

these matrices can then be utilized in the subsequent steps. Here, $\tilde{\boldsymbol{s}}_d(t)$ represents the dominant independent components, whereas $\tilde{\boldsymbol{s}}_e(t)$ corresponds to the residual independent components.

Step 2. Reference Signal Modeling for Normal Condition

A key task in SPC is to define appropriate monitoring statistics. Analogous to Hotelling's T^2 and SPE statistics used in PCA-based monitoring, the temporal I^2 statistic (T-I^2) and the spatial mean squared prediction error (SM-SPE) statistic are introduced to quantify signal variation in the temporal and spatial domains, respectively. Specifically, the T-I^2 statistic at time t is computed as the sum of squared independent component scores:

$$\text{T-}I^2(t) = \text{T-}I_d^2(t) + \text{T-}I_e^2(t) = \tilde{\boldsymbol{s}}_d{}^{\mathrm{T}}(t)\tilde{\boldsymbol{s}}_d(t) + \tilde{\boldsymbol{s}}_e{}^{\mathrm{T}}(t)\tilde{\boldsymbol{s}}_e(t). \tag{10.8}$$

Since $\tilde{\boldsymbol{s}}_d(t)$ is the dominant independent components, here we use the T-$I_d^2(t)$ statistic for the ICA monitoring model. The T-I_d^2 statistic reflects the systematic part for the process variation. In addition, the SM-SPE statistic is defined by:

$$\begin{aligned}\text{SM-SPE}(t) &= \|e(z,t)\|_2^2,\\ &= \left\|\hat{y}(z,t) - \tilde{y}(z,t)\right\|_2^2.\end{aligned} \tag{10.9}$$

$\hat{y}(z,t) - \tilde{y}(z,t)$ can be obtained by time/space synthesis in Eq. 10.7:

$$\hat{y}(z,t) - \tilde{y}(z,t) = \phi^{\mathrm{T}}(z)(\boldsymbol{a}(t) - \tilde{\boldsymbol{a}}(t)), \tag{10.10}$$

where

$$\begin{aligned}\tilde{\boldsymbol{a}}(t) &= \boldsymbol{Q}^{-1}\tilde{\boldsymbol{A}}_d\tilde{\boldsymbol{s}}_d(t)\\ &= \boldsymbol{Q}^{-1}\tilde{\boldsymbol{A}}_d\boldsymbol{W}_d\boldsymbol{a}(t).\end{aligned} \tag{10.11}$$

Combining Eqs. 10.9, 10.10 and 10.11 together, we have

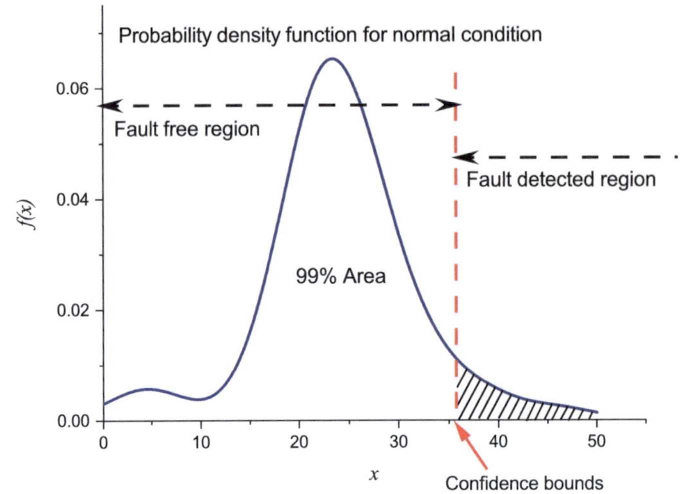

Fig. 10.3 Schematic graph of KDE in normal condition

$$\text{SM-SPE}(t) = \left\| \phi^{\mathrm{T}}(z)(\boldsymbol{I} - \boldsymbol{Q}^{-1}\tilde{\mathbf{A}}_d \boldsymbol{W}_d)\boldsymbol{a}(t) \right\|_2^2, \tag{10.12}$$

where $\boldsymbol{I}$ denotes the identity matrix.

In PCA-based monitoring, the SPE statistic captures the residual variation of the process. In contrast, taking into account the S-T dynamics of DPSs, the proposed SM-SPE statistic quantifies the residual component across the entire spatial domain at each sampling instant t.

Step 3. Confidence Bounds Estimation

Once the T-I_d^2 and SM-SPE statistics are computed, the next step is to determine their corresponding confidence bounds, denoted as T-I_R^2 and SM-SPE$_R$, respectively. In PCA-based monitoring, confidence bounds are typically derived under the assumption that the PDF follow a multivariate Gaussian distribution [34, 35]. However, this assumption is often violated in practical industrial processes. Moreover, since the latent variables in ICA monitoring do not necessarily follow a multivariate Gaussian distribution, a non-parametric approach based on KDE [33] is employed to estimate the PDFs of the monitoring statistics. Additional details of the KDE procedure are provided in Sect. A.2 of the appendix.

Specifically, after obtaining the T-I_d^2 and SM-SPE statistics under normal operating conditions, a univariate kernel density estimator is applied to derive the confidence bounds for both statistics. Figure 10.3 illustrates the KDE process. Here, x represents the monitoring statistic (T-I_d^2 or SM-SPE), i.e., $x \in$ T-I_d^2, SM-SPE, and $f(x)$ denotes the estimated PDF under normal conditions. The confidence bound for each statistic is defined as the value covering 99% of the density function [33]. A schematic of the reference signal construction is presented in Fig. 10.4.

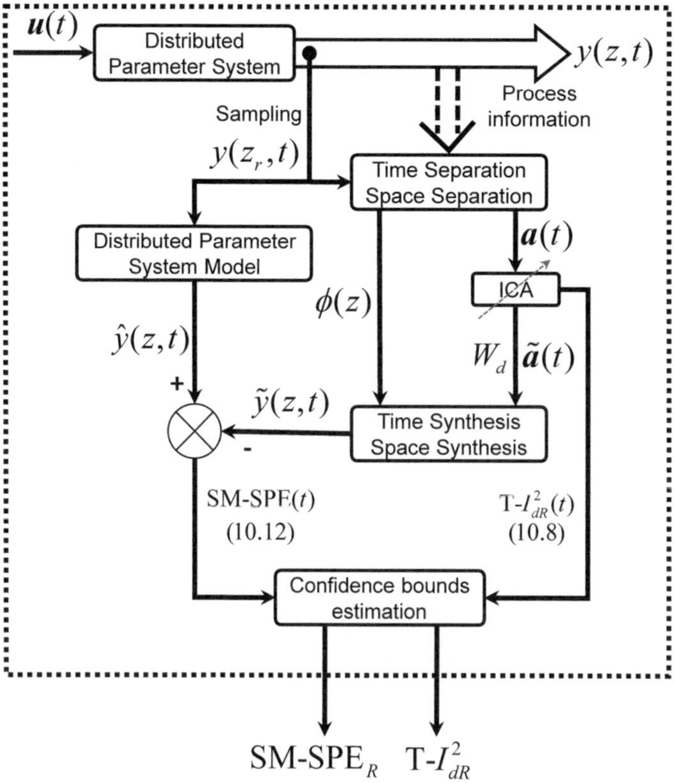

Fig. 10.4 Reference signal construction

10.3.2 Reference-Based Fault Detection and Spatial Identification

Step 1. States Extraction in Abnormal Condition
When the new data are collected, firstly obtain the new temporal coefficients $\boldsymbol{a}_{\text{new}}(t)$ with the same spatial basis functions $\phi(z)$:

$$\boldsymbol{a}_{\text{new}}(t) = \boldsymbol{\varphi}^{\dagger}\boldsymbol{y}_{\text{new}m}(t),$$

where $\boldsymbol{y}_{\text{new}m}(t) = [y_{\text{new}}(z_1, t), y_{\text{new}}(z_2, t), \cdots, y_{\text{new}}(z_N, t)]^{\text{T}}$ denotes the new measurement when a fault occurs; $\boldsymbol{\varphi}^{\dagger}$ denotes the Moore-Penrose inverse of $\boldsymbol{\varphi}$.

Then we calculate the new independent components by using the same de-mixing matrix $\boldsymbol{W}_d$ obtained from reference signal construction,

$$\tilde{\boldsymbol{s}}_{\text{newd}}(t) = \boldsymbol{W}_d\boldsymbol{a}_{\text{new}}(t). \tag{10.13}$$

Step 2. Monitoring Statistics Construction
The T-I_d^2 statistic of the new data at time t can be obtained by

$$\text{T-}I_{\text{newd}}^2(t) = \tilde{\boldsymbol{s}}_{\text{newd}}^{\text{T}}(t)\tilde{\boldsymbol{s}}_{\text{newd}}(t). \tag{10.14}$$

Then the SM-SPE statistic is calculated as

$$\begin{aligned}\text{SM-SPE}_{\text{new}}(t) &= \|e_{\text{new}}(z,t)\|_2^2, \\ &= \left\| \hat{y}_{\text{new}}(z,t) - \tilde{y}_{\text{new}}(z,t) \right\|_2^2,\end{aligned} \tag{10.15}$$

where $\hat{y}_{\text{new}}(z,t) - \tilde{y}_{\text{new}}(z,t)$ is obtained by time/space synthesis:

$$\hat{y}_{\text{new}}(z,t) - \tilde{y}_{\text{new}}(z,t) = \phi^{\text{T}}(z)(\boldsymbol{a}_{\text{new}}(t) - \tilde{\boldsymbol{a}}_{\text{new}}(t)), \tag{10.16}$$

where

$$\begin{aligned}\tilde{\boldsymbol{a}}_{\text{new}}(t) &= \boldsymbol{Q}^{-1}\tilde{\boldsymbol{A}}_d\tilde{\boldsymbol{s}}_{\text{newd}}(t) \\ &= \boldsymbol{Q}^{-1}\tilde{\boldsymbol{A}}_d\boldsymbol{W}_d\boldsymbol{a}_{\text{new}}(t).\end{aligned} \tag{10.17}$$

Combining Eqs. 10.15, 10.16 and 10.17 together, we have

$$\text{SM-SPE}_{\text{new}}(t) = \left\| \phi^{\text{T}}(z)(\boldsymbol{I} - \boldsymbol{Q}^{-1}\tilde{\boldsymbol{A}}_d\boldsymbol{W}_d)\boldsymbol{a}_{\text{new}}(t) \right\|_2^2. \tag{10.18}$$

Step 3. Fault Detection in Temporal Domain
By comparing Eqs. 10.8 with 10.14 and Eqs. 10.12 with 10.18, it is evident that the T-I_d^2 and SM-SPE statistics effectively capture the deviations between normal and faulty conditions. Once T-$I_{\text{new}}^2(t)$ and SM-SPE$_{\text{new}}(t)$ are computed, fault detection can be performed by referencing the previously established confidence bounds T-I_R^2 and SM-SPE$_R$. The sets of detected fault samples, corresponding to instances exceeding the reference thresholds, are denoted by $\boldsymbol{t}_f^{\text{T}}$ and $\boldsymbol{t}_f^{\text{SM}}$, respectively.

The principles of fault detection are as follows:

$$t \in \begin{cases} \boldsymbol{t}_f^{\text{T}}, & \text{if T-}I_{\text{new}}^2(t) > \text{T-}I_R^2, \\ \boldsymbol{t}_f^{SM}, & \text{if SM-SPE}_{\text{new}}(t) > \text{SM-SPE}_R. \end{cases} \tag{10.19}$$

Step 4. Spatial Identification of Fault
Following fault detection in the temporal domain, the spatial locations of the detected faults can be determined by computing the contribution plots at the fault sample sets $\boldsymbol{t}_f^{\text{T}}$ and $\boldsymbol{t}_f^{\text{SM}}$, respectively. In the context of process monitoring for dynamic systems governed by ODEs, contribution plots quantify the impact of individual variables on the T^2 and SPE statistics. Here, this concept is extended to evaluate the location-specific contributions to the T-I_{new}^2 and SM-SPE$_{\text{new}}$ statistics at the detected fault samples, thereby identifying the spatial positions of faults. The mean location contribution for T-$I_{\text{new}}^2(\boldsymbol{t}_f^{\text{T}})$ and SM-SPE$_{\text{new}}(\boldsymbol{t}_f^{SM})$ is computed as follows:

$$
\begin{aligned}
\hat{y}_{c\mathrm{T}}(z,t) &= \frac{\phi^{\mathrm{T}}(z)\boldsymbol{Q}^{-1}\tilde{\boldsymbol{A}}_d\hat{\boldsymbol{s}}_{\mathrm{newd}}(t)}{\left\|\phi^{\mathrm{T}}(z)\boldsymbol{Q}^{-1}\tilde{\boldsymbol{A}}_d\hat{\boldsymbol{s}}_{\mathrm{newd}}(t)\right\|_2}\left\|\hat{\boldsymbol{s}}_{\mathrm{newd}}(t)\right\|, \\
\hat{y}_{c\mathrm{T}}(z,\boldsymbol{t}_f^{\mathrm{T}}) &= \frac{1}{M}\sum\nolimits_{i=1}^{M}\left|\hat{y}_{c\mathrm{T}}(z,t_i)\right|,
\end{aligned}
\tag{10.20}
$$

and

$$
\begin{aligned}
\hat{y}_{c\text{SM-SPE}}(z,t) &= \hat{y}_{\mathrm{new}}(z,t) - \tilde{y}_{\mathrm{new}}(z,t), \\
\hat{y}_{c\text{SM-SPE}}(z,\boldsymbol{t}_f^{\mathrm{SM}}) &= \frac{1}{N}\sum\nolimits_{i=1}^{N}\left|\hat{y}_{c\text{SM-SPE}}(z,t_i)\right|,
\end{aligned}
\tag{10.21}
$$

respectively. Here, $\|\cdot\|$ represents the Euclidean norm, and $|\cdot|$ denotes the absolute value. The quantities M and N correspond to the cardinalities of the sets $\boldsymbol{t}_f^{\mathrm{T}}$ and $\boldsymbol{t}_f^{\mathrm{SM}}$, respectively.$\hat{y}_{c\mathrm{T}}(z,\boldsymbol{t}_f^{\mathrm{T}})$ indicates the location contribution to the T-I^2 statistic at the detected fault sample set $\boldsymbol{t}_f^{\mathrm{T}}$, whereas $\hat{y}_{c\text{SM-SPE}}(z,\boldsymbol{t}_f^{\mathrm{SM}})$represents the location contribution to the SM-SPE statistic at $\boldsymbol{t}_f^{\mathrm{SM}}$. The fault spatial locations, $\boldsymbol{z}_f^{\mathrm{T}}$ and $\boldsymbol{z}_f^{\mathrm{SM}}$, are then determined based on the location contributions at the corresponding fault sample sets. In the contribution plots, the spatial position exhibiting the maximum contribution is identified as the fault location. A schematic overview of the reference-based fault detection and spatial identification procedure is illustrated in Fig. 10.5.

10.4 Illustrative Examples

10.4.1 Catalytic Rod

In this section, we studied one typical DPS-catalytic rod to evaluate the performance of the proposed fault detection and spatial identification method.

As illustrated in Fig. 10.6, a long, slender rod within a reactor supplied with pure species A represents a typical transport–reaction process encountered in the chemical industry [2]. The rod undergoes a zero-order exothermic catalytic reaction of the form $A \rightarrow B$. Cooling is achieved by a medium in contact with the rod. The model is based on standard assumptions: the rod has constant density and heat capacity, uniform thermal conductivity, fixed temperatures at both ends, and an excess of species A in the furnace. Under these conditions, the temperature distribution along the rod can be described by the following parabolic PDE:

$$
\begin{aligned}
\frac{\partial y(z,t)}{\partial t} &= \frac{\partial^2 y(z,t)}{\partial z^2} + \beta_T\left(e^{-\frac{\gamma}{1+y(z,t)}} - e^{-\gamma}\right) + \beta_U\left(b_u(z)u(t) - y(z,t)\right) + H(t-t_f)d(z,t), \\
y_m(z_r,t) &= y(z_r,t) + w(z_r,t), \\
\boldsymbol{y}_m(t) &= [y_m(z_1,t),\cdots,y_m(z_N,t)]^{\mathrm{T}},
\end{aligned}
\tag{10.22}
$$

subject to the Dirichlet boundary and initial conditions:

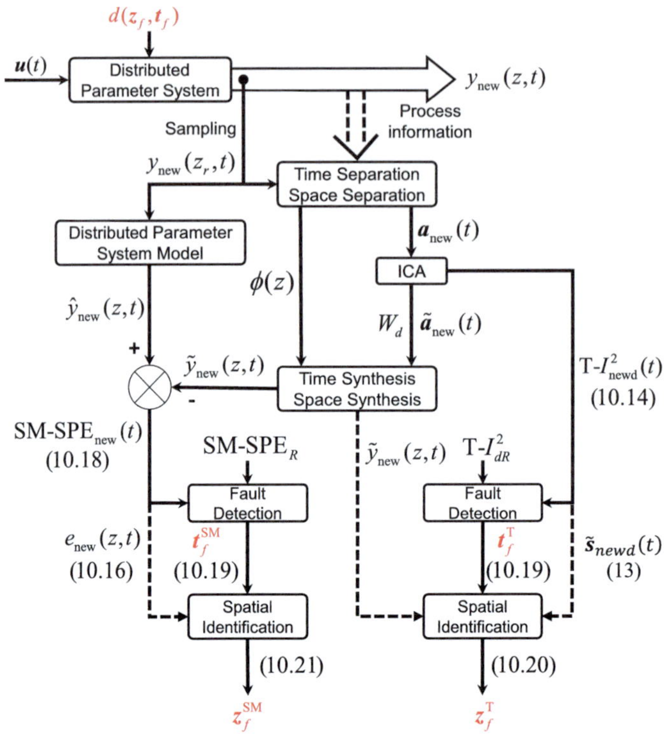

Fig. 10.5 Fault detection and spatial identification

Fig. 10.6 Catalytic rod

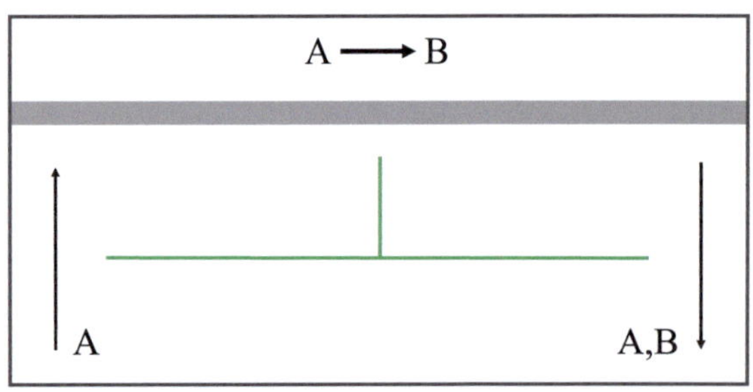

$$y(0, t) = 0, \;\; y(\pi, t) = 0, \;\; y(z, 0) = 0,$$

where $y(z, t)$ represents the temperature at spatial location z and time t; β_T is the heat of reaction; γ denotes the activation energy; β_U is the heat transfer coefficient; $u(t)$ represents the manipulated input corresponding to the cooling medium temperature, and $b_u(z)$ describes the spatial distribution of the actuator. The measured outputs and measurement noise at sensor locations z_1 to z_N are denoted by $y_m(z_r, t)$ and $w(z_r, t)$, respectively. Typical values of the process parameters are given as follows:

Table 10.1 Settings of faults for catalytic rod

Fault	Type	Fault occurring time and spatial location
1	$b_f(\pi/8) = 8 * b_u(\pi/8)$	$t_f = 350, z_f = \pi/8$
2	$b_f(7\pi/8) = 9 * b_u(7\pi/8)$	$t_f = 350, z_f = 7\pi/8$

$$\beta_T = 50, \ \beta_U = 2, \ \gamma = 4.$$

The eigenvalue problem for the spatial operator of the process:

$$\mathcal{A}y = \frac{\partial^2 y}{\partial z^2},$$
$$y \in \mathcal{S}(\mathcal{A}) = \{y \in \mathcal{L}_2([0, \pi]; \mathbb{R}); y(0, t) = 0, y(\pi, t) = 0\}$$

can be solved theoretically, and the solution is of the following form:

$$\lambda_j = -j^2, \ \phi_j(z) = \sqrt{\frac{2}{\pi}} \sin(jz), \ j = 1, \cdots, \infty.$$

We select the first nine eigenvalues as dominant, yielding $\varepsilon = |\lambda_1| / |\lambda_{10}| = 0.01 \ll 1$. Accordingly, the temporal coefficients $\boldsymbol{a}(t)$ are determined by solving Eq. 10.5.

The system includes a single actuator $u(t)$ with spatial profile $b_u(z)$. The temporal input $u(t)$ is uniformly distributed over $(-3, 6)$, and the spatial distribution $b_u(z)$ is uniformly distributed within (0, 1). Eight uniformly spaced sensors are deployed for measurement, with spacing set to $\pi/7$. Eq. 10.22 is solved using the FDM [36]. The sampling step is $\Delta t = 0.01$, s, and the random process noise $w(z_r, t)$ is characterized by the signal-to-noise ratio (SNR) between the measured output and the noise, fixed here at 30 dB.

For the analysis, a total of 750 samples were collected. The first 300 samples were utilized to construct the reference signals, while the remaining samples were employed for fault detection and spatial localization. The fault scenarios considered are specified in Table 10.1, with the mathematical formulations of Fault 1 and Fault 2 given as:

$$d(z, t) = \beta_U b_f(z) f(t),$$

where $f(t)$ follows a uniform distribution over $(-3, 6)$, and $b_f(z) \equiv 0$ when $z \neq z_f$. For Fault 1, the occurrence time is $t_f = 350$ with its location at $z_f = \pi/8$. The corresponding fault detection and spatial identification results for Faults 1 and 2 are shown in Figs. 10.7 and 10.8, respectively. The blue dashed lines denote the confidence limits of the monitoring statistics under fault-free operation, i.e., T-I_R^2 and SM-SPE$_R$, while the orange solid circles on the spatial axis mark the actual

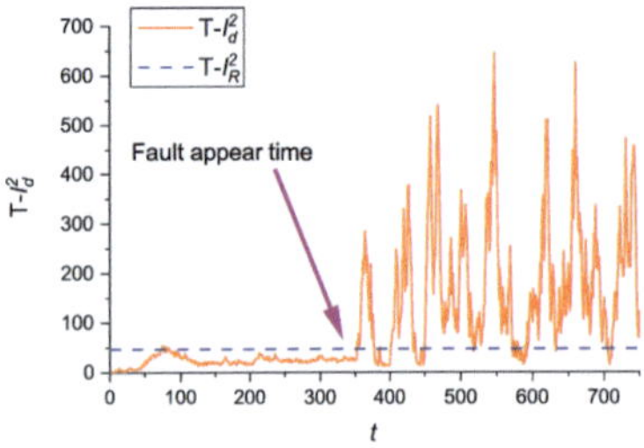

(a) T-I_d^2 statistic based fault detection

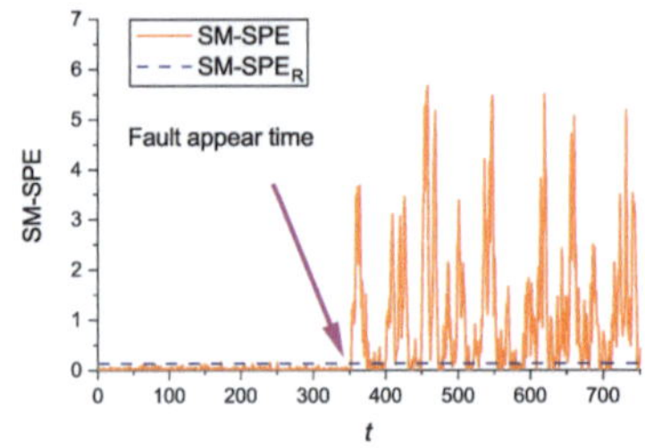

(b) SM-SPE statistic based fault detection

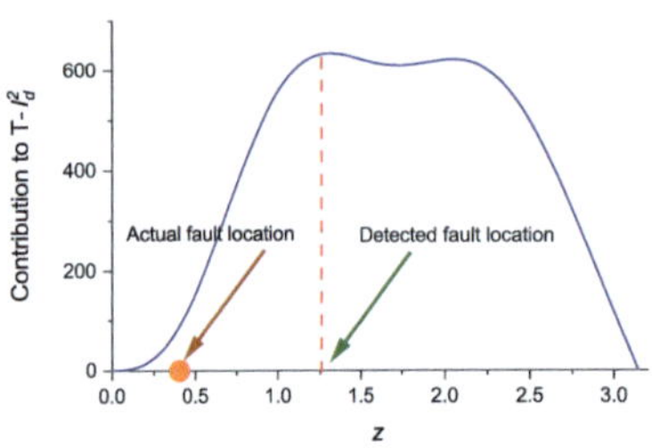

(c) T-I_d^2 statistic based spatial identification

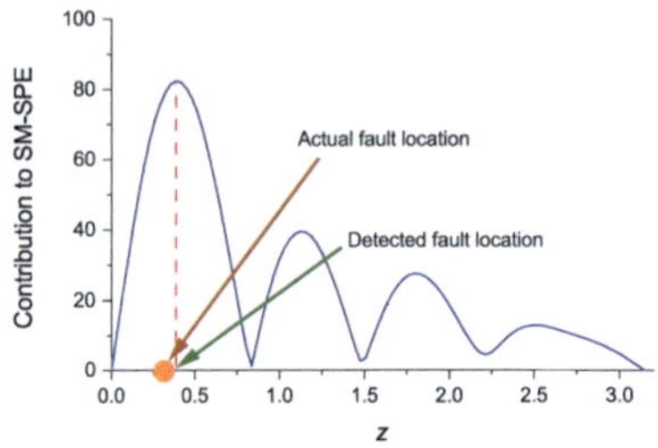

(d) SM-SPE statistic based spatial identification

Fig. 10.7 Detection and spatial identification results of Fault 1

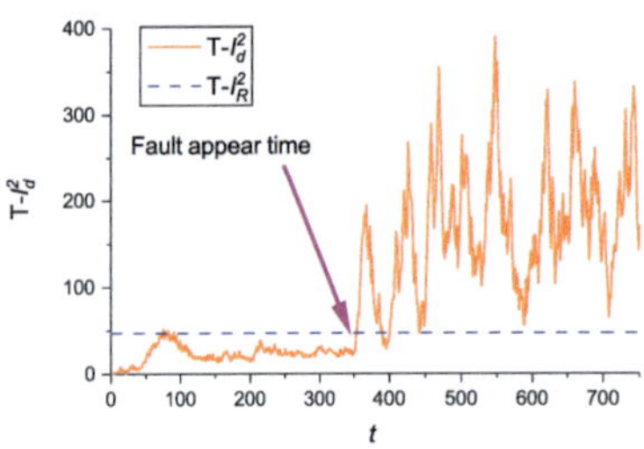

(a) T-I_d^2 statistic based fault detection

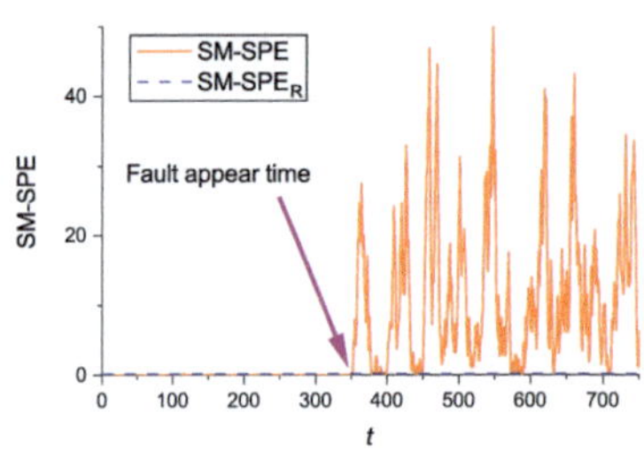

(b) SM-SPE statistic based fault detection

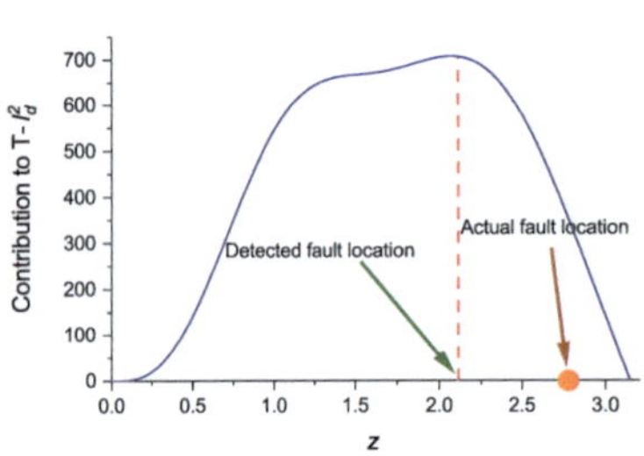

(c) T-I_d^2 statistic based spatial identification

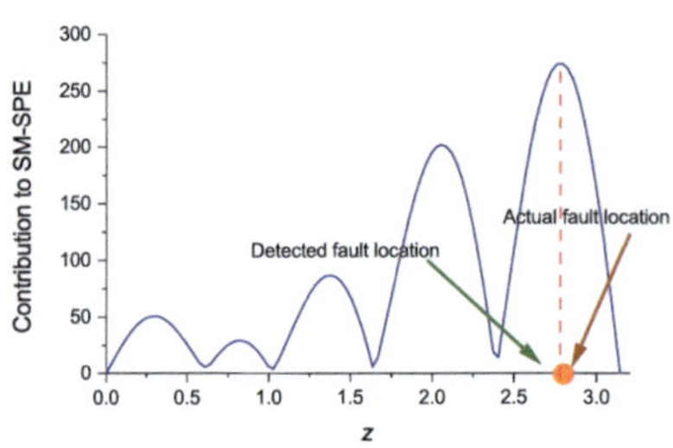

(d) SM-SPE statistic based spatial identification

Fig. 10.8 Detection and spatial identification results of Fault 2

Table 10.2 FAR and FDR by the proposed methods on catalytic rod

Fault		1	2
FAR(%)	T-I_d^2	2.59	1.28
	SM-SPE	1.60	0.52
FDR(%)	T-I_d^2	**86.53**	**97.75**
	SM-SPE	78.56	96.26

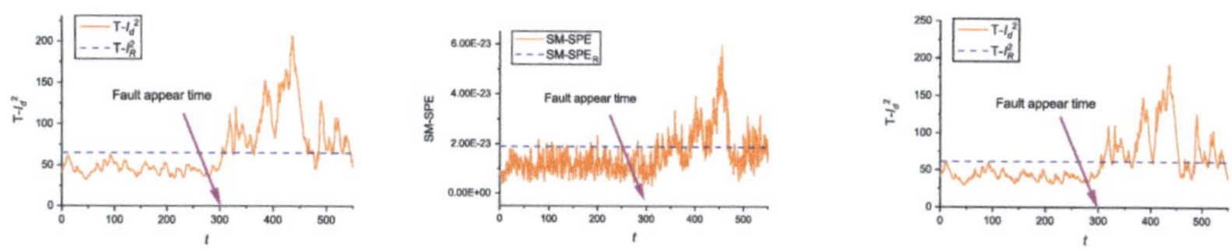

(a) T-I_d^2 statistic based fault detection with no disturbance (b) SM-SPE statistic based fault detection with no disturbance (c) T-I_d^2 statistic based fault detection with SNR=100

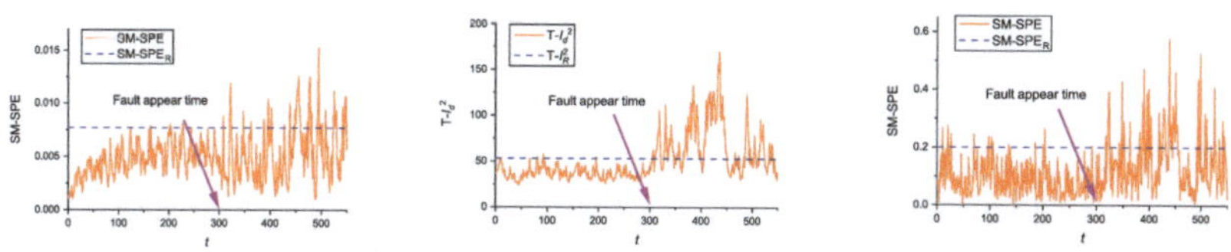

(d) SM-SPE statistic based fault detection with SNR=100 (e) T-I_d^2 statistic based fault detection with SNR=20 (f) SM-SPE statistic based fault detection with SNR=20

Fig. 10.9 Detection and spatial identification results of fault under different disturbances

fault locations. The position with the maximum contribution value is identified as the detected fault location.

In statistical process monitoring, the "FDR" is adopted to assess the monitoring performance. Specifically, the "detection rate" is defined as the proportion of samples that exceed the confidence bounds after a fault occurs. The FAR together with FDR results are summarized in Table 10.2, where the FAR values for both T-I_d^2 and SM-SPE statistics remain within acceptable levels.

To further validate the robustness of fault detection, additional simulations were carried out, with results presented in Fig. 10.9 and Table 10.3. In these tests, the faults were assumed to share the same temporal signal and spatial distribution as the input, introduced between samples 300 and 550. To evaluate detection performance under disturbances, three cases were examined: no disturbance, disturbance with SNR=100, and disturbance with SNR=20. As illustrated in Fig. 10.9, faults were reliably detected in all scenarios. The FAR and FDR results provided in Table 10.3 further indicate that stronger disturbances tend to increase FAR or reduce FDR.

Table 10.3 FAR and FDR by the proposed methods on catalytic rod with different disturbances

Disturbance		No disturbance	SNR=100	SNR=20
FAR(%)	T-I_d^2	**0**	**0**	**1.6**
	SM-SPE	5.17	2.79	5.58
FDR(%)	T-I_d^2	**86.06**	**74.10**	**62.95**
	SM-SPE	51.79	22.31	25.9

10.4.2 Snap Curing Oven

In this section, the curing thermal process in a snap oven [3, 37, 38] is employed to validate the effectiveness of the proposed fault detection and spatial identification approach. The snap curing oven is a key facility in industrial curing, widely used in chip packaging, where achieving the desired S-T temperature distribution is crucial for product quality. As illustrated in Fig. 10.10, the oven cures chips placed on the leadframe under a prescribed temperature profile. Temperature regulation is achieved through four identical heaters $h_1 - h_4$ (700 W each), while 16 thermocouples are installed for temperature measurement, with sensor placement on the leadframe shown in Fig. 10.11. Each heater is driven by a pulse-width modulation ratio and a power amplifier, and the resulting temperature signals are acquired via the dSPACE real-time simulation platform (see Fig. 10.10). A total of 1000 experimental samples were collected from eight sensors (s1, s4, s5, s8, s9, s12, s13, s16) at a 10 s sampling interval. In this study, only half of the available sensors were used to evaluate the performance of the proposed fault detection and spatial identification method.

The fundamental heat transfer equation of the oven can be expressed as follows:

$$
\begin{aligned}
\rho(T)c(T)\frac{\partial T}{\partial t} &= \frac{\partial}{\partial z_1}(k(T)\frac{\partial T}{\partial z_1}) + \frac{\partial}{\partial z_2}(k(T)\frac{\partial T}{\partial z_2}) + \frac{\partial}{\partial z_3}(k(T)\frac{\partial T}{\partial z_3}) + f_c(T) + f_r(T) \\
&\quad + \boldsymbol{Q}\boldsymbol{p}_0\boldsymbol{f}_u(\boldsymbol{u}(t)) + H(t - t_f)d(z, t), \\
\boldsymbol{T}_m(t) &= [T(z_{11}, z_{21}, z_{31}, t), \cdots, T(z_{1N}, z_{2N}, z_{3N}, t)]^{\mathrm{T}},
\end{aligned}
\tag{10.23}
$$

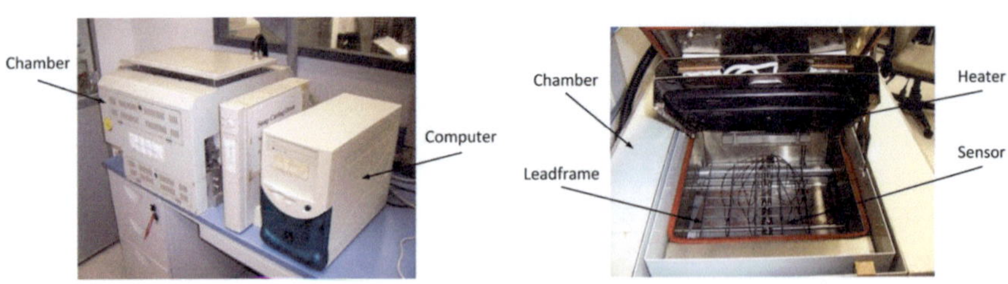

(a) Curing thermal system (b) Curing oven

Fig. 10.10 Snap curing oven system

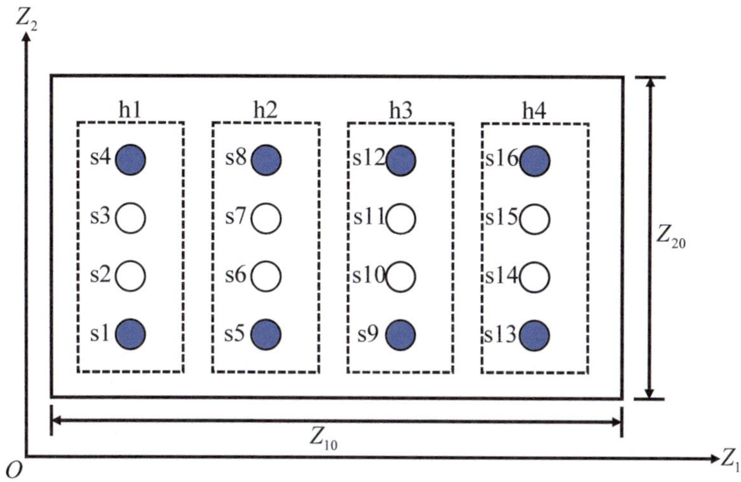

Fig. 10.11 Sensor placement for fault detection and localization

subject to the following boundary conditions:

$$k(T)\left.\frac{\partial T}{\partial z_i}\right|_{z_i=0} = f_{i1}(z_1, z_2, z_3, T, T_a)|_{z_i=0},$$

$$k(T)\left.\frac{\partial T}{\partial z_i}\right|_{z_i=z_{i0}} = f_{i2}(z_1, z_2, z_3, T, T_a)|_{z_i=z_{i0}}, i = 1, 2, 3$$

and the initial condition:

$$T_0 = T(z_1, z_2, z_3, 0),$$

where $\rho(T)$, $c(T)$, and $k(T)$ represent the temperature-dependent density, specific heat, and thermal conductivity, respectively. $T = T(z, t)$ denotes the temperature at spatial location $z = (z_1, z_2, z_3)$ and time t. The terms $f_c(T)$ and $f_r(T)$ correspond to thermal convection and radiation. The input PWM signals of the four heaters are denoted as $\boldsymbol{u}(t) = [u_1(t), \cdots, u_4(t)]^{\mathrm{T}}$, with $\boldsymbol{Q} = [q_1(z), \cdots, q_4(z)]$ describing their spatial distributions. The diagonal matrix $\boldsymbol{p}0 = \mathrm{diag}[p01, p_{02}, p_{03}, p_{04}]$ characterizes the efficiency and power of each heater, while f_{i1} and f_{i2} denote unknown nonlinear functions. T_0 represents the initial temperature distribution, and T_a is the ambient temperature. Typically, the oven operates around a nominal working point. To apply the proposed fault detection method, several assumptions and modifications are adopted following [30], with detailed derivations provided in Sect. A.3 of the appendix. As highlighted in Remark 10.1, once the slow dynamics of the snap curing oven can be represented in the form of Eq. 10.4, the proposed approach becomes applicable.

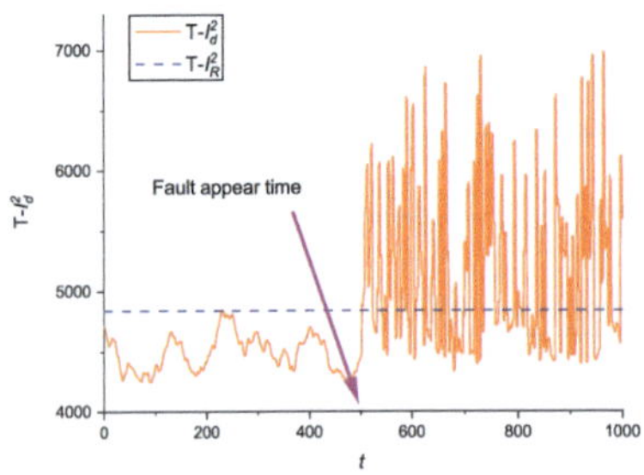

(a) T-I_d^2 statistic based fault detection

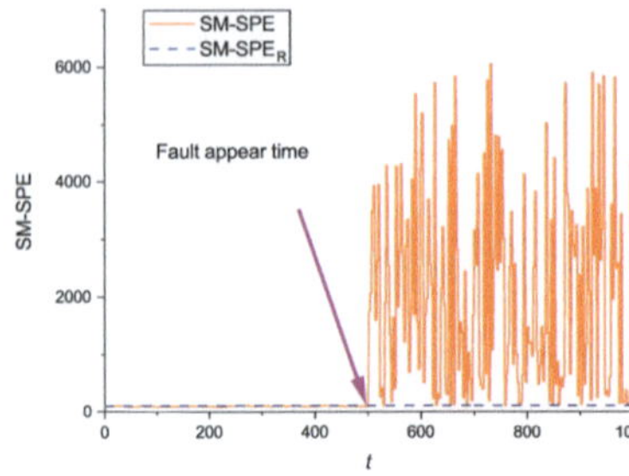

(b) SM-SPE statistic based fault detection

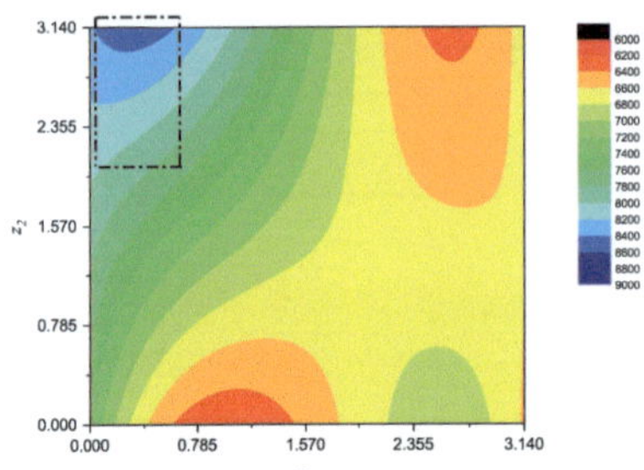

(c) T-I_d^2 statistic based spatial identification

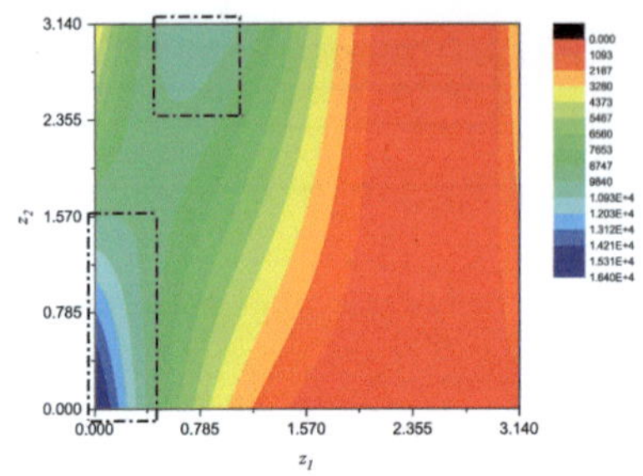

(d) SM-SPE statistic based spatial identification

Fig. 10.12 Detection and spatial identification results of Fault 1

For analysis, 1000 samples were collected, with the first 500 used for reference signal modeling and the remaining for fault detection and spatial identification. The four inputs $\boldsymbol{u}(t)$ follow a uniform distribution over $(0, 0.2)$. Fault settings are summarized in Table 10.3. Both Fault 1 and Fault 2 correspond to actuator faults. Specifically, Fault 1 shares the same temporal input and spatial distribution as heater 1, while Fault 2 is associated with heater 4, where its input exhibits a mean shift of 0.3. These faults are introduced from sample 500 onward, and their spatial locations correspond to heaters 1 and 4 in the normalized coordinate system. For instance, as shown in Fig. 10.11, heater 1 is located within $z_1 \in (0, z_{10}/4)$ and $z_2 \in (0, z_{20})$. Without loss of generality, the original coordinates (z_1, z_2) are mapped to the normalized domain $(\frac{z_1}{z_{10}}\pi, \frac{z_2}{z_{20}}\pi)$. The fault detection and spatial identification results for Faults 1 and 2 are illustrated in Figs. 10.12 and 10.13, respectively, where the mean location contribution defined in Eqs. 10.20 and 10.21 is employed. The identified fault locations are marked with black rectangles, with coordinates normalized within $z_1 \in (0, \pi)$, $z_2 \in (0, \pi)$. Furthermore, Table 10.5 reports the FAR and FDR results, showing that the FAR values for both T-I_d^2 and SM-SPE statistics are within acceptable limits (Table 10.4).

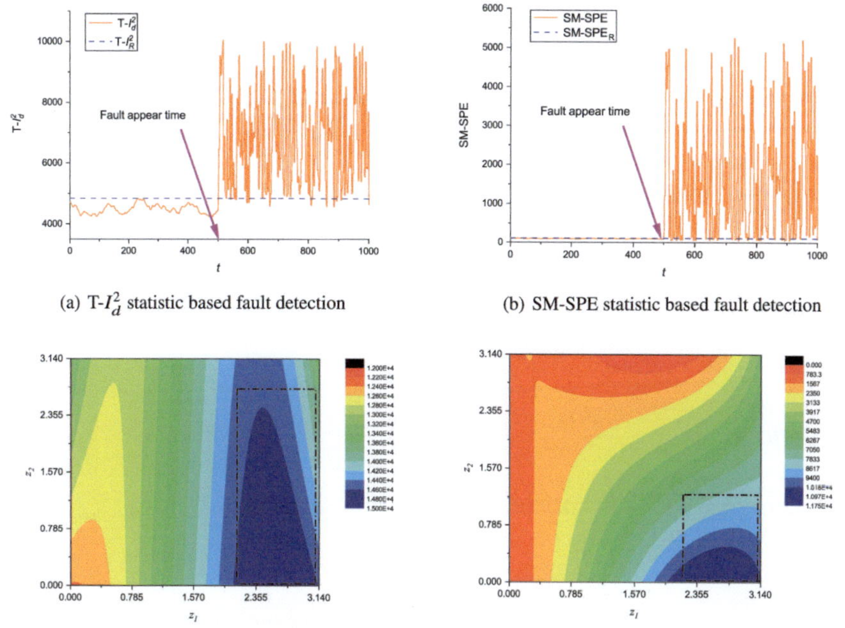

(a) T-I_d^2 statistic based fault detection

(b) SM-SPE statistic based fault detection

(c) T-I_d^2 statistic based spatial identification

(d) SM-SPE statistic based spatial identification

Fig. 10.13 Detection and spatial identification results of Fault 2

Table 10.4 Settings of faults for snap curing oven

Fault	Type	Fault occurring time and spatial location
1	$u_1 \to 2 * u_1$	$t_f = 500$, h1 $(z_1 \in (0, \pi/4), z_2 \in (0, \pi))$
2	$u_4 \to u_4 + 0.3$	$t_f = 500$, h4 $(z_1 \in (0, 3\pi/4), z_2 \in (0, \pi))$

Table 10.5 FAR and FDR by the proposed methods on snap curing oven

Fault		1	2
FAR(%)	T-I_d^2	0.37	0.22
	SM-SPE	0.41	0.48
FDR(%)	T-I_d^2	53.50	**93.00**
	SM-SPE	**98.40**	83.80

10.5 Conclusion

In this chapter, a fault detection and spatial identification method for parabolic DPSs with limited sensors has been developed. The framework consists of two stages: offline reference signal modeling and real-time fault estimation. In the modeling stage, normal operating data are employed to establish confidence bounds for monitoring statistics, which subsequently serve as reference signals during real-time fault estimation. Experimental results verify the effectiveness of the proposed approach in both fault detection and spatial localization. Extending this method to online spatial fault identification for DPSs[39–41]is a worthwhile research direction.

Appendix

A.1 Time/Space Separation and Model Reduction

$$\mathcal{A}y = a_1\frac{\partial y}{\partial z} + a_2\frac{\partial^2 y}{\partial z^2},\ y \in \mathcal{S}(\mathcal{A}) \triangleq \{y \in \mathcal{H}, \text{and the conditions in Eq. 10.2 hold}\}.$$

For the operator $\mathcal{A}$, the eigenvalue problem is defined as:

$$\mathcal{A}\phi_j(z) = \lambda_j\phi_j(z),\ \ j = 1, \cdots, \infty,$$

where λ_j represents an eigenvalue and ϕ_j its associated eigenfunction. The eigenspectrum of $\mathcal{A}$, denoted as $\sigma(\mathcal{A})$, is defined as the collection of all eigenvalues, i.e., $\sigma(\mathcal{A}) = \lambda_1, \cdots$. We assume that $\sigma(\mathcal{A})$ can be decomposed into two parts: a finite-dimensional subset containing m slow eigenvalues, and an infinite-dimensional stable subset consisting of the remaining fast eigenvalues. Moreover, it is required that the gap between the slow and fast eigenvalues of $\mathcal{A}$ is sufficiently large.

Assumption 10.1
1.$Re\{\lambda_1\} \geqslant Re\{\lambda_2\} \geqslant \cdots \geqslant Re\{\lambda_j\} \geqslant \cdots$, where $Re\{\lambda_j\}$ denotes the real part of λ_j.
2. $\sigma(\mathcal{A})$ can be partitioned as $\sigma(\mathcal{A}) = \sigma_1(\mathcal{A}) + \sigma_2(\mathcal{A})$, where $\sigma_1(\mathcal{A})$ consists of the first m (with m finite) eigenvalues, that is $\sigma_1(\mathcal{A}) = \{\lambda_1, \cdots, \lambda_m\}$, and $|Re\{\lambda_1\}|/|Re\{\lambda_m\}| = O(1)$.
3. $Re\{\lambda_{m+1}\} < 0$ and $|Re\{\lambda_m\}|/|Re\{\lambda_{m+1}\}| = O(\varepsilon)$, where $\varepsilon < 1$ is a small positive number.

Based on the theorem of separation of variables [42], the solution of system Eqs. (10.1)–(10.3) can be expressed by the following time/space decoupled form:

$$y(z,t) = \sum_{i=1}^{\infty} a_i(t)\phi_i(z) = \phi^{\mathrm{T}}(z)\boldsymbol{a}(t), \tag{10.24}$$

where $\phi(z) = [\phi_1(z), \cdots, \phi_\infty(z)]^{\mathrm{T}} \in \mathbb{R}^\infty$ denotes the spatial basis functions and $a(t) = [a_1(t), \cdots, a_\infty(t)]^{\mathrm{T}}$ denotes the temporal coefficients of this sum of products. Substituting Eqs. 10.24 into 10.1 yields:

$$\begin{aligned} \phi^{\mathrm{T}}(z)\dot{\boldsymbol{a}}(t) &= \mathcal{A}\phi^{\mathrm{T}}(z)\boldsymbol{a}(t) + f(\phi^{\mathrm{T}}(z)\boldsymbol{a}(t), \boldsymbol{u}(t)) \\ &\quad + k_w \boldsymbol{b}_w^{\mathrm{T}}(z)\boldsymbol{w}(t) + k_u \boldsymbol{b}_u^{\mathrm{T}}(z)\boldsymbol{u}(t) + H(t-t_f)d(z,t), \\ y(z_r,t) &= \phi^{\mathrm{T}}(z_r)\boldsymbol{a}(t), r = 1, 2, \cdots, N. \end{aligned} \tag{10.25}$$

Based on Assumption 10.1 and applying Galerkin's method, we can define $\mathcal{H}_s \triangleq span\{\phi_1, \cdots, \phi_m\}$ and $\mathcal{H}_f \triangleq span\{\phi_{m+1}, \cdots, \phi_\infty\}$ as the modal subspaces of $\mathcal{A}$. Also, the orthogonal projection operators $\mathcal{P}_s$ and $\mathcal{P}_f$ are defined as:

$$\begin{aligned} \boldsymbol{a}_s(t) = \mathcal{P}_s y(z,t) &\triangleq \begin{pmatrix} <\phi_1(z), y(z,t)> \\ \vdots \\ <\phi_m(z), y(z,t)> \end{pmatrix}, \\ \boldsymbol{a}_f(t) = \mathcal{P}_f y(z,t) &\triangleq \begin{pmatrix} <\phi_{m+1}(z), y(z,t)> \\ \vdots \\ <\phi_\infty(z), y(z,t)> \end{pmatrix}. \end{aligned} \tag{10.26}$$

Define $\phi_s(z) = [\phi_1(z), \cdots, \phi_m(z)]^{\mathrm{T}} \in \mathbb{R}^m$, $\phi_f(z) = [\phi_{m+1}(z), \cdots, \phi_\infty(z)]^{\mathrm{T}}$. Due to the fact that the spatial basis functions $\phi_i(z), i = 1, \cdots, \infty$ are orthonormal, we can rewrite $\boldsymbol{a}_s(t)$ and $\boldsymbol{a}_f(t)$ as $\boldsymbol{a}_s(t) = [a_1(t), \cdots, a_m(t)]^{\mathrm{T}} \in \mathbb{R}^m$, and $\boldsymbol{a}_f(t) = [a_{m+1}(t), \cdots, a_\infty(t)]^{\mathrm{T}} \in \mathbb{R}^\infty$. Then, Eq. 10.24 can be rewritten as:

$$y(z,t) = \phi^{\mathrm{T}}(z)\boldsymbol{a}(t) = \phi_s^{\mathrm{T}}(z)\boldsymbol{a}_s(t) + \phi_f^{\mathrm{T}}(z)\boldsymbol{a}_f(t). \tag{10.27}$$

By applying $\mathcal{P}_s$ and $\mathcal{P}_f$ to Eq. 10.24 and combining with Eq. 10.25, we obtain the following infinite-dimensional ODE system:

$$\begin{aligned} \dot{\boldsymbol{a}}_s(t) &= \boldsymbol{A}_s\boldsymbol{a}_s(t) + \boldsymbol{f}_s(\boldsymbol{a}_s(t), \boldsymbol{a}_f(t), \boldsymbol{u}(t)) + \boldsymbol{B}_{w,s}\boldsymbol{w}(t) \\ &\quad + \boldsymbol{B}_{u,s}\boldsymbol{u}(t) + H(t-t_f)\boldsymbol{d}_s(t), \\ \dot{\boldsymbol{a}}_f(t) &= \boldsymbol{A}_f\boldsymbol{a}_f(t) + \boldsymbol{f}_f(\boldsymbol{a}_s(t), \boldsymbol{a}_f(t), \boldsymbol{u}(t)) + \boldsymbol{B}_{w,f}\boldsymbol{w}(t) \\ &\quad + \boldsymbol{B}_{u,f}\boldsymbol{u}(t) + H(t-t_f)\boldsymbol{d}_f(t), \\ y(z_r,t) &= {\phi_s}^{\mathrm{T}}(z_r)\boldsymbol{a}_s(t) + {\phi_f}^{\mathrm{T}}(z_r)\boldsymbol{a}_f(t), \end{aligned} \tag{10.28}$$

with the initial conditions

$$\boldsymbol{a}_s(0) = \mathcal{P}_s\boldsymbol{a}(0), \boldsymbol{a}_f(0) = \mathcal{P}_f\boldsymbol{a}(0),$$

where

$$\begin{aligned}
&\boldsymbol{A}_s = \mathcal{P}_s\mathcal{A}\phi^{\mathrm{T}}(z) = diag\{\lambda_1, \cdots, \lambda_m\},\\
&\boldsymbol{A}_f = \mathcal{P}_f\mathcal{A}\phi^{\mathrm{T}}(z) = diag\{\lambda_{m+1}, \cdots, \lambda_\infty\},\\
&\boldsymbol{f}_s(\boldsymbol{a}_s(t), \boldsymbol{a}_f(t), \boldsymbol{u}(t)) = \mathcal{P}_s f(\phi_s^{\mathrm{T}}(z)\boldsymbol{a}_s + \phi_f^{\mathrm{T}}(z)\boldsymbol{a}_f, \boldsymbol{u}(t)),\\
&\boldsymbol{f}_f(\boldsymbol{a}_s(t), \boldsymbol{a}_f(t), \boldsymbol{u}(t)) = \mathcal{P}_f f(\phi_s^{\mathrm{T}}(z)\boldsymbol{a}_s + \phi_f^{\mathrm{T}}(z)\boldsymbol{a}_f, \boldsymbol{u}(t)),\\
&\boldsymbol{B}_{w,s} = \mathcal{P}_s k_w \boldsymbol{b}_w^{\mathrm{T}}(z),\ \ \boldsymbol{B}_{w,f} = \mathcal{P}_f k_w \boldsymbol{b}_w^{\mathrm{T}}(z),\\
&\boldsymbol{B}_{u,s} = \mathcal{P}_s k_u \boldsymbol{b}_u^{\mathrm{T}}(z),\ \ \boldsymbol{B}_{u,f} = \mathcal{P}_f k_u \boldsymbol{b}_u^{\mathrm{T}}(z),\\
&\boldsymbol{d}_s(t) = \mathcal{P}_s d(z,t),\ \ \boldsymbol{d}_f(t) = \mathcal{P}_f d(z,t).
\end{aligned}$$

Using $\varepsilon = \frac{|\mathrm{Re}\{\lambda_1\}|}{|\mathrm{Re}\{\lambda_{m+1}\}|}$ and multiplying the $\boldsymbol{a}_f$-subsystem by ε yields the singular perturbation model:

$$\begin{aligned}
\dot{\boldsymbol{a}}_s(t) &= \boldsymbol{A}_s\boldsymbol{a}_s(t) + \boldsymbol{f}_s(\boldsymbol{a}_s(t), \boldsymbol{a}_f(t), \boldsymbol{u}(t)) + \boldsymbol{B}_{w,s}\boldsymbol{w}(t)\\
&\quad + \boldsymbol{B}_{u,s}\boldsymbol{u}(t) + H(t - t_f)\boldsymbol{d}_s(t),\\
\varepsilon\dot{\boldsymbol{a}}_f(t) &= \boldsymbol{A}_{f\varepsilon}\boldsymbol{a}_f(t) + \varepsilon\boldsymbol{f}_f(\boldsymbol{a}_s(t), \boldsymbol{a}_f(t), \boldsymbol{u}(t)) + \varepsilon\boldsymbol{B}_{w,f}\boldsymbol{w}(t)\\
&\quad + \varepsilon\boldsymbol{B}_{u,f}\boldsymbol{u}(t) + \varepsilon H(t - t_f)\boldsymbol{d}_f(t),
\end{aligned} \tag{10.29}$$

where $\boldsymbol{A}_{f\varepsilon} = \varepsilon\boldsymbol{A}_f$.

Then the singular perturbation theory[43] can be applied here. Introducing the fast time-scale $\tau = t/\varepsilon$ and setting $\varepsilon = 0$, the following infinite dimensional fast subsystem can be obtained from model 10.29:

$$\frac{d\boldsymbol{a}_f(\tau)}{d\tau} = \boldsymbol{A}_{f\varepsilon}\boldsymbol{a}_f(\tau). \tag{10.30}$$

According to the fact that $\lambda_{m+1} < 0$ and the definition of ε, system 10.30 is globally exponentially stable. Setting $\varepsilon = 0$ in Eq. 10.29 and considering the non-singularity of $\boldsymbol{A}_{f\varepsilon}$, we have that $\boldsymbol{a}_f(t) = 0$. Substituting $\boldsymbol{a}_f(t) = 0$ into Eq. 10.29, we have the finite-dimensional ODE system presented in Eqs. 10.4–10.6.

A.2 KDE

Here a univariate kernel density estimator with kernel K is used:

$$\hat{f}(x) = \frac{1}{nh}\sum_{i=1}^{n} K\left(\frac{x - X_i}{h}\right),$$

where x denotes the data point of interest, X_i represents an observation from the sample, h is the bandwidth (smoothing parameter), n is the sample size, and K denotes the kernel function. The kernel estimator can thus be interpreted as a superposition of "bumps" centered at the observations. The form of these "bumps" is determined by the kernel function K, which must satisfy the condition:

$$\int_{-\infty}^{\infty} K(x)\mathrm{d}x = 1.$$

Various kernel functions can be employed, including the Biweight, Triangular, Gaussian, and Rectangular kernels, among others [44]. In this work, we utilize the widely adopted Gaussian kernel:

$$K(t) = \frac{1}{\sqrt{2\pi}} e^{-(1/2)t^2}.$$

A key step in KDE is to determine the optimal smoothing parameter h, denoted as h_{opt}. For this purpose, we employ the least-squares cross-validation (LSCV) method [44], which offers a fully automatic procedure for selecting h.

A.3 Derivations of Snap Curing Oven for the Proposed Approach

Around a specific working point, the specific heat c of dry air at atmospheric pressure may be regarded as constant, while the thermal conductivity $k(T)$ and density $\rho(T)$ remain temperature-dependent and can be expressed as:

$$k(T) = k_0 + \bar{k}(T) \text{ and } \rho(T) = \frac{\rho_0}{1 + \bar{\rho}(T)},$$

where k_0 and ρ_0 denote the nominal values at the operating point, while $\bar{k}(T)$ and $\bar{\rho}(T)$ are expressed as functions of T. The term $\boldsymbol{f_u}(\boldsymbol{u}(t))$ characterizes the nonlinear mapping between the input pulse width modulation (PWM) ratio $\boldsymbol{u}(t)$ and the power supplied to the chamber. This mapping can be decomposed into a linear component $\boldsymbol{u}(t)$ and a residual nonlinear component $\boldsymbol{f_{u0}}(\boldsymbol{u}(t))$, yielding $\boldsymbol{f_u}(\boldsymbol{u}(t)) = \boldsymbol{u}(t) + \boldsymbol{f_{u0}}(\boldsymbol{u}(t))$. Accordingly, the system Eq. 10.23 is reformulated as:

$$\frac{\partial T}{\partial t} = k_1 \nabla^2 T + F_0(T) + G(T, \boldsymbol{u}(t)) + k_2 \boldsymbol{Q}\boldsymbol{p}_0 \boldsymbol{u}(t) + H(t - t_f)D(z, t), \tag{10.31}$$

where $k_1 = (k_0/\rho_0 c)$ and $k_2 = (1/\rho_0 c)$ are constants. $\nabla^2 = \frac{\partial^2}{\partial {z_1}^2} + \frac{\partial^2}{\partial {z_2}^2} + \frac{\partial^2}{\partial {z_3}^2}$ is the Laplacian operator,

$$F_0(T) = \frac{k_0\bar{\rho}(T)}{\rho_0 c}\nabla^2 T + \frac{1+\bar{\rho}(T)}{\rho_0 c}(\bar{k}(T)\nabla^2 T + \frac{\partial \bar{k}(T)}{\partial z_1}\frac{\partial T}{\partial z_1} + \frac{\partial \bar{k}(T)}{\partial z_2}\frac{\partial T}{\partial z_2} + \frac{\partial \bar{k}(T)}{\partial z_3}\frac{\partial T}{\partial z_3} + f_c(T) + f_r(T)),$$
$$G(T, \boldsymbol{u}(t)) = \boldsymbol{Qp}_0(\frac{1+\bar{\rho}(T)}{\rho_0 c}\boldsymbol{f_{u0}}(\boldsymbol{u}(t)) + \frac{\bar{\rho}(T)}{\rho_0 c}\boldsymbol{u}(t)),$$
$$D(z, t) = \frac{1+\bar{\rho}(T)}{\rho_0 c}d(z, t).$$

Due to the lack of exact boundary conditions, the eigenfunctions cannot be obtained directly. Consequently, the model in (10.31) is reformulated into the following equivalent representation:

$$\begin{aligned} &\frac{\partial T}{\partial t} = k_1\nabla^2 T + F(T, \boldsymbol{u}(t)) + k_2\boldsymbol{Qp}_0\boldsymbol{u}(t) + H(t-t_f)D(z, t), \\ &\left.\frac{\partial T}{\partial z_i}\right|_{z_i=0} = 0, \ \left.\frac{\partial T}{\partial z_i}\right|_{z_i=z_{i0}} = 0, \ i = 1, 2, 3. \end{aligned} \tag{10.32}$$

where

$$F(T, \boldsymbol{u}(t)) = G(T, \boldsymbol{u}(t)) + F_0(T) - \delta(z_1 - 0)\frac{k_1}{k(T)}f_{11} + \delta(z_1 - z_{10})\frac{k_1}{k(T)}f_{12} - \delta(z_2 - 0)\frac{k_1}{k(T)}f_{21}$$
$$+ \delta(z_2 - z_{20})\frac{k_1}{k(T)}f_{22} - \delta(z_3 - 0)\frac{k_1}{k(T)}f_{31} + \delta(z_3 - z_{30})\frac{k_1}{k(T)}f_{32},$$

where $\delta(\cdot)$ denotes the Dirac delta function. The proof of this transformation is provided in the Appendix of [30]. In this formulation, the eigenvalue problem can be addressed as follows:

$$\lambda^2_{z_1 i} = \frac{i^2\pi^2}{z^2_{10}}, \lambda^2_{z_2 j} = \frac{j^2\pi^2}{z^2_{20}}, \lambda^2_{z_3 k} = \frac{k^2\pi^2}{z^2_{30}},$$
$$i = 0, 1, \cdots, \infty; \ j = 0, 1, \cdots, \infty; k = 0, 1, \cdots, \infty;$$

$$\phi_i(z_1) = \begin{cases} 1, \ i = 0 \\ \cos(i\pi z_1/z_{10}), i \neq 0 \end{cases}$$
$$\phi_j(z_2) = \begin{cases} 1, \ j = 0 \\ \cos(j\pi z_2/z_{20}), \ j \neq 0 \end{cases}$$
$$\phi_k(z_3) = \begin{cases} 1, \ k = 0 \\ \cos(k\pi z_3/z_{30}), k \neq 0 \end{cases}$$

The solution of the equivalent model (10.32) can thus be expressed as an orthogonally decoupled series:

$$T(z_1, z_2, z_3, t) = \sum_{i=0}^{\infty}\sum_{j=0}^{\infty}\sum_{k=0}^{\infty}\tilde{a}_{ijk}(t)\phi_i(z_1)\phi_j(z_2)\phi_k(z_3). \tag{10.33}$$

Substitute Eqs. 10.33 into 10.32 and do the following integration on both sides after multiplying the basis functions

$$\iiint\limits_{\Omega} \frac{\partial T}{\partial t} \phi_i(z_1)\phi_j(z_2)\phi_k(z_3)\mathrm{d}z_1\,\mathrm{d}z_2\,\mathrm{d}z_3 =$$

$$\iiint\limits_{\Omega} (k_1\nabla^2 T + F(T, \boldsymbol{u}(t)) + k_2 \boldsymbol{Q}\boldsymbol{p}_0\boldsymbol{u}(t) + H(t - t_f)D(z, t))\phi_i(z_1)\phi_j(z_2)\phi_k(z_3)\mathrm{d}z_1\,\mathrm{d}z_2\,\mathrm{d}z_3. \tag{10.34}$$

Due to the orthogonality of the eigenfunctions, Eq. 10.34 can be rewritten as

$$\begin{aligned} \dot{\tilde{a}}_{ijk}(t) &= k_1\bar{\lambda}_{ijk}\tilde{a}_{ijk}(t) + f_{ijk}(\tilde{\boldsymbol{a}}, \boldsymbol{u}) + k_2\bar{\boldsymbol{b}}_{ijk}\boldsymbol{p}_0\boldsymbol{u}(t), \\ T(z_{1r}, z_{2r}, z_{3r}, t) &= \sum \tilde{a}_{ijk}(t)\phi_i(z_{1r})\phi_j(z_{2r})\phi_k(z_{3r}), \\ r &= 1, \cdots, N \end{aligned} \tag{10.35}$$

where

$$\begin{aligned} \bar{\lambda}_{ijk} &= -(\lambda_{z_1 i}^2 + \lambda_{z_2 j}^2 + \lambda_{z_3 k}^2), \\ \bar{\boldsymbol{b}}_{ijk} &= [\bar{b}_{ijk}^1, \bar{b}_{ijk}^2, \bar{b}_{ijk}^3, \bar{b}_{ijk}^4], \\ \bar{b}_{ijk}^l &= \iiint\limits_{\Omega} q_l(z_1, z_2, z_3)\phi_i(z_1)\phi_j(z_2)\phi_k(z_3)\mathrm{d}z_1\,\mathrm{d}z_2\,\mathrm{d}z_3, \\ l &= 1, 2, 3, 4 \\ f_{ijk}(\tilde{\boldsymbol{a}}, \boldsymbol{u}) &= \iiint\limits_{\Omega} (F(T, \boldsymbol{u}(t)) + H(t - t_f)D(z, t)) \\ &\phi_i(z_1)\phi_j(z_2)\phi_k(z_3)\mathrm{d}z_1\,\mathrm{d}z_2\,\mathrm{d}z_3, \end{aligned}$$

in which $\bar{\lambda}_{ijk}$ is rearranged in order of magnitude and then denoted as a new sequence of λ_n with $\phi_n(z_1, z_2, z_3)$ being its corresponding eigenfunction, as follows:

$$\begin{aligned} \lambda_n &= \left\{\bar{\lambda}_{ijk}\right\}, \text{ with } \lambda_1 > \lambda_2 > \cdots > \lambda_n > \cdots \\ \phi_n(z_1, z_2, z_3) &= \{\phi_i(z_1)\phi_j(z_2)\phi_k(z_3)\} \\ a_n &= \{\tilde{a}_{ijk}\} \\ b_n^l &= \{\bar{b}_{ijk}^l\} \\ c_{rn} &= \phi_n(z_{1r}, z_{2r}, z_{3r}) \\ f_n(\boldsymbol{a}, \boldsymbol{u}) &= \{f_{ijk}(\boldsymbol{a}, \boldsymbol{u})\} \end{aligned}$$

Then Eq. 10.35 can be rewritten in a general form as follows:

$$\begin{aligned} \dot{\boldsymbol{a}}(t) &= k_1\boldsymbol{A}_0\boldsymbol{a}(t) + k_2\boldsymbol{B}_0\boldsymbol{p}_0\boldsymbol{u}(t) + \boldsymbol{f}_0(\boldsymbol{a}, \boldsymbol{u}), \\ \boldsymbol{T}_m &= \boldsymbol{C}_0\boldsymbol{a}(t), \end{aligned} \tag{10.36}$$

where

$$\boldsymbol{a}(t) = [a_1, \cdots a_n, \cdots]^{\mathrm{T}}$$
$$\boldsymbol{A}_0 = diag(\lambda_1, \cdots \lambda_n, \cdots)$$
$$\boldsymbol{B}_0 = \begin{pmatrix} b_1^1 \cdots b_n^1 \cdots \\ b_1^2 \cdots b_n^2 \cdots \\ b_1^3 \cdots b_n^3 \cdots \\ b_1^4 \cdots b_n^4 \cdots \end{pmatrix}$$
$$\boldsymbol{C}_0 = \begin{pmatrix} c_{11} & \cdots & c_{1n} & \cdots \\ \vdots & & \vdots & \\ c_{m1} & \cdots & c_{mn} & \cdots \end{pmatrix}$$
$$\boldsymbol{f}_0(\boldsymbol{a}, \boldsymbol{u}) = [f_1(\boldsymbol{a}, \boldsymbol{u}), \cdots f_n(\boldsymbol{a}, \boldsymbol{u}), \cdots]^{\mathrm{T}}$$

Then the the slow mode of the snap curing oven's dynamics can be expressed by a system in the form of Eq. 10.4 as follows:

$$\begin{aligned} \dot{\boldsymbol{a}}_s(t) &= k_1 \boldsymbol{A}_s \boldsymbol{a}_s(t) + k_2 \boldsymbol{B}_s \boldsymbol{p}_0 \boldsymbol{u}(t) + \boldsymbol{f}_s(\boldsymbol{a}_s, 0, \boldsymbol{u}), \\ \boldsymbol{T}_m &= \boldsymbol{C}_s \boldsymbol{a}_s(t), \end{aligned} \tag{10.37}$$

where

$$\boldsymbol{A}_s = \mathcal{P}_s \boldsymbol{A}_0, \; \boldsymbol{B}_s = \mathcal{P}_s \boldsymbol{B}_0, \; \boldsymbol{f}_s(\boldsymbol{a}_s, 0, \boldsymbol{u}) = \mathcal{P}_s \boldsymbol{f}_0(\boldsymbol{a}, \boldsymbol{u}).$$

Then the proposed approach can be used.

References

1. Li H-X, Qi C (2010) Modeling of distributed parameter systems for applications-synthesized review from time-space separation. J Process Control 20(8):891–901
2. Christofides PD (2012) Nonlinear and robust control of PDE systems: Methods and applications to transport-reaction processes. Springer Science+Business Media, Berlin
3. Lu X, Zou W, Huang M (2017) Robust spatiotemporal LS-SVM modeling for nonlinear distributed parameter system with disturbance. IEEE Trans Ind Electron 64(10):8003–8012
4. Li H-X, Lu X (2014) System design and control integration for advanced manufacturing. John Wiley & Sons, Chichester
5. Li H-X, Qi C (2011) Spatio-temporal modeling of nonlinear distributed parameter systems: a time/space separation based approach. Springer Science+Business Media, Berlin
6. Xiao T, Li H-X (2017) Eigenspectrum-based iterative learning control for a class of distributed parameter system. IEEE Trans Autom Control 62(2):824–836
7. Wang J-W, Li H-X, Wu H-N (2017) A membership-function-dependent approach to design fuzzy pointwise state feedback controller for nonlinear parabolic distributed parameter systems with spatially discrete actuators. IEEE Trans Syst Man Cybern Syst 47(7):1486–1499
8. Wang J-W, Wu H-N, Li H-X (2012) Stochastically exponential stability and stabilization of uncertain linear hyperbolic PDE systems with Markov jumping parameters. Automatica 48(3):569–576
9. Deutscher J, Kerschbaum S (2018) Backstepping control of coupled linear parabolic PIDEs with spatially-varying coefficients. IEEE Trans Autom Control 63(12):4218–4233

10. Chiang LH, Russell EL, Braatz RD (2000) Fault detection and diagnosis in industrial systems. Springer Science+Business Media, Berlin
11. Vachtsevanos GJ, Lewis F, Hess A et al (2006) Intelligent fault diagnosis and prognosis for engineering systems. John Wiley & Sons, Chichester
12. Isermann R (2006) Fault-diagnosis systems: an introduction from fault detection to fault tolerance. Springer Science+Business Media, Berlin
13. Ding SX (2014) Data-driven design of fault diagnosis and fault-tolerant control systems. Springer Science+Business Media, Berlin
14. Ding SX (2008) Model-based fault diagnosis techniques: design schemes, algorithms, and tools. Springer Science+Business Media, Berlin
15. Jackson EJ (1985) Multivariate quality control. Commun Stat Theory Methods 14(11):2657–2688
16. Yi J, Huang D, He H et al (2017) A novel framework for fault diagnosis using kernel partial least squares based on an optimal preference matrix. IEEE Trans Ind Electron 64(5):4315–4324
17. Hyvärinen A, Karhunen J, Oja E (2004) Independent component analysis, vol 46. John Wiley & Sons, Chichester
18. Rashid MM, Yu J (2012) A new dissimilarity method integrating multidimensional mutual information and independent component analysis for non-Gaussian dynamic process monitoring. Chemom Intell Lab Syst 115:44–58
19. Yin S, Ding SX, Xie X et al (2014) A review on basic data-driven approaches for industrial process monitoring. IEEE Trans Ind Electron 61(11):6418–6428
20. Dobson J, Cawley P (2016) Independent component analysis for improved defect detection in guided wave monitoring. Proc IEEE 104(8):1620–1631
21. Fan J, Wang Y (2014) Fault detection and diagnosis of non-linear non-Gaussian dynamic processes using kernel dynamic independent component analysis. Inf Sci 259:369–379
22. Cai L, Tian X, Chen S (2017) Monitoring nonlinear and non-Gaussian processes using Gaussian mixture model-based weighted kernel independent component analysis. IEEE Trans Neural Netw Learn Syst 28(1):122–135
23. Demetriou MA (2002) A model-based fault detection and diagnosis scheme for distributed parameter systems: A learning systems approach. ESAIM Control Optim Calc Var 7:43–67
24. El-Farra NH (2006) Integrated fault detection and fault-tolerant control architectures for distributed processes. Ind Eng Chem Res 45(25):8338–8351
25. Deutscher J (2016) Fault detection for linear distributed-parameter systems using finite-dimensional functional observers. Int J Control 89(3):550–563
26. Fischer F, Deutscher J (2017) Fault detection for parabolic systems with distributed inputs and outputs using the modulation function approach. IFAC-PapersOnLine 50(1):6774–6779
27. Fischer F, Deutscher J (2016) Algebraic fault detection and isolation for parabolic distributed-parameter systems using modulation functions. IFAC-PapersOnLine 49(8):162–167
28. Cai J, Ferdowsi H, Sarangapani J (2016) Model-based fault detection, estimation, and prediction for a class of linear distributed parameter systems. Automatica 66:122–131
29. Asiri S, Laleg-Kirati T-M (2017) Modulating functions-based method for parameters and source estimation in one-dimensional partial differential equations. Inverse Probl Sci Eng 25(8):1191–1215
30. Deng H, Li H-X, Chen G (2005) Spectral-approximation-based intelligent modeling for distributed thermal processes. IEEE Trans Control Syst Technol 13(5):686–700
31. Ray WH (1981) Advanced process control. McGraw-Hill, Columbus
32. Hyvarinen A (1999) Fast and robust fixed-point algorithms for independent component analysis. IEEE Trans Neural Netw 10(3):626–634
33. Lee J-M, Yoo CK, Lee I-B (2004) Statistical process monitoring with independent component analysis. J Process Control 14(5):467–485
34. Martin EB, Morris AJ (1996) Non-parametric confidence bounds for process performance monitoring charts. J Process Control 6(6):349–358
35. Besterfield DH (2004) Quality control. Pearson Education India

36. Strikwerda JC (2004) Finite difference schemes and partial differential equations. SIAM, Philadelphia
37. Xu K-K, Li H-X, Yang H-D (2017) Kernel-based random vector functional-link network for fast learning of spatiotemporal dynamic processes. IEEE Trans Syst Man Cybern Syst 99:1–11
38. Xu K-K, Li H-X, Yang H-D (2017) Dual least squares support vector machines based spatiotemporal modeling for nonlinear distributed thermal processes. J Process Control 54:81–89
39. Wang Z, Li H-X (2017) Incremental spatio-temporal learning for online modeling of distributed parameter systems. IEEE Trans Syst Man Cybern Syst 49(12):2612–2622
40. Wang B-C, Li H-X (2018) A sliding window based dynamic spatiotemporal modeling for distributed parameter systems with time-dependent boundary conditions. IEEE Trans Ind Informat 15(4):2044–2053
41. Meng X-B, Li H-X, Yang H-D (2018) Evolutionary design of spatio-temporal learning model for thermal distribution in lithium-ion batteries. IEEE Trans Ind Informat 15(5):2838–2848
42. MacCluer CR (1994) Boundary value problems and orthogonal expansions: physical problems from a Sobolev viewpoint. IEEE Press, Piscataway
43. Khalil HK, Grizzle JW (2002) Nonlinear systems. Prentice Hall, Upper Saddle River
44. Silverman BW (1986) Density estimation for statistics and data analysis. CRC Press, Boca Raton

Part III
Purely Data-Driven Modeling and Diagnosis for Completely Unknown DPSs

Chapter 11
From On-Line Systems Modeling to Fault Detection for A Class of Unknown High-Dimensional DPSs

Abstract Fault detection for distributed-parameter systems has predominantly been model-based, with performance heavily contingent on prior model information, which limits applicability in industrial settings. This chapter makes an initial attempt to establish a novel framework that integrates on-line systems modeling with fault detection for unknown high-dimensional distributed-parameter systems. The two components are coupled: the systems modeling error is repurposed as the residual signal for fault detection, while the on-line modeling is switched to off-line mode conditional on the detection outcomes. The high-dimensional distributed-parameter systems are first decomposed into spatial features and temporal sequences. A receding-horizon scheme is then employed to learn the temporal dynamics, and the residual signal is constructed from the temporal validation error. Experimental studies on sensor-fault diagnosis for the thermal process of a two-dimension battery cell are conducted to validate the method.

11.1 Introduction

DPSs are classes of dynamical systems typically represented by PDEs. Exemplars encompass flexible mechanical systems [1–3], battery thermal processes [4, 5], and chemical processes [6, 7]. Fault diagnosis of DPSs is fundamental to ensuring safe and efficient operation; however, systematic investigation in the open literature remains limited. For illustration, battery thermal faults [8, 9] may escalate to thermal runaway and fires, thereby constraining the safety of electric vehicles and hybrid electric vehicles; furthermore, actuator failures—widespread in flexible manipulators—can engender adverse responses and forced shutdowns of controlled systems, with the potential to culminate in production accidents [10].

In contrast with the extensive literature on fault diagnosis of LPS governed by ODE—from both model-based [11] and data-driven [12] perspectives—the body of work on DPSs is comparatively limited. To the best of our knowledge, most reported studies remain model-based and can be classified according to whether *early-lumping* is employed. In *early-lumping* approaches, the original PDE system is first reduced to a set of ODEs, after which established methods for LPSs are applied [13, 14]. While

© The Author(s), under exclusive license to Springer Nature Singapore Pte Ltd. 2026
Y. Feng et al., *Spatio-Temporal Abnormality Diagnosis for Industrial Distributed Parameter Systems*, https://doi.org/10.1007/978-981-95-3750-1_11

this reduction lowers analytical and implementation complexity, it introduces observation spillover [15], whereby residual dynamics excite the measurement channel and deteriorate diagnostic performance. By contrast, *late-lumping* methods typically construct PDE observers directly via PDE backstepping [16]; representative contributions include [17–20]. A more recent *late-lumping* alternative is the modulation-functions-based approach, originally proposed for parameter estimation in LPSs [21], later adapted to DPS fault diagnosis [22, 23] by transforming the governing PDEs into algebraic equations that can be solved directly.

An indispensable premise in the foregoing works is the prior availability of explicit DPS model information. Under this premise, both ODE and PDE observers can be synthesized, and the algebraic equations arising from modulation-functions methods can be solved. This requirement is, however, restrictive: model uncertainty is ubiquitous in industrial applications, and in some cases first-principles models are inadequate. As in LPSs, system I/O data can be exploited to develop data-driven fault diagnosis methods [24]. In the same spirit, DPS fault diagnosis can be pursued beyond the well-studied observer-based paradigm. For example, [25] proposes a statistical approach based on ICA for fault detection and localization in partially known DPSs; yet, akin to certain SPC techniques, its performance degrades on highly dynamic systems. From a systems-modeling perspective, a potential remedy is to maintain an on-line updating model and conduct diagnosis on this data-driven surrogate. This consideration motivates a systematic framework that integrates on-line systems modeling with fault diagnosis of DPSs, wherein modeling errors are explicitly routed into residuals for detection. Furthermore, a switching mechanism is required to prevent the data-driven learner from absorbing the fault, thereby preserving detectability.

Motivated by the above considerations, an integrated methodological framework for joint systems modeling and fault detection in a class of high-dimensional DPSs is developed. To be more specific, we first decompose the original system into temporal sequences and spatial features for the ease of model learning. Temporal sequences are learned within a receding-horizon on-line learning scheme. The residual signal is then constructed and evaluated using the validation error of the temporal sequences. To further enhance detection performance, an on-line to off-line switching scheme is introduced.

The remainder of this chapter is structured as follows. Section 11.2 formulates the problem statement. Section 11.3 develops the methodology framework, comprising on-line systems modeling and fault detection. Experimental validation on the thermal process of a 2-D battery cell is presented. Finally, Sect. 11.5 concludes the chapter.

11.2 Problem Statement

Notations: $\mathbb{R}$ and $\mathbb{R}^n$ denote the set of all real numbers and n-dimensional Euclidean space with the norm $|\cdot|_2$, respectively. Moreover, $|\cdot|_1$ denotes the 1-norm in the n-dimensional Euclidean space. $(\phi(\cdot,t),\varphi(\cdot,t)) \triangleq \int_\Omega \phi(\cdot,t)\varphi(\cdot,t)\mathrm{d}x\mathrm{d}y\mathrm{d}z$ denotes the inner product and $||u(\cdot, t)||_2$ denotes the norm of the real Hilbert space $\mathcal{L}^2(\Omega) \triangleq \mathcal{L}^2(\Omega; \mathbb{R})$. $< a(k) >_1^N = \frac{1}{N}\sum_{k=1}^{N} a(k)$ denotes the ensemble average of the sequence.

In this chapter, we consider a general class of unknown DPSs in 3-D spatial domain $\Omega \triangleq (0, \bar{x}) \times (0, \bar{y}) \times (0, \bar{z}) \subset \mathbb{R}^3$ as follows:

$$\begin{aligned} \frac{\partial u}{\partial t} &= g(\boldsymbol{q}(t), \boldsymbol{\rho}(x, y, z)) + \boldsymbol{\lambda}^{\mathrm{T}}(x, y, z)\boldsymbol{f}_p(t) \\ &+ \mathcal{L}\left(u, \frac{\partial u}{\partial x}, \frac{\partial u}{\partial y}, \frac{\partial u}{\partial z}, \frac{\partial^2 u}{\partial x \partial y}, \frac{\partial^2 u}{\partial x \partial z}, \frac{\partial^2 u}{\partial y \partial z}, \frac{\partial^2 u}{\partial x^2}, \frac{\partial^2 u}{\partial y^2}, \frac{\partial^2 u}{\partial z^2}\right), \end{aligned} \tag{11.1a}$$

$$\boldsymbol{p}(t) = \int_{\Omega} \boldsymbol{c}(x, y, z)u(x, y, z, t)\mathrm{d}x\mathrm{d}y\mathrm{d}z + \boldsymbol{f}_s(t), \tag{11.1b}$$

subject to the following boundary conditions:

$$h_1\left(u, \frac{\partial u}{\partial x}, \frac{\partial u}{\partial y}, \frac{\partial u}{\partial z}\right)\bigg|_{x=0} = 0, \tag{11.2a}$$

$$h_2\left(u, \frac{\partial u}{\partial x}, \frac{\partial u}{\partial y}, \frac{\partial u}{\partial z}\right)\bigg|_{x=\bar{x}} = 0, \tag{11.2b}$$

$$h_3\left(u, \frac{\partial u}{\partial x}, \frac{\partial u}{\partial y}, \frac{\partial u}{\partial z}\right)\bigg|_{y=0} = 0, \tag{11.2c}$$

$$h_4\left(u, \frac{\partial u}{\partial x}, \frac{\partial u}{\partial y}, \frac{\partial u}{\partial z}\right)\bigg|_{y=\bar{y}} = 0, \tag{11.2d}$$

$$h_5\left(u, \frac{\partial u}{\partial x}, \frac{\partial u}{\partial y}, \frac{\partial u}{\partial z}\right)\bigg|_{z=0} = 0, \tag{11.2e}$$

$$h_6\left(u, \frac{\partial u}{\partial x}, \frac{\partial u}{\partial y}, \frac{\partial u}{\partial z}\right)\bigg|_{z=\bar{z}} = 0, \tag{11.2f}$$

and with the following initial condition:

$$u(x, y, z, 0) = u_0(x, y, z), \tag{11.3}$$

where $u(x, y, z, t)$ denotes the state variable, $x \in [0, \bar{x}]$, $y \in [0, \bar{y}]$ and $z \in [0, \bar{z}]$ are the spatial coordinates, $t \geq 0$ is the time variable. $\mathcal{L}(\cdot) \in \mathbb{R}$ is an *unknown nonlinear* function of the spatial operators. $\boldsymbol{q}(t) \triangleq [q_1(t)\ q_2(t)\ \cdots\ q_n(t)]^{\mathrm{T}} \in \mathbb{R}^n$ denotes the exogenous input of the system while $\boldsymbol{\rho}(x, y, z) \triangleq [\rho_1(x, y, z)\ \rho_2(x, y, z)\ \cdots\ \rho_n(x, y, z)]^{\mathrm{T}} \in \mathbb{R}^n$ denotes its *unknown* shape function, $g(\cdot)$ is an unknown nonlinear function. $\boldsymbol{f}_p(t) \triangleq [f_{p1}(t)\ f_{p2}(t)\ \cdots\ f_{ps}(t)]^{\mathrm{T}} \in \mathbb{R}^s$ denotes the additive distributed process fault with $\boldsymbol{\lambda}(x, y, z) \triangleq [\lambda_1(x, y, z)\ \lambda_2(x, y, z)\ \cdots\ \lambda_s(x, y, z)]^{\mathrm{T}} \in \mathbb{R}^s$ denotes its shape function. It includes both actuator faults with known shape function (the same as that of the input $\boldsymbol{q}(t)$) and other process faults with unknown shape function. $\boldsymbol{c}(x, y, z) \triangleq [c_1(x, y, z)\ c_2(x, y, z)\ \cdots\ c_M(x, y, z)]^{\mathrm{T}} \in \mathbb{R}^M$ denotes the output function of the measuring sensor and $\boldsymbol{p}(t) \triangleq [p_1(t)\ p_2(t)\ \cdots\ p_M(t)]^{\mathrm{T}} \in \mathbb{R}^M$

is the output measurement. For simplicity, only point-wise measurement sensors [26] are considered in this chapter. $h_1(\cdot)$, $h_2(\cdot)$, $h_3(\cdot)$, $h_4(\cdot)$, $h_5(\cdot)$, and $h_6(\cdot)$ are *unknown nonlinear* functions describing the boundary conditions. $\boldsymbol{f}_s(t) \triangleq [f_{s1}(t)\ f_{s2}(t)\ \cdots\ f_{sM}(t)]^{\mathrm{T}} \in \mathbb{R}^M$ denotes the sensor fault that affects the measurement sensors physically, whose space locations are specifically given.

Remark 11.1 It is worth noting that the considered system class also includes passive systems when $\boldsymbol{q}(t) = 0$ holds. Moreover, for the following systems modeling and fault detection, the measured input $\bar{\boldsymbol{q}}(t)$ rather than the true value $\boldsymbol{q}(t)$ is adopted.

Remark 11.2 The considered model (11.1a)–(11.3) covers a wide range of industrial applications. For example, the thermal process of 1-D cylindrical Li-ion battery cells [27], 2-D [28] and even the 3-D temperature distribution [29] in Li-ion pouch cells, the 1-D vibrating model of flexible string [30], and the 2-D gas or chemical dispersion in ambient atmosphere [31].

The problem investigated in this chapter can be summarized as follows.

Use the measured I/O data $(\bar{\boldsymbol{q}}(t),\ \boldsymbol{p}(t))$ to design a *purely data-driven* systems modeling and fault detection framework for the considered unknown S-T systems in 3-D spatial domain (11.1a)–(11.3).

11.3 Temporal Dynamics Learning-Based Detection

The methodology framework depicted in Fig. 11.1 consists of three components. First, to address the S–T distribution of the system, a high-dimensional S–T decomposition is employed to partition the original dynamics into temporal sequences and spatial features. Second, leveraging the extracted temporal sequences, a receding-horizon on-line learning scheme is formulated to track time-varying temporal dynamics driven by variations in operating conditions or system parameters. In addition, the residual signal is constructed from modeling error rather than the observer error assumed in classical model-based paradigm. Third, because the learning model is updated on-line, faults may be assimilated by the learning model and thereby degrade detection performance; accordingly, a *fault-holder* mechanism is instituted to switch the learning model to an off-line mode in which updates are suspended.

11.3.1 High-Dimensional S-T Dynamics Decomposition

The principal challenge in systems modeling stems from the mismatch between the S-T dynamics decomposition method and the limited spatial sensing afforded by a finite number of sensors. To address this, and following the *pseudo-modal methods* in [32], the solution of (11.1a)–(11.3) is expanded in a finite set of spatial basis functions, $\phi_i(x, y, z)$ $(i = 1, \cdots, N)$, as follows:

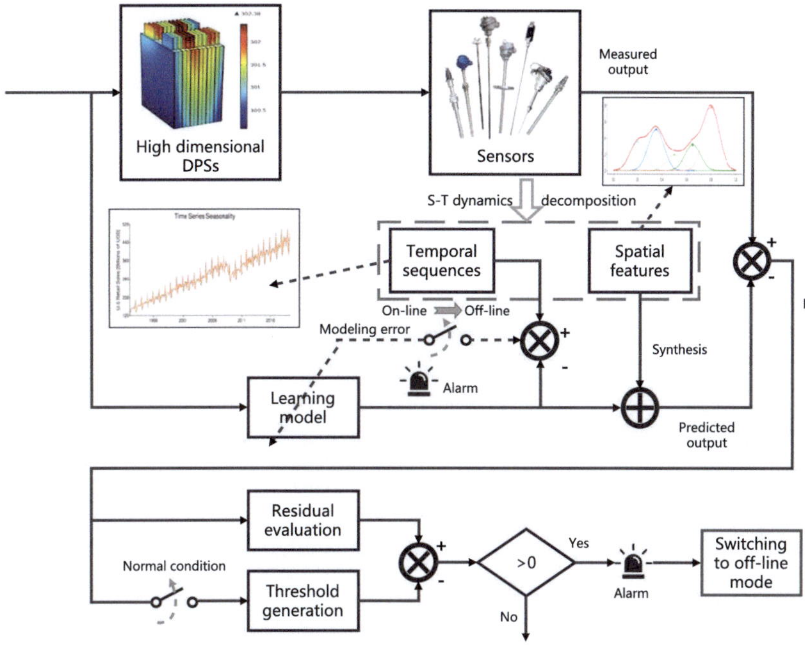

Fig. 11.1 Methodology framework

$$u_N(x, y, z, t) = \sum_{i=1}^{N} \phi_i(x, y, z) a_i(t), \tag{11.4}$$

where $u_N(x, y, z, t)$ denotes the N-th order approximation of $u(x, y, z, t)$, $\phi_i(x, y, z)$ and $a_i(t)$ denote the i-th spatial basis function and the corresponding temporal variable, respectively. The major difference between the modal approximation methods and the *pseudo-modal methods* is that the latter can use any set of known spatial basis functions while the former utilizes the eigenfunctions of the spatial operator. In this manner, we can use some goodness-of-fit criterion to determine the temporal variable $a_i(t)(i = 1, \cdots, N)$ in such a way that $u_N(x, y, z, t)$ is a good approximation to the solution of (11.1a)–(11.3).

To minimize the truncation error, the following optimization problem is formulated:

$$\begin{aligned} &\min_{\{\phi_i(x,y,z)\}_{i=1}^{N}} \left\langle \| u(x, y, z, t) - u_N(x, y, z, t) \|_2^2 \right\rangle_1^{N_t} \\ &\text{s.t.} \quad \big(\phi_i(x, y, z), \phi_j(x, y, z)\big) = \begin{cases} 1, & i = j \\ 0, & i \neq j \end{cases} \\ &i, j = 1, \cdots, N, \end{aligned} \tag{11.5}$$

which gives N spatial basis functions $\{\phi_i(x, y, z)\}_{i=1}^N$ and the corresponding temporal variables $a_i(t) = (\phi_i(\cdot), u_N(\cdot, t))$.

In practice, the methods of snapshots [33] can be used to solve the above optimization problem.

Similarly, we can also obtain the N-th order approximation of the input term $g(\cdot)$ as:

$$g_N(x, y, z, t) = \sum_{i=1}^{N} \phi_i(x, y, z) b_i(t), \tag{11.6}$$

where the temporal term $b_i(t)$ is related to the exogenous input $\boldsymbol{q}(t)$.

11.3.2 Receding Horizon Temporal Dynamics Learning

After decomposing the state variable $u(x, y, z, t)$ into spatial features and temporal variables, the temporal dynamics can be obtained by the orthogonal projection method. Define the orthogonal operator $\mathcal{P}$ so that

$$\boldsymbol{a}(t) \triangleq \mathcal{P} u_N(x, y, z, t) = \begin{bmatrix} (\phi_1, u_N) \\ \vdots \\ (\phi_N, u_N) \end{bmatrix} \in \mathbb{R}^N.$$

Applying the operator $\mathcal{P}$ to both sides of (11.1a) and employing Galerkin's method yields

$$\dot{a}_i(t) = f_i(a_i(t)) + b_i(t) + \boldsymbol{B}_i \boldsymbol{f}_p(t), i = 1, \cdots, N, \tag{11.7}$$

where $f_i(\cdot)$ is an unknown nonlinear scalar function and $\boldsymbol{B}_i \triangleq [(\phi_1, \lambda_1)\ (\phi_1, \lambda_2)\ \cdots\ (\phi_1, \lambda_k)] \in \mathbb{R}^{1\times k}$ is an unknown vector. The aim of on-line learning is to identify the unknown function and parameters of model (11.7) from the measured I/O data $(\bar{\boldsymbol{q}}(t), \boldsymbol{p}(t))$. Since $b_i(t)$ depends on the exogenous input $\boldsymbol{q}(t)$, the relation is described by unknown nonlinear functions $\delta_i(\cdot), i = 1, \cdots, N$:

$$b_i(t) = \delta_i(\boldsymbol{q}(t)).$$

Then (11.7) can be rewritten as

$$\dot{a}_i(t) = f_i(a_i(t)) + \delta_i(\boldsymbol{q}(t)) + \boldsymbol{B}_i \boldsymbol{f}_p(t), i = 1, \cdots, N. \tag{11.8}$$

The detailed procedures for the learning implementation are outlined as follows. The first step is to obtain the temporal sequences $\boldsymbol{a}(t)$ via the decomposition technique. By combining the decomposition form (11.4) with the measurement equation (11.1b), it follows that

$$\boldsymbol{p}(t) = \boldsymbol{\Phi}\boldsymbol{a}(t) + \boldsymbol{\Delta}(t) + \boldsymbol{f}_s(t), \tag{11.9}$$

where

$$\boldsymbol{\Phi} \triangleq \begin{pmatrix} (c_1(\cdot), \phi_1(\cdot)) & \cdots & (c_1(\cdot), \phi_N(\cdot)) \\ \vdots & \ddots & \vdots \\ (c_M(\cdot), \phi_1(\cdot)) & \cdots & (c_M(\cdot), \phi_N(\cdot)) \end{pmatrix} \in \mathbb{R}^{M \times N},$$

where $\boldsymbol{\Delta}(t) \in \mathbb{R}^N$ denotes the truncation error arising from the omission of higher-order terms. The effect of $\boldsymbol{\Delta}(t)$ can be neglected with a suitable choice of the order N. Furthermore, by an appropriate selection of the number and placement of sensors, the following compatibility condition can be satisfied:

$$\operatorname{rank} \boldsymbol{\Phi} = N,$$

such that one can obtain the approximation of $\boldsymbol{a}(t)$ in the *fault-free* scenario as

$$\bar{\boldsymbol{a}}(t) \triangleq \boldsymbol{\Phi}^+ \boldsymbol{p}(t) = [\bar{a}_1(t)\ \bar{a}_2(t)\ \cdots\ \bar{a}_N(t)]^{\mathrm{T}} \in \mathbb{R}^N, \tag{11.10}$$

where $\boldsymbol{\Phi}^+ = (\boldsymbol{\Phi}\boldsymbol{\Phi}^{\mathrm{T}})^{-1}\boldsymbol{\Phi}^{\mathrm{T}}$ denotes the Moore-Penrose inverse of matrix $\boldsymbol{\Phi}$.

$$\boldsymbol{\omega}_i(t) = \begin{pmatrix} \bar{a}_i(t) & \bar{a}_i(t+1) & \cdots & \bar{a}_i(t+l-1) & \bar{q}_1(t+l-1) & \cdots & \bar{q}_n(t+l-1) \\ \bar{a}_i(t+1) & \bar{a}_i(t+2) & \cdots & \bar{a}_i(t+l) & \bar{q}_1(t+l) & \cdots & \bar{q}_n(t+l) \\ \vdots & \vdots & \ddots & \vdots & \vdots & \ddots & \vdots \\ \bar{a}_i(t+m-1) & \bar{a}_i(t+m) & \cdots & \bar{a}_i(t+m+l-2) & \bar{q}_1(t+m+l-2) & \cdots & \bar{q}_n(t+m+l-2) \end{pmatrix}$$
$$\in \mathbb{R}^{m \times (n+l)}, i = 1, \cdots, N, \tag{11.11}$$

$$\begin{aligned} \boldsymbol{\xi}_i(t) = [&\bar{a}_i(t+m)\ \bar{a}_i(t+m+1)\ \cdots \\ &\bar{a}_i(t+m+l-1)\ \bar{q}_1(t+m+l-1)\ \cdots\ \bar{q}_n(t+m+l-1)] \in \mathbb{R}^{1 \times (n+l)}. \end{aligned} \tag{11.12}$$

After obtaining the temporal sequence data $\bar{\boldsymbol{a}}(t)$, the measured exogenous input data $\bar{\boldsymbol{q}}(t)$ are used for model training. Specifically, the current and previous temporal sequence data are employed to produce the prediction $\hat{\boldsymbol{a}}(t)$ for subsequent fault diagnosis.

First, a receding-horizon learning scheme is adopted for the on-line learning implementation. The input matrix for the i-th temporal variable $\bar{a}_i$ in the *training window* t is defined as $\boldsymbol{\omega}_i(t)$ in (11.11), where m denotes the length of the training window and l is the auto-regression (AR) order of the temporal sequences. In this way, the effect of previous temporal data is incorporated into model training, which is consistent with the model equation in (11.7).

A schematic figure of the proposed receding horizon learning scheme is shown in Fig. 11.2. The method trains from "window" to "window". Specifically, each

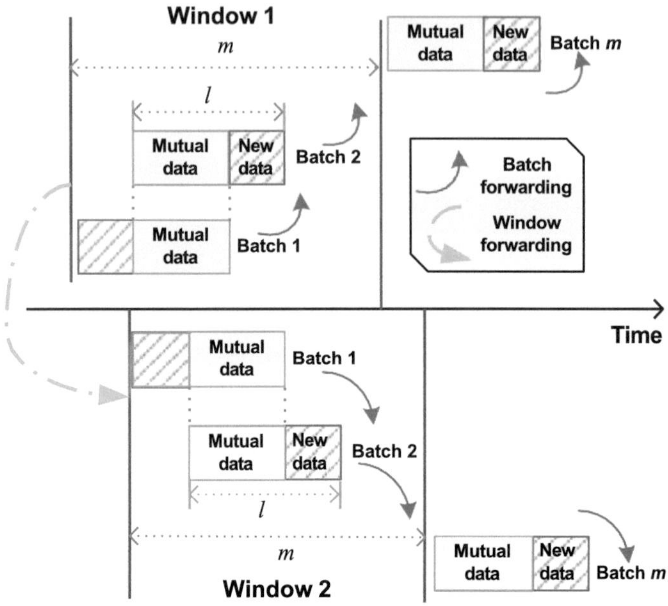

Fig. 11.2 Receding horizon learning scheme

window contains m data, on which a learning model is trained. Within each window, training proceeds "batch" by "batch". For example, in the first batch of window t, the input to the training model is $[\bar{a}_i(t)\ \bar{a}_i(t+1)\ \cdots\ \bar{a}_i(t+l-1)\ \bar{q}_1(t+l-1)\ \cdots\ \bar{q}_n(t+l-1)] \in \mathbb{R}^{1\times(n+l)}$, which means that the previous data with time lag l and the measured input at time $t+l-1$ are used to predict the output at time $t+l$. In the second batch, the input is $[\bar{a}_i(t+1)\ \bar{a}_i(t+2)\ \cdots\ \bar{a}_i(t+l)\ \bar{q}_1(t+l)\ \cdots\ \bar{q}_n(t+l)] \in \mathbb{R}^{1\times(n+l)}$, which shares some data with batch 1 and also adds new data. The procedure then continues to batch 3 and onward. After batch m is trained, the scheme shifts from window t to window $t+1$ and again trains from batch 1 to batch m.

In this chapter, Lasso (least absolute shrinkage and selection operator) is selected for the model training, the details of Lasso can be referred to [34]. The objective function for parameter estimation of batch $k = 1, 2, \cdots, m$ in window t is

$$\hat{\boldsymbol{\theta}}_{tk} = \arg\min_{\boldsymbol{\theta}_{tk}} \frac{1}{2} \left|\bar{a}_i(t+l+k-1) - \boldsymbol{\varphi}^{\mathrm{T}} \left(\boldsymbol{\omega}_{ik}(t)\right) \boldsymbol{\theta}_{tk}\right|_2^2 + \lambda \left|\boldsymbol{\theta}_{tk}\right|_1, \qquad (11.13)$$

where $\hat{\boldsymbol{\theta}}_{tk} \in \mathbb{R}^{n+l}$ are the estimated parameters, $\boldsymbol{\omega}_{ik}(t) \in \mathbb{R}^{1\times(n+l)}$ denotes the k-th row of $\boldsymbol{\omega}_i(t)$. $\boldsymbol{\varphi}$ and λ are the basis functions and regularization coefficient of Lasso, respectively.

Upon completion of model training in window t, that is, after the parameter estimate $\hat{\boldsymbol{\theta}}_{tk}$ has been obtained, the one-step-ahead validation output is computed using the most recent estimate $\hat{\boldsymbol{\theta}}_{tm}$ as follows:

$$\hat{a}_i(t+m+l) = \boldsymbol{\varphi}^{\mathrm{T}}\left(\boldsymbol{\xi}_i(t)\right)\hat{\boldsymbol{\theta}}_{tm}, \tag{11.14}$$

where $\boldsymbol{\xi}_i(t)$ is defined in (11.12).

11.3.3 Residual Generation and Evaluation Based on Modeling Error

Residual signal generation is a fundamental step in model-based fault diagnosis [11]. In most cases, the output estimation error between the sensor measurements and the observer output is used to construct the residual signal. In the same paradigm, the systems modeling error, in particular the validation error, is adopted for residual signal generation. To this end, the system output is first synthesized from the low-dimensional learning model.

Motivated by the transformed model output equation in (11.9), the following estimated output is obtained:

$$\hat{\boldsymbol{p}}(t+m+l) \triangleq \boldsymbol{\Phi}\hat{\boldsymbol{a}}(t+m+l). \tag{11.15}$$

Then we can construct the residual signal for window t as

$$\boldsymbol{r}(t+m+l) \triangleq \boldsymbol{p}(t+m+l) - \hat{\boldsymbol{p}}(t+m+l) = \boldsymbol{p}(t+m+l) - \boldsymbol{\Phi}\hat{\boldsymbol{a}}(t+m+l), \tag{11.16}$$

where $\hat{\boldsymbol{a}}(t+m+l) \triangleq [\hat{a}_1(t+m+l) \ \cdots \ \hat{a}_N(t+m+l)] \in \mathbb{R}^N$ denotes the one-step ahead validation output for the temporal sequences.

By a simple transformation $\tau \triangleq t+m+l$, we have that

$$\boldsymbol{r}(\tau) = \boldsymbol{p}(\tau) - \boldsymbol{\Phi}\hat{\boldsymbol{a}}(\tau). \tag{11.17}$$

Combining (11.9) with (11.17), we have that

$$\boldsymbol{r}(\tau) = \boldsymbol{\Phi}(\boldsymbol{a}(\tau) - \hat{\boldsymbol{a}}(\tau)) + \boldsymbol{\Delta}(\tau) + \boldsymbol{f}_s(\tau). \tag{11.18}$$

Since the influence of truncation error is negligible, the sensor fault $\boldsymbol{f}_s(\tau)$ or the input measured fault at time step τ perturbs $\boldsymbol{r}(\tau)$ and is therefore potentially observable. Moreover, because a process fault $\boldsymbol{f}_p(\tau)$ alters the state variable $u(x,\tau)$ and, consequently, $\boldsymbol{a}(\tau)$, the quantity $\boldsymbol{r}(\tau)$ deviates from its normal range and can be detected. Hence, the proposed residual signal $\boldsymbol{r}(\tau)$ can be employed for fault detection.

Algorithm 11.1: Fault detection framework

Input: Measured I/O data $(\bar{\boldsymbol{q}}(t), \boldsymbol{p}(t))$
Output: Faulty time set t_f
/****Initialization****/
  Select m (training window length), l (AR order) and α (significance level);
/****Systems modeling****/
  Calculate the spatial basis functions $\{\phi_i(x, y, z)\}_{i=1}^{N}$ and matrix $\boldsymbol{\Phi} \leftarrow$ (11.5) and (11.9);
for *temporal variable* $i \in [1, N]$ **do**
  for *window* $t \in [t_{\min}, t_{\max}]$ **do**
    for *batch* $k \in [1, m]$ **do**
      Model training: $\hat{\boldsymbol{\theta}}_{tk} \leftarrow$ (11.13);
  Calculate one step ahead validation output $\hat{a}_i(t + m + l)$ and $\hat{\boldsymbol{p}}(t + m + l) \leftarrow$ (11.14) and (11.15) ;
  Calculate residual signal $\boldsymbol{r}(\tau)$, detection index $J(\tau)$ and threshold $J_{\text{th}} \leftarrow$ (11.18) and (11.19);
/****Fault detection****/
for $\tau \in [\tau_{\min}, \tau_{\max}]$ **do**
  Residual evaluation: $J(\tau) \leftarrow$ (11.19);
  if $J(\tau) > J_{\text{th}}$ **then**
    $t \in t_f$
    if $J(\tau) > \sup_{\text{fault-free}} |\boldsymbol{r}(\tau)|_{\text{peak}}^2$ **then**
      Switch to off-line mode
    else
      Constant on-line learning
  else
    $t \notin t_f$

In addition, the following residual evaluation criterion for fault detection is defined as:

$$J(\tau) \triangleq |\boldsymbol{r}(\tau)|_2^2, \tag{11.19}$$

where $J(\tau)$ denotes the residual evaluation statistic and J_{th} is the corresponding threshold.

For threshold computation, the kernel density estimation technique [35] is applied to approximate the PDF of $J(\tau)$ and to determine the threshold J_{th} at a prescribed significance level α (typically 0.05 or 0.01).

After the detection threshold J_{th} is obtained, the following decision logic is used for fault detection and mode switching:

$$\begin{cases} J(\tau) > J_{\text{th}} \Longrightarrow \text{faulty at time } \tau \\ \text{Otherwise} \Longrightarrow \text{fault-free at time } \tau, \end{cases} \tag{11.20}$$

$$\begin{cases} J(\tau) > \sup\limits_{\text{fault-free}} |\boldsymbol{r}(\tau)|_{\text{peak}}^2 \Longrightarrow \text{Switch to off-line mode} \\ \text{Otherwise} \Longrightarrow \text{Constant on-line learning}, \end{cases} \tag{11.21}$$

where $|\cdot|_{\text{peak}}$ denotes the peak norm [11, 36].

A summary of the proposed systems modeling and fault detection framework by pseudo code is shown in Algorithm 11.1.

Remark 11.3 *Since the proposed learning model is updated in an on-line manner, both process and sensor faults may be "learned" by the model, thereby compromising the effectiveness of the fault detection scheme. To mitigate this issue, the above switching mechanism is introduced: once a fault is initially detected and certain additional conditions are satisfied, the on-line learning is switched to an off-line mode in which the model is no longer updated. In this way, the detection outcomes are preserved until further actions are undertaken.*

11.4 Experiment Studies on Li-Ion Batteries

For validation of the proposed fault detection method, a thermal model of a battery cell with negligible thickness [4] is adopted. The thermal model is governed by the following 2-D PDE:

$$\rho C_p \frac{\partial T}{\partial t} = \frac{\partial}{\partial x}\left(\lambda_x \frac{\partial T}{\partial x}\right) + \frac{\partial}{\partial y}\left(\lambda_y \frac{\partial T}{\partial y}\right) + Q(x, y, t), \tag{11.22}$$

subject to the boundary conditions:

$$\begin{aligned} \left.\frac{\partial T}{\partial x}\right|_{x=0} &= \frac{h}{\lambda_x}(T - T_\infty), & \left.\frac{\partial T}{\partial x}\right|_{x=\bar{x}} &= \frac{h}{\lambda_x}(T_\infty - T), \\ \left.\frac{\partial T}{\partial y}\right|_{y=0} &= \frac{h}{\lambda_y}(T - T_\infty), & \left.\frac{\partial T}{\partial y}\right|_{y=\bar{y}} &= \frac{h}{\lambda_y}(T_\infty - T), \end{aligned} \tag{11.23}$$

where the state variable $T(x, y, t)$ represents the temperature distribution on the surface of the battery cell at time $t \geq 0$, with $x \in [0, \bar{x}]$ and $y \in [0, \bar{y}]$ denoting the spatial coordinates. ρ, C_p, and h denote the density, heat capacity, and convective coefficient of the battery cell, respectively. λ_x and λ_y are the thermal conductivities along the x- and y-directions, respectively. T_∞ denotes the ambient temperature. $Q(x, y, t)$ is the heat generation term, which is treated as an unknown nonlinear function of the current $I(t)$, the terminal voltage $V(t)$, and their shape functions. Hence, the system input is defined as $\boldsymbol{q}(t) \triangleq [I(t)\ V(t)]$. Moreover, twenty thermocouples are affixed to the surface of the battery cell for temperature measurement, as presented in Fig. 11.3.

To evaluate the performance of the proposed fault detection framework, a pouch type Li-ion battery with LiFePO4/graphite chemistry is employed. The battery

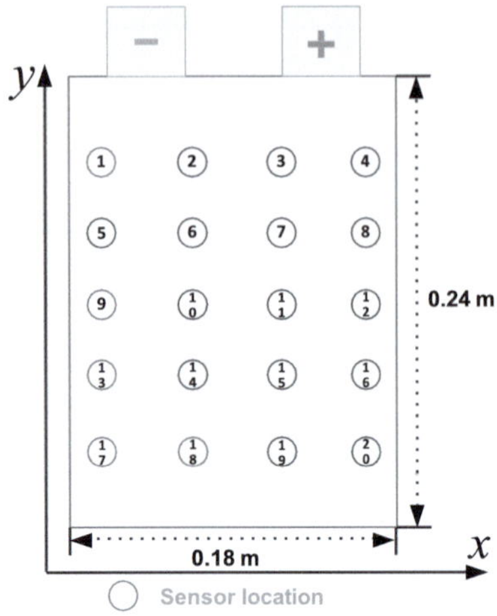

Fig. 11.3 Schematic plot of sensor placement

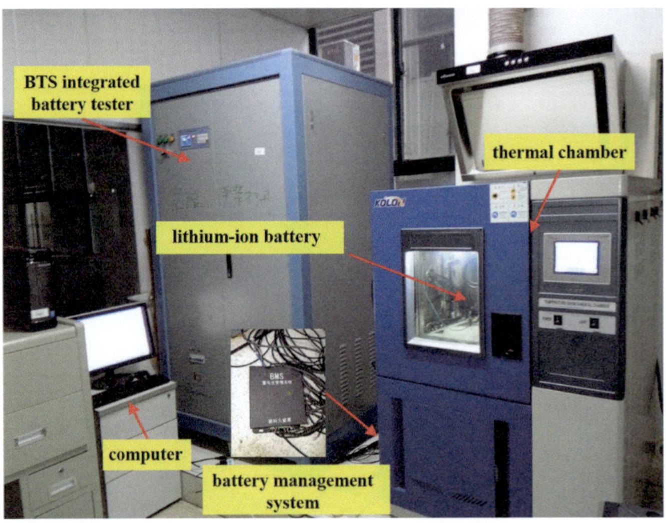

Fig. 11.4 Experiment setup

dimensions (length, width, and thickness) are 0.24 m, 0.18 m, and 7.83E-3 m, respectively. The nominal capacity, nominal voltage, charge cut-off voltage, and discharge cut-off voltage are 60 Ah, 3.2 V, 3.65 V, and 2 V, respectively. Since the thickness is small relative to the length and width, it is neglected in the physical model (11.22). The experiment platform is shown in Fig. 11.4 and comprises a BTS integrated battery tester, a thermal chamber, a BMS, and a host computer. The BTS integrated battery tester generates the charge/discharge current for experiment validation, while the thermal chamber maintains a constant ambient temperature of 25°C. The BMS

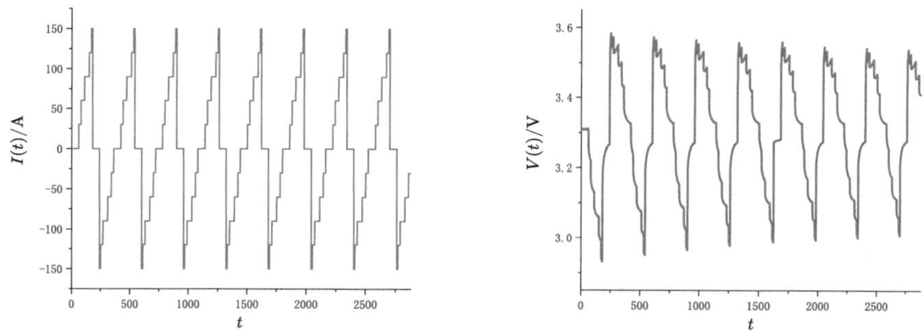

(a) Input current $I(t)$ (b) Terminal voltage $V(t)$

Fig. 11.5 Input current and measured terminal voltage for systems modeling and fault detection

and host computer are responsible for temperature data collection and implementation of the fault detection algorithm, respectively. Specifically, temperature signals are collected from twenty thermocouples affixed to the cell surface with a resolution of 0.01°C; the signals are first transmitted to the BMS and subsequently to the host computer.

The input current used for experiment validation and the measured terminal voltage are shown in Fig. 11.5a, b, respectively. The detection threshold is computed at a significance level of $\alpha = 0.05$. For illustration, two types of sensor faults are considered, following the fault settings in [37].

Case 1: The temperature sensor bias fault of 1°C from $t = 2500$ sec for sensors No. 10-No. 15 in Fig. 11.3.

Case 2: The voltage sensor bias fault of 3V from $t = 2500$ sec.

With the training window length $m = 70$ and AR order $l = 5$, the on-line systems modeling results for Case 1 are shown in Fig. 11.6a, d. The validation error in the fault-free scenario is acceptable. However, once the fault occurs at $\tau = 2500$ sec, the validation error becomes relatively large for the corresponding faulty sensors.

Moreover, to further compare the proposed switching scheme with the constant on-line learning scheme, the systems modeling results at $\tau = 2502$ sec under both temperature and voltage sensor fault are presented in Fig. 11.7a, b, c, d. It is observed that shortly after the fault occurs at $\tau = 2500$ sec, the validation error under the constant on-line learning scheme decreases rapidly, whereas that under the proposed switching scheme remains comparatively larger. This behavior arises because, within the proposed switching scheme, the learning model ceases updating once certain conditions are satisfied; consequently, discrepancies persist between the previously learned model and the incoming measurements.

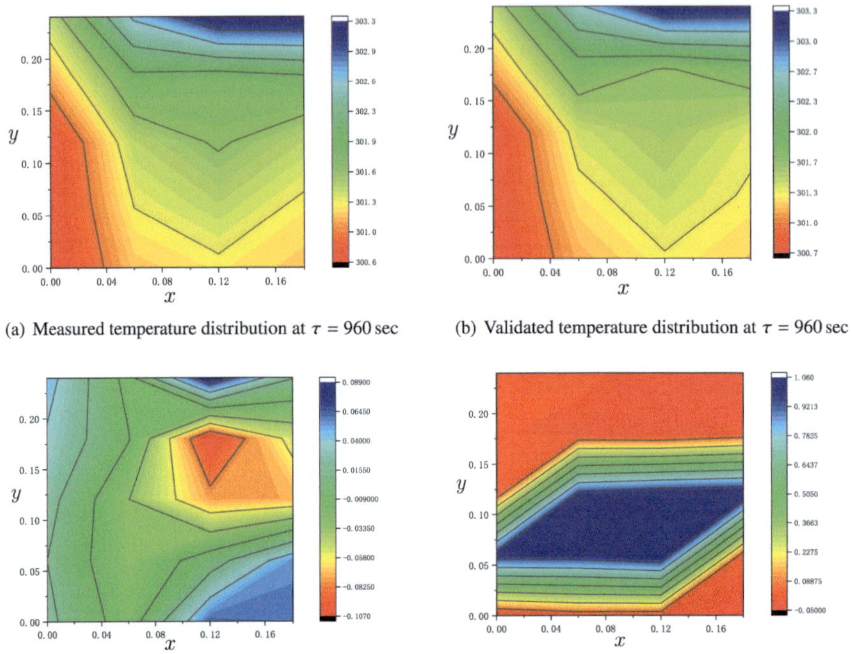

(a) Measured temperature distribution at τ = 960 sec

(b) Validated temperature distribution at τ = 960 sec

(c) Validation error at τ = 960 sec

(d) Validation error at τ = 2500 sec

Fig. 11.6 On-line systems modeling results under Case 1

The fault detection results under the decision logic (11.20) are shown in Fig. 11.8a–d, demonstrating that the proposed method effectively identifies both faults using only the measured I/O data, without requiring explicit model knowledge. For quantitative evaluation, the FDR and FAR are defined as follows:

$$\begin{aligned} \text{FDR} &= \text{prob}\left(J > J_{\text{th}} \mid f \neq 0\right), \\ \text{FAR} &= \text{prob}\left(J > J_{\text{th}} \mid f = 0\right). \end{aligned} \tag{11.24}$$

Table 11.1 summarizes the indices for both cases, from which it follows that the proposed method attains strong detection performance and surpasses the constant on-line learning scheme.

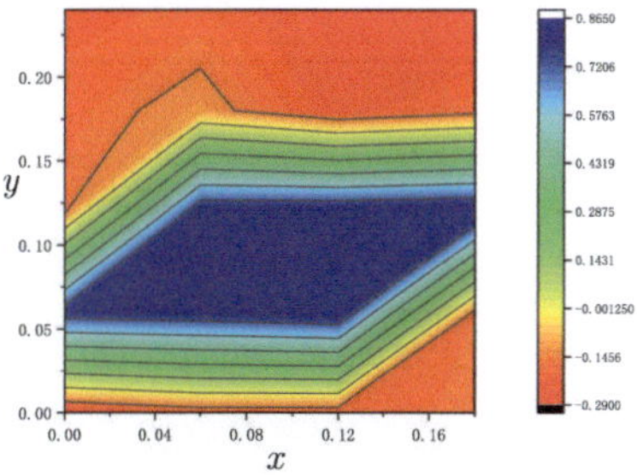

(a) Validation error with the proposed switching scheme (Case 1)

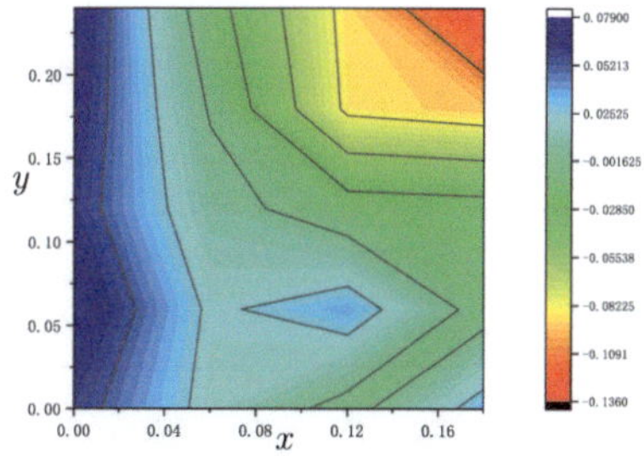

(b) Validation error with the constant on-line learning scheme (Case 1)

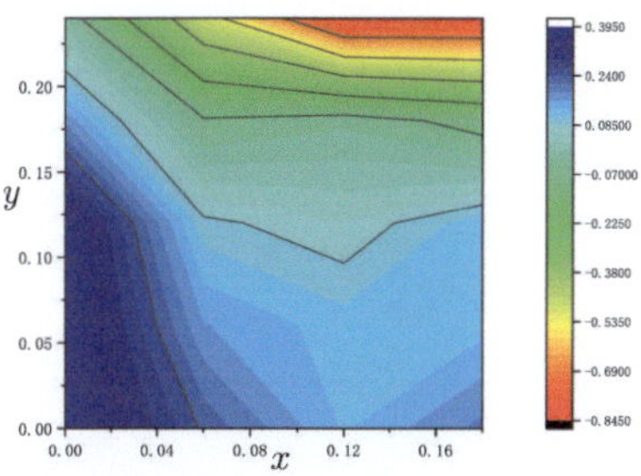

(c) Validation error with the proposed switching scheme (Case 2)

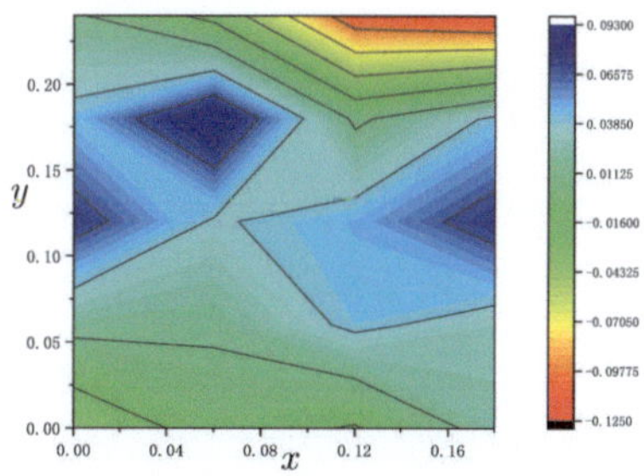

(d) Validation error with the constant on-line learning scheme (Case 2)

Fig. 11.7 Systems modeling results based on the proposed switching scheme and constant on-line learning scheme under faults. ($\tau = 2502$ sec)

11.5 Conclusion

In this chapter, a novel systems modeling and fault detection framework for a class of high-dimensional DPSs is introduced by recasting modeling error as residuals generation for fault detection. Furthermore, the fault detection module modulates the systems modeling by switching the learning mode from on-line to off-line, thereby preserving the detection outcomes. Experimental results on the thermal process of a battery cell corroborate the effectiveness of the proposed method. This chapter focuses exclusively on the fault detection problem for high-dimensional DPSs.

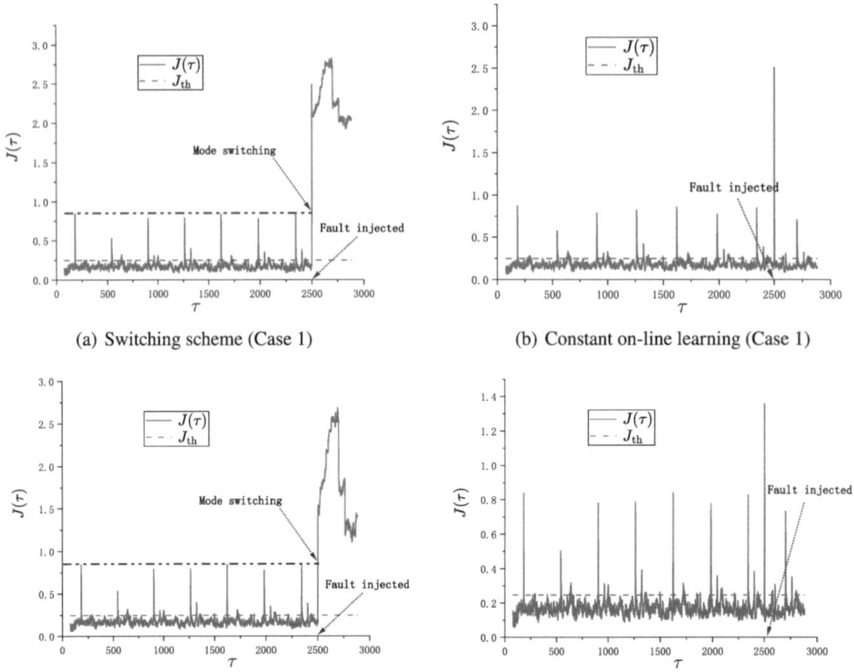

(a) Switching scheme (Case 1)
(b) Constant on-line learning (Case 1)
(c) Switching scheme (Case 2)
(d) Constant on-line learning (Case 2)

Fig. 11.8 Fault detection results based on the proposed switching scheme and constant on-line learning scheme

Table 11.1 Fault detection indexes under both cases

	Switching	Constant on-line
Case 1: FDR(%)	**100**	8.68
Case 1: FAR(%)	**4.29**	7.22
Case 2: FDR(%)	**100**	7.63
Case 2: FAR(%)	4.41	**4.37**

References

1. Liu Z, Liu J, He W (2017) Modeling and vibration control of a flexible aerial refueling hose with variable lengths and input constraint. Automatica 77:302–310
2. Liu Z, He X, Zhao Z et al (2020) Vibration control for spatial aerial refueling hoses with bounded actuators. IEEE Trans Ind Electron 68(5):4209–4217
3. Zhao X, Liu Z, Zhang S et al (2022) Time-varying trajectory tracking boundary control of a flexible rotation beam based on servomechanism. IEEE Trans Ind Electron 69(9):9185–9195
4. Wang B-C, Li H-X, Yang H-D (2019) Spatial correlation-based incremental learning for spatiotemporal modeling of battery thermal process. IEEE Trans Ind Electron 67(4):2885–2893

5. Xu K, Fan B, Yang H et al (2022) Locally weighted principal component analysis-based multimode modeling for complex distributed parameter systems. IEEE Trans Cybernetics 52(10):10504–10514
6. Christofides PD (2012) Nonlinear and robust control of PDE systems: Methods and applications to transport-reaction processes. Springer Science+Business Media, Berlin
7. Feng Y, Wang Y, Wang B-C et al (2022) Spatial decomposition-based fault detection framework for parabolic-distributed parameter processes. IEEE Trans Cybernetics 52(8):7319–7327
8. Doughty DH, Roth EP (2012) A general discussion of Li ion battery safety. Electrochem Soc Interface 21(2):37–44
9. Dey S, Perez HE, Moura SJ (2017) Model-based battery thermal fault diagnostics: Algorithms, analysis, and experiments. IEEE Trans Control Syst Technol 27(2):576–587
10. Zhao Z, Liu Z, He W et al (2021) Boundary adaptive fault-tolerant control for a flexible Timoshenko arm with backlash-like hysteresis. Automatica 130:109690
11. Ding SX (2013) Model-based fault diagnosis techniques: design schemes, algorithms, and tools, 2nd edn. Springer-Verlag, Berlin
12. Ding SX (2014) Data-driven design of fault diagnosis and fault-tolerant control systems. Springer, Berlin
13. Ghantasala S, El-Farra NH (2009) Robust actuator fault isolation and management in constrained uncertain parabolic PDE systems. Automatica 45(10):2368–2373
14. Feng Y, Wang Y, Wang J-W et al (2021) Spatio-temporal fault localization for nonlinear spatially distributed processes: A spatial mapping filter-based framework. Int J Robust Nonlinear Control 31(14):6953–6971
15. Harkort C (2014) Early-lumping based controller synthesis for linear infinite-dimensional systems. FAU Forschungen, Reihe B, Medizin, Naturwissenschaft, Technik 1. FAU University Press, Erlangen
16. Krstic M, Smyshlyaev A (2008) Boundary control of PDEs: A course on backstepping designs. SIAM, Philadelphia
17. Cai J, Ferdowsi H, Sarangapani J (2016) Model-based fault detection, estimation, and prediction for a class of linear distributed parameter systems. Automatica 66:122–131
18. Dey S, Perez HE, Moura SJ (2019) Robust fault detection of a class of uncertain linear parabolic PDEs. Automatica 107:502–510
19. Feng Y, Wang Y, Wang J-W et al (2022) Backstepping-based distributed abnormality localization for linear parabolic distributed parameter systems. Automatica 135:109930
20. Xu X, Yuan Y, Wang M et al (2019) Fault detection and estimation for a class of PIDE systems based on boundary observers. Int J Robust Nonlinear Control 29(17):5867–5885
21. Shinbrot M (1954) On the analysis of linear and nonlinear dynamical systems from transient-response data. National Advisory Committee for Aeronautics, Washington
22. Fischer F, Deutscher J (2020) Flatness-based algebraic fault diagnosis for distributed-parameter systems. Automatica 117:108987
23. Fischer F, Deutscher J (2022) Fault diagnosis for linear heterodirectional hyperbolic ODE-PDE systems using backstepping-based trajectory planning. Automatica 135:109952
24. Yin S, Ding SX, Xie X et al (2014) A review on basic data-driven approaches for industrial process monitoring. IEEE Trans Ind Electron 61(11):6418–6428
25. Feng Y, Li H-X (2018) Detection and spatial identification of fault for parabolic distributed parameter systems. IEEE Trans Ind Electron 66(9):7300–7309
26. Wang J-W, Liu Y-Q, Sun C-Y (2018) Pointwise exponential stabilization of a linear parabolic PDE system using non-collocated pointwise observation. Automatica 93:197–210
27. Kim Y, Mohan S, Siegel JB et al (2014) The estimation of temperature distribution in cylindrical battery cells under unknown cooling conditions. IEEE Trans Control Syst Technol 22(6):2277–2286
28. Sattarzadeh S, Roy T, Dey S (2021) Real-time estimation of 2-D temperature distribution in lithium-ion pouch cells. IEEE Trans Transport Electrification 7(4):2249–2259
29. Ghalkhani M, Bahiraei F, Nazri GA et al (2017) Electrochemical-thermal model of pouch-type lithium-ion batteries. Electrochim Acta 247:569–587

30. Ren Y, Zhu P, Zhao Z et al (2022) Adaptive fault-tolerant boundary control for a flexible string with unknown dead zone and actuator fault. IEEE Trans Cybern 52(7):7084–7093
31. Demetriou MA, Gatsonis NA, Court JR (2013) Coupled controls-computational fluids approach for the estimation of the concentration from a moving gaseous source in a 2-D domain with a Lyapunov-guided sensing aerial vehicle. IEEE Trans Control Syst Technol 22(3):853–867
32. Ray WH (1981) Advanced process control. McGraw-Hill, Columbus
33. Qi C, Li H-X (2008) A Karhunen-Loève decomposition-based Wiener modeling approach for nonlinear distributed parameter processes. Ind Eng Chem Res 47(12):4184–4192
34. Tibshirani R (1996) Regression shrinkage and selection via the lasso. J R Stat Soc B 58(1):267–288
35. Silverman BW (2018) Density estimation for statistics and data analysis. Routledge, Abingdon
36. Feng Y, Li H-X, Yang H-D (2021) Abnormal source identification for parabolic distributed parameter systems. IEEE Trans Syst Man Cybern Syst 51(9):5698–5707
37. Dey S, Mohon S, Pisu P et al (2016) Sensor fault detection, isolation, and estimation in lithium-ion batteries. IEEE Trans Control Syst Technol 24(6):2141–2149

Chapter 12
Dynamic Spatial-Independent-Component-Analysis-Based Abnormality Localization for DPSs

Abstract This chapter presents a data-driven method for abnormality localization in distributed parameter systems. The cross-correlation order in the spatial domain is first obtained using a cumulants-based identification technique. A spatial augmented matrix is then constructed from the spatio-temporal distribution data, and a dynamic spatial independent component analysis is introduced for independent decomposition. The dominant spatial independent components are extracted, and spatial residuals are generated as reference statistics. Using kernel density estimation, confidence intervals of these statistics under normal (abnormality-free) conditions are established as spatial references. These two references enable reliable spatial localization of abnormalities. Unlike model-based approaches that depend on explicit process models, the proposed method is model-free and relies solely on recorded process data. Experimental studies on two representative DPSs confirm the effectiveness of the approach.

12.1 Introduction

DPSs are commonly encountered in industrial processes, such as transport-reaction, thermal, and fluid processes. A distinctive property of DPSs is their S-T dynamics [1], meaning that the system's inputs, outputs, and even parameters may vary across both space and time. In recent years, extensive research has been devoted to the modeling [2–6] and control [7] of DPSs.

With the increasing demand for high-quality production and cost-effective operation in complex industrial manufacturing systems, system safety and reliability have become increasingly critical. Consequently, FD [8, 9] of such processes is of great importance. Existing fault diagnosis approaches can generally be divided into model-based [8] and data-driven [9] categories. Model-based methods rely on system modeling within modern control theory, whereas data-driven methods exploit large amounts of process data together with statistical algorithms. Among data-driven approaches, the most widely used include PCA [10, 11], PLS [12, 13], and their variants [14, 15]. These methods are regarded as powerful tools in SPC owing to their efficiency and relatively low design complexity compared with model-based

© The Author(s), under exclusive license to Springer Nature Singapore Pte Ltd. 2026
Y. Feng et al., *Spatio-Temporal Abnormality Diagnosis for Industrial Distributed Parameter Systems*, https://doi.org/10.1007/978-981-95-3750-1_12

approaches. For instance, a sparse exponential discriminant analysis algorithm was proposed in [15] to separate multiple faults and identify the key variables responsible for fault differences. ICA [16], originally developed for blind source separation in signal processing, has also been applied to fault diagnosis [17, 18], showing advantages over PCA-based methods. Specifically, ICA exploits higher-order statistics, while PCA is limited to second-order statistics, and ICA can handle non-Gaussian distributions that frequently occur in industrial systems. Furthermore, as an extension of ICA, spatial independent component analysis (SICA) has been applied in functional magnetic resonance imaging (fMRI) [19, 20] to extract spatially independent components.

In recent years, increasing attention has been devoted to fault and anomaly detection in DPSs [21–28]. For instance, [22] introduced a framework that integrates fault detection and fault-tolerant control for spatially distributed processes modeled by quasi-linear parabolic PDE with control constraints and actuator faults. In [24], a modulation function approach was developed for fault detection in parabolic systems with distributed inputs and outputs. Meanwhile, [27] constructed a fault detection observer directly from the PDE representation of linear DPSs, thereby avoiding the transformation into a set of ODEs. Although these works have demonstrated novelty, the spatial localization of faults and abnormalities has not been thoroughly addressed, despite its importance in anomaly elimination given the S-T dynamics of DPSs. More recently, [29] proposed a spatial identification method for faults in parabolic DPSs. However, its localization accuracy strongly depends on the reliability of time-domain detection, and there remains significant potential for enhancing spatial localization performance.

Building on the above insights, we aim to design a data-driven method for abnormality localization in DPSs. The proposed approach consists of two main steps: offline construction of spatial references and online localization of abnormalities.

The main advantages of this approach are threefold:

- The cross-correlation in the space domain is considered.
- Abnormality localization can be directly achieved without a prior step of abnormality detection in the time domain.
- Explicit system model or prior information of the process is not needed.

The structure of this chapter is arranged as follows: Sect. 12.4 presents the overall framework of the dynamic SICA method. Section 12.5 reports experimental results to demonstrate its effectiveness. Section 12.6 concludes the chapter.

12.2 Problem Formulation

In Fig. 12.1, a typical structure of DPSs is illustrated. The temporal system input is represented as $\boldsymbol{u}(t) \in \mathbb{R}^{n_u}$, while $y(z, t)$ and $d(z_f, \boldsymbol{t}_f)$ denote the spatio-temporal system output and the abnormality occurring in the system, respectively. Specifically,

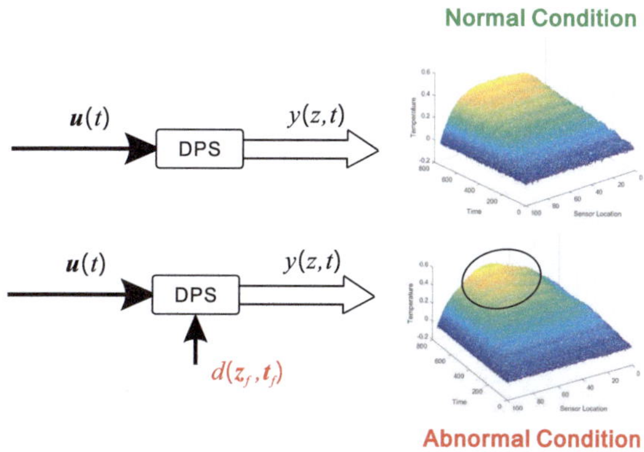

Fig. 12.1 Problem framework

$t_s \in \boldsymbol{t}_f$ indicates the occurrence of an abnormality at time t_s, and $z_s \in \boldsymbol{z}_f$ implies that the abnormality affects the spatial location z_s.

The abnormality localization problem considered in this chapter can be summarized as follows:

- Determine the abnormality location $\boldsymbol{z}_f$ in DPSs based on the sampled output data $\{y(z_s, t_s)\}_{z_s=1,t_s=1}^{N,L}$, without relying on any prior system information. Here, N and L represent the number of samples in the spatial and temporal domains, respectively.

12.3 ICA Fundamentals: From TICA to SICA

This chapter introduces a data-driven method, namely dynamic SICA, which enables accurate localization of abnormalities. The proposed approach is inspired by SICA [19], a technique widely applied in fMRI studies.

To provide a clear comparison between SICA and temporal independent component analysis (TICA), a schematic illustration is shown in Fig. 12.2. As depicted, TICA extracts temporally independent components ($\boldsymbol{IC_T}$), whose dimensionality in the time domain matches that of the data $\boldsymbol{Y}$. In contrast, SICA aims to obtain spatially independent components ($\boldsymbol{IC_S}$), with their spatial dimensionality consistent with the data $\boldsymbol{Y}^{\mathrm{T}}$.

The overall framework of the proposed method is illustrated in Fig. 12.3, consisting of two main phases: offline training and online localization. During offline training, spatial references are generated using the dynamic SICA together with KDE. In the subsequent online phase, these references serve as thresholds for abnormality localization.

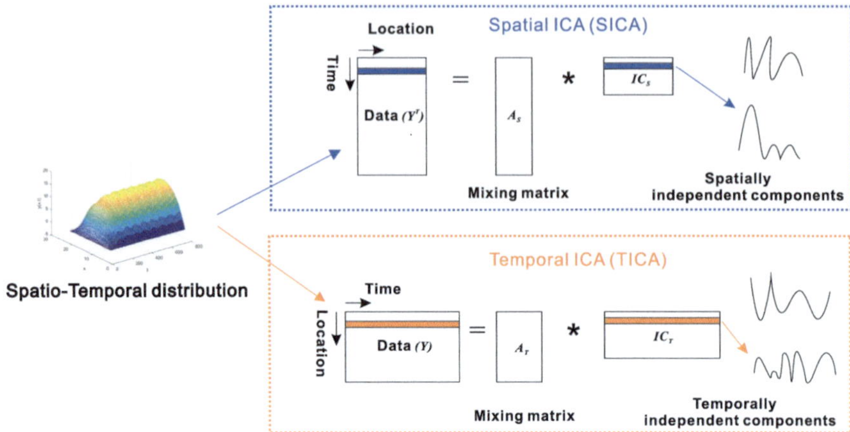

Fig. 12.2 Schematic graph of SICA and TICA

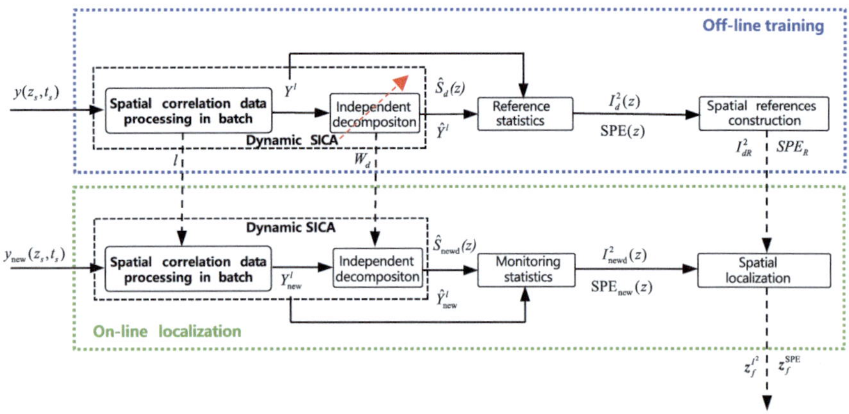

Fig. 12.3 Dynamic SICA based abnormality localization

12.4 DSICA-Based Abnormality Localization

12.4.1 Dynamic SICA Formulation

12.4.1.1 Spatial Correlation Data Formulation

The S-T nature of DPSs implies that the output measurements $y(z_s, t_s)_{z_s=1,t_s=1}^{N,L}$ depend on both time and space, resulting in potential temporal auto-correlation and spatial cross-correlation. The sampled output of the DPSs can be expressed as follows:

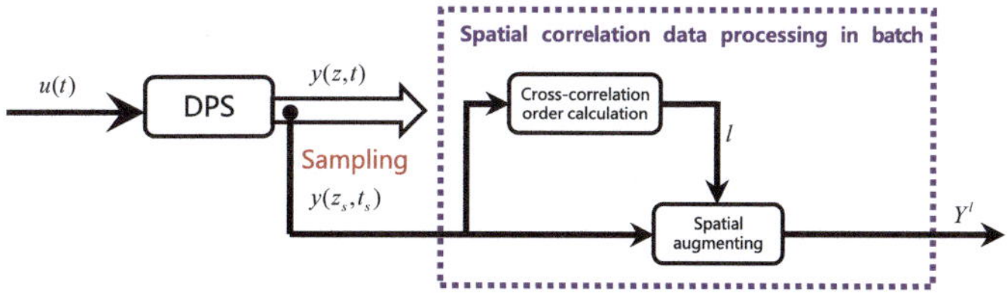

Fig. 12.4 Spatial correlation data processing in batch

$$\boldsymbol{Y}^{\mathrm{T}} = \begin{pmatrix} y(1,1) & y(2,1) & \cdots & y(N,1) \\ y(1,2) & y(2,2) & \cdots & y(N,2) \\ \vdots & \vdots & \vdots & \vdots \\ y(1,L) & y(2,L) & \cdots & y(N,L) \end{pmatrix} \in \mathbb{R}^{L\times N}. \tag{12.1}$$

Here, a "sample" refers to the system output measured at a specific spatial location $z_s(z_s = 1, 2, \cdots, N)$ and at a specific time $t_s(t_s = 1, 2, \cdots, L)$. All such samples together form a batch of S-T data, where "batch" represents the data spanning from $t_s = 1$ to $t_s = L$, remaining consistent across both offline training and online localization stages.

As shown in Eq. 12.1, the k-th column of $\boldsymbol{Y}^{\mathrm{T}}$ represents the system output at the k-th spatial location across the time points from $t_s = 1$ to $t_s = L$. Denoting $\boldsymbol{y}_k \in \mathbb{R}^{L\times 1}$ as the k-th column of $\boldsymbol{Y}^{\mathrm{T}}$ $(k = 1, 2, \cdots, N)$, $\boldsymbol{Y}^{\mathrm{T}}$ can then be expressed as $\boldsymbol{Y}^{\mathrm{T}} = [\boldsymbol{y}_1, \boldsymbol{y}_2, \cdots, \boldsymbol{y}_N]$.

For abnormality localization, it is essential to account for spatial cross-correlation. Accordingly, the first step of dynamic SICA involves determining the cross-correlation order of the output data in the spatial domain. Here, the cumulants-based identification approach [30] is employed to estimate the cross-correlation order l. Inspired by dynamic principal component analysis (DPCA) [31] and dynamic independent component analysis (DICA) [18], which incorporate dynamic behavior in modeling, a spatially augmented matrix is constructed as follows:

$$\boldsymbol{Y}^{l} = \begin{pmatrix} \boldsymbol{y}_m & \boldsymbol{y}_{m+1} & \cdots & \boldsymbol{y}_{m+N_s} \\ \boldsymbol{y}_{m-1} & \boldsymbol{y}_m & \cdots & \boldsymbol{y}_{m+N_s-1} \\ \vdots & \vdots & \vdots & \vdots \\ \boldsymbol{y}_{m-l} & \boldsymbol{y}_{m+1-l} & \cdots & \boldsymbol{y}_{m+N_s-l} \end{pmatrix}, \tag{12.2}$$

here, $m \in [l+1, N-N_s]$ and $N_s + 1 \leqslant N$ denotes the number of spatial samples utilized. It is evident that $\boldsymbol{Y}^{l} \in \mathbb{R}^{L(l+1)\times(N_s+1)}$ resembles the Hankel matrix commonly employed in system identification [32]. This spatial correlation processing allows $\boldsymbol{Y}^{l}$ to effectively capture the spatial dynamics.

A schematic illustration of the spatial correlation data construction summarizing this step, is shown in Fig. 12.4.

12.4.1.2 Independent Decomposition for Spatial Correlation Data

The next stage of dynamic SICA treats the spatial correlation data $\boldsymbol{Y}^l$ as *mixed signals* in the blind source separation framework [33], and extracts the *spatial independent components* $\hat{\boldsymbol{s}}(z)$, which are then used to construct reference statistics:

$$\hat{\boldsymbol{s}}(z) = \tilde{\boldsymbol{A}}^{\mathrm{T}}\boldsymbol{Q}\boldsymbol{y}^l(z) = \boldsymbol{W}\boldsymbol{y}^l(z),$$

where $\boldsymbol{y}^l(z)$ denotes the z-th column of the augmented matrix $\boldsymbol{Y}^l$, and $\tilde{\boldsymbol{A}}$ and $\boldsymbol{Q}$ represent the orthogonal mixing and whitening matrices, respectively. To reduce dimensionality, a few rows of $\boldsymbol{W}$ with the largest sum of squares are selected, denoted as $\boldsymbol{W}_d$ [17]. Similarly, by choosing the columns of $\tilde{\boldsymbol{A}}$ corresponding to the selected rows of $\boldsymbol{W}$, a matrix $\tilde{\boldsymbol{A}}_d$ is formed. This allows extraction of the dominant spatial independent components as follows:

$$\hat{\boldsymbol{s}}_d(z) = \boldsymbol{W}_d\boldsymbol{y}^l(z). \tag{12.3}$$

The ICA technique is applied for independent decomposition, and further details can be found in Sect. A.1 of the appendix.

12.4.2 *Off-Line Training for Spatial References*

The goal of offline training is to characterize the system behavior under normal conditions. Specifically, it involves constructing spatial reference statistics and determining their confidence bounds under normal operation, which serve as spatial references for abnormality localization in new data.

Step 1. Spatial Reference Statistics Construction
Inspired by the temporal I^2 and squared prediction error (SPE) statistics in [18], the spatial I_d^2 statistic at spatial location z is defined as follows:

$$I_d^2(z) = \hat{\boldsymbol{s}}_d^{\mathrm{T}}(z)\hat{\boldsymbol{s}}_d(z). \tag{12.4}$$

And the spatial SPE statistic can be formulated as:

$$\mathrm{SPE}(z) = \boldsymbol{e}^{\mathrm{T}}(z)\boldsymbol{e}(z) = (\boldsymbol{y}^l(z) - \hat{\boldsymbol{y}}^l(z))^{\mathrm{T}}(\boldsymbol{y}^l(z) - \hat{\boldsymbol{y}}^l(z)), \tag{12.5}$$

where $\hat{\boldsymbol{y}}^l(z)$ can be obtained by:

$$\hat{\boldsymbol{y}}^l(z) = \boldsymbol{Q}^{-1}\tilde{\boldsymbol{A}}_d\hat{\boldsymbol{s}}_d(z) = \boldsymbol{Q}^{-1}\tilde{\boldsymbol{A}}_d\boldsymbol{W}_d\boldsymbol{y}^l(z). \tag{12.6}$$

Remark 12.1 *To be noticed, for the abnormality localization problem, traditional TICA-based methods decompose the data into temporally independent components and form the* I_d^2 *and SPE statistics in the time domain. Abnormality localization is conducted by the contribution plots of each location point to the detected abnormal samples in the time domain. Compared with the proposed method, the disadvantages of the traditional methods are threefold:*

- *The localization accuracy relies on the detection accuracy in the time domain.*
- *The localization principle is based on the contributions of location points to each detected abnormal sample in the time domain. However, as it can be seen from Eqs. 12.4 and 12.5, the proposed* I_d^2 *and SPE statistics can reflect variations on all temporal samples for each location point when abnormality exist.*
- *When using the contribution plot for abnormality localization, the contributions of abnormal location points should be far larger than that of the normal location points, which makes it not sensitive to minor abnormalities.*

Therefore, the proposed method is inherently superior over the traditional TICA-based methods on abnormality localization problem.

Step 2. Probability based Spatial References Construction
Once a model capturing the normal operating region of the system is established, it is necessary to detect any deviations from standard behavior, which requires calculating confidence bounds to determine if the system remains in control. Specifically, after obtaining the spatial I_d^2 and SPE reference statistics, their confidence bounds, denoted as I_{dR}^2 and SPE_R, must be determined and used as spatial references for abnormality localization. For PCA-based approaches, confidence bounds are typically derived by assuming that the PDFs follow a multivariate Gaussian distribution [34], an assumption that does not hold for many industrial manufacturing systems. For ICA-based approaches, since the independent components do not follow a multivariate Gaussian distribution, a non-parametric KDE method [16, 35] is employed to overcome this limitation.

Thus, after obtaining the two spatial reference statistics, a univariate kernel density estimator is applied to estimate the PDFs for both. A schematic of the KDE is shown in Fig. 12.5, where the horizontal axis represents the reference statistic (I_d^2 or SPE) and the vertical axis represents the estimated PDF. Notably, the critical point encompassing 99% of the estimated PDF area is considered the confidence bound of these spatial reference statistics [17, 18]. Further details on KDE can be found in Sect. A.2 of the appendix.

A key advantage of confidence regions obtained via KDE is that they more closely follow the actual data and are less likely to include regions corresponding to unknown operating conditions compared to confidence regions derived from Hotelling's T^2 statistics.

In summary, the confidence bounds in the proposed dynamic-SICA method are determined as follows. First, I_d^2 or SPE values from normal operating data are collected. Next, a univariate kernel density estimator is applied to estimate the PDF of

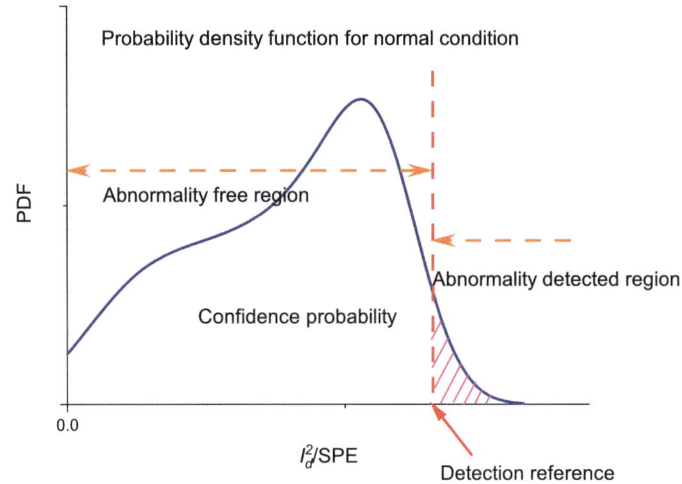

Fig. 12.5 Schematic graph of KDE

these normal values. The point covering 99% of the estimated PDF is then selected as the confidence bound for the normal operating data (I_d^2 or SPE values).

12.4.3 On-Line Reference-Based Abnormality Localization

Step 1. Spatial Monitoring Statistics Construction
When a new batch of data is collected, the first step is to construct the augmented matrix $\boldsymbol{Y}^l_{\text{new}}$ using the previously determined cross-correlation order l. Then, the new dominant spatial independent components can be computed using the same de-mixing matrix $\boldsymbol{W}_d$ obtained from the offline training:

$$\hat{\boldsymbol{s}}_{\text{new}d}(z) = \boldsymbol{W}_d \boldsymbol{y}^l_{\text{new}}(z). \tag{12.7}$$

And the spatial $I^2_{\text{new}d}$ statistic is obtained as:

$$I^2_{\text{new}d}(z) = \hat{\boldsymbol{s}}^{\text{T}}_{\text{new}d}(z)\hat{\boldsymbol{s}}_{\text{new}d}(z). \tag{12.8}$$

Then the spatial SPE_{new} statistic can be calculated as:

$$\text{SPE}_{\text{new}}(z) = (\boldsymbol{y}^l_{\text{new}}(z) - \hat{\boldsymbol{y}}^l_{\text{new}}(z))^{\text{T}}(\boldsymbol{y}^l_{\text{new}}(z) - \hat{\boldsymbol{y}}^l_{\text{new}}(z)), \tag{12.9}$$

where $\hat{\boldsymbol{y}}^l_{\text{new}}(z)$ can be obtained by:

$$\hat{\boldsymbol{y}}^l_{\text{new}}(z) = \boldsymbol{Q}^{-1}\tilde{\boldsymbol{A}}_d\hat{\boldsymbol{s}}_{\text{new}d}(z) = \boldsymbol{Q}^{-1}\tilde{\boldsymbol{A}}_d\boldsymbol{W}_d\boldsymbol{y}^l_{\text{new}}(z). \tag{12.10}$$

Step 2. Abnormality Localization Principles Formulation
By comparing Eqs. 12.4 with 12.8 and Eqs. 12.5 with 12.9, the differences between normal and abnormal conditions can be clearly observed using the spatial I_d^2 and SPE statistics. Furthermore, once the $I_{\text{new}d}^2$ and SPE_{new} statistics are obtained, abnormality localization can be performed by comparing them with the spatial references I_{dR}^2 and SPE_R. The sets of detected abnormal location points for these two statistics are denoted as $z_f^{I^2}$ and z_f^{SPE}, respectively.

Thus, the underlying principles of abnormality localization can be summarized as follows:

$$z \in \begin{cases} z_f^{I^2}, & \text{if } I_{\text{new}d}^2(z) > I_{dR}^2, \\ z_f^{\text{SPE}}, & \text{if } \text{SPE}_{\text{new}}(z) > \text{SPE}_R. \end{cases} \tag{12.11}$$

The framework of the proposed abnormality localization method is summarized in Algorithm 12.1.

Algorithm 12.1: Framework of dynamic SICA-based spatial localization method.

1: Calculate the cross-correlation order l of data $y(z_s, t_s)$ in the normal condition and formulate the spatial augmented matrix $\boldsymbol{Y}^l$.
2: Use the ICA method to find the independent components $\hat{\boldsymbol{s}}(z)$ and the de-mixing matrix $\boldsymbol{W}$.
3: Find the dominant parts of the de-mixing matrix $\boldsymbol{W}$, denoted by $\boldsymbol{W}_d$. And calculate the dominant spatial independent components by $\hat{\boldsymbol{s}}_d(z) = \boldsymbol{W}_d \boldsymbol{y}^l(z)$.
4: Calculate the spatial I_d^2 and SPE reference statistics by $I^2(z) = \hat{\boldsymbol{s}}_d^{\text{T}}(z)\hat{\boldsymbol{s}}_d(z)$, $\text{SPE}(z) = (\boldsymbol{y}^l(z) - \hat{\boldsymbol{y}}^l(z))^{\text{T}}(\boldsymbol{y}^l(z) - \hat{\boldsymbol{y}}^l(z))$, where $\hat{\boldsymbol{y}}^l(z) = \boldsymbol{Q}^{-1}\tilde{\boldsymbol{A}}_d \boldsymbol{W}_d \boldsymbol{y}^l(z)$.
5: Calculate the confidence bounds of I_d^2 and SPE statistics by the KDE method.
6: For new data $y_{\text{new}}(z, t)$, formulate the spatial augmented matrix $\boldsymbol{Y}_{\text{new}}^l$ using the same cross-correlation order l.
7: Calculate the dominant spatial independent components with the same de-mixing matrix $\boldsymbol{W}_d$. $\hat{\boldsymbol{s}}_{\text{new}d}(z) = \boldsymbol{W}_d \boldsymbol{y}_{\text{new}}^l(z)$.
8: Calculate the spatial $I_{\text{new}d}^2$ and SPE_{new} monitoring statistics as it is shown in Line 4.
9: Compare the $I_{\text{new}d}^2$ and SPE_{new} monitoring statistics with the confidence bounds I_{dR}^2 and SPE_R determined in Line 5.
10: **RETURN** the location $z_f^{I^2}$ and z_f^{SPE} if the corresponding $I_{\text{new}d}^2$ and SPE_{new} are beyond the confidence bounds.

12.5 Illustrative Examples

In this section, two typical DPSs are used to evaluate the proposed abnormality localization method.

Fig. 12.6 Catalytic rod

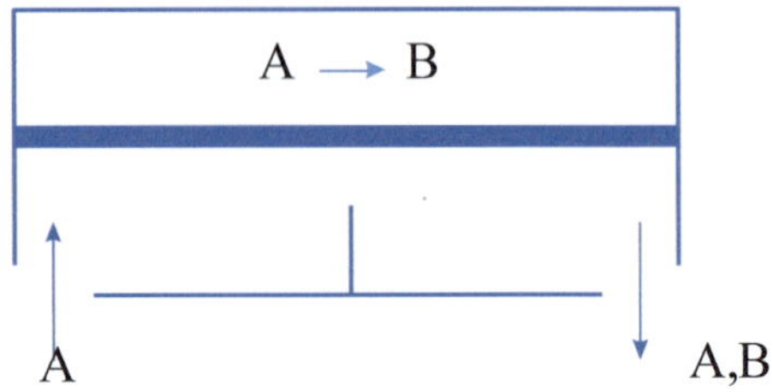

12.5.1 Catalytic Rod

As illustrated in Fig. 12.6, a long, slender rod inside a reactor is considered [1]. The reactor is supplied with pure species A, along which a zero-order exothermic catalytic reaction $A \rightarrow B$ takes place. In addition, a cooling medium in contact with the rod provides temperature control. For the analysis, several assumptions are made: an excess of species A in the furnace, constant rod conductivity, constant density and heat capacity, and fixed temperatures at both ends of the rod. Under these assumptions, the temperature distribution along the catalytic rod can be described by the following parabolic PDE:

$$\frac{\partial y(z,t)}{\partial t} = \frac{\partial^2 y(z,t)}{\partial z^2} + \beta_T (e^{-\frac{\gamma}{1+y(z,t)}} - e^{-\gamma}) + \beta_U (\boldsymbol{b}^{\mathrm{T}}(z)\boldsymbol{u}(t) - y(z,t)), \tag{12.12}$$

subject to the following Dirichlet boundary and initial conditions:

$$y(0,t) = 0, \quad y(\pi,t) = 0, \quad y(z,0) = y_0(z),$$

where $y(z,t)$ denotes the temperature at location z and time t, β_T represents the heat of reaction, β_U is the heat transfer coefficient, γ is the activation energy, $\boldsymbol{u}(t)$ is the manipulated input (temperature of the cooling medium), and $\boldsymbol{b}(z)$ represents the actuator distribution across the spatial domain. The typical values of the system parameters are:

$$\beta_T = 50, \ \beta_U = 2, \ \gamma = 4.$$

There are four actuators, $\boldsymbol{u}(t) = [u_1(t), \cdots, u_4(t)]$, with spatial distributions $\boldsymbol{b}(z) = [b_1(z), \cdots, b_4(z)]$, where $b_i(z) = H(z - (i-1)\pi/4) - H(z - i\pi/4)$ for $i = 1, \cdots, 4$, and $H(\cdot)$ denotes the standard Heaviside function. To fully excite the system, the temporal inputs are set as $u_i(t) = 1.1 + 5\sin[(t/10) + (i/10)]$ for $i = 1, \cdots, 4$. Twenty sensors are uniformly distributed in space to collect measurements. The finite-difference method (FDM) [36], a commonly used approach for solving PDEs, is employed to solve Eq. 12.12. The sampling interval Δt is 0.01 s, and the random measurement noise is determined by the SNR between the measurements and noise, with SNR set to 20 dB in this experiment.

Normal operating data consisting of 750 temporal samples are used for the analysis, with the cross-correlation order calculated as $l = 2$. Details on the computation of the cross-correlation order can be found in Sect. A.3 of the appendix. Abnormalities 1, 3, and 7 correspond to variations in the temporal inputs of the actuators, while abnormality 2 corresponds to changes in the spatial distribution of an actuator. Abnormalities 4, 5, and 6 correspond to variations in the parameters β_U, β_T, and γ, respectively. Notably, abnormality 7 represents an impulsive change in the actuator spatial distribution at temporal sample 350.

In conventional SPC methods, the "FDR" and "FAR" are commonly used to assess the performance of diagnostic methods. FDR represents the percentage of samples exceeding the 99% confidence bounds after the occurrence of an abnormality, while FAR indicates the proportion of samples that exceed the detection threshold in the absence of any fault. Inspired by these two metrics, the "detection rate" and "misdetection rate" are defined to evaluate the proposed abnormality localization approach as follows:

$$\text{Detection rate} = \frac{\text{Detected abnormal region}}{\text{Actual abnormal region}} - \frac{\text{Misdetected abnormal region}}{\text{Actual abnormal region}},$$

$$\text{Misdetection rate} = \frac{\text{Misdetected abnormal region}}{\text{Detected abnormal region}}.$$

The SICA method is used as a benchmark for comparison with the proposed approach. Additionally, the performance of the proposed method under different cross-correlation orders ($l = 2$ and $l = 1$) is also evaluated. The localization results for abnormality 1 are shown in Fig. 12.7a–f, and Table 12.2 presents the detection and misdetection rates for all abnormalities. Overall, the proposed method achieves high detection rates, although some abnormalities (2 and 3) exhibit relatively high misdetection rates. This is mainly because the action areas of neighboring actuators overlap at their junctions, so a change in one actuator can also affect its neighbors, making precise localization—i.e., achieving both high detection and low misdetection rates—challenging. Nevertheless, considering the detection and misdetection rates of both I_d^2 and SPE statistics, the proposed dynamic SICA method outperforms the traditional SICA method across all abnormalities. For abnormalities 1, 2, 3, and 7, dynamic SICA with $l = 2$ performs better than with $l = 1$, while for abnormalities 4, 5, and 6, the performance is worse with $l = 2$. This indicates that a lower cross-correlation order may be more suitable for abnormalities caused by variations in system parameters. Moreover, robustness analysis of the proposed method on the catalytic rod can be found in Sect. A.3 of the appendix. The settings for abnormalities are defined in Table 12.1.

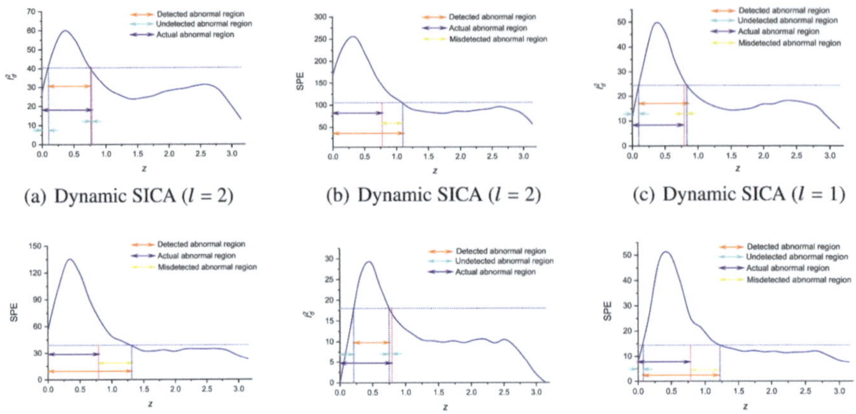

(a) Dynamic SICA (l = 2) (b) Dynamic SICA (l = 2) (c) Dynamic SICA (l = 1)

(d) Dynamic SICA (l = 1) (e) SICA (f) SICA

Fig. 12.7 Localization results in the case of Abnormality 1

Table 12.1 Settings of Abnormalities for catalytic rod

Abnormality	Type	Samples in the time domain
0	Normal condition	$1 \rightarrow 750$
1	$\boldsymbol{u}_1 \rightarrow 2 * \boldsymbol{u}_1$	$350 \rightarrow 750$
2	$\boldsymbol{b}_1 \rightarrow 3 * \boldsymbol{b}_1$	$350 \rightarrow 750$
3	$\boldsymbol{u}_2 \rightarrow 1.8 * \boldsymbol{u}_2, \boldsymbol{u}_3 \rightarrow 3 * \boldsymbol{u}_3$	$350 \rightarrow 750$
4	$\beta_U \rightarrow 2 * \beta_U$	$350 \rightarrow 750$
5	$\beta_T \rightarrow 10 + \beta_T$	$350 \rightarrow 750$
6	$\gamma \rightarrow 2.5 * \gamma$	$350 \rightarrow 750$
7	$\boldsymbol{u}_3 \rightarrow 8 * \boldsymbol{u}_3$	350 (Impulsive)

12.5.2 *Tubular Reactor*

The tubular reactor shown in Fig. 12.8 [37] represents another typical convection–diffusion–reaction process in chemical engineering. It can be modeled using a dimensionless formulation that incorporates both diffusion and convection effects, along with a nonlinear heat generation term, as follows:

$$\begin{cases} \dfrac{\partial T}{\partial t} = \dfrac{1}{Pe_h}\dfrac{\partial^2 T}{\partial z^2} - \dfrac{1}{Le}\dfrac{\partial T}{\partial z} + \eta C \exp[\gamma(1-\dfrac{1}{T})] + \mu[T_W(z,t) - T(z,t)], \\ \dfrac{\partial C}{\partial t} = \dfrac{1}{Pe_m}\dfrac{\partial^2 C}{\partial z^2} - \dfrac{\partial C}{\partial z} - Da \cdot C \exp[\gamma(1-\dfrac{1}{T})], \end{cases} \tag{12.13}$$

with the following boundary conditions:

Table 12.2 Abnormality localization results on catalytic rod

Method		Index/Abnormality	1	2	3	4	5	6	7
Dynamic SICA($l = 2$)	Detection rate(%)	I_d^2	83.93	**100**	**100**	78.29	81.76	66.00	0
		SPE	**100**	**100**	**100**	**100**	**100**	**100**	**100**
Dynamic SICA($l = 1$)	Detection rate(%)	I_d^2	88.92	98.47	**100**	88.35	93.73	94.37	0
		SPE	**100**	**100**	**100**	**100**	**100**	**100**	65.35
SICA	Detection rate(%)	I_d^2	67.01	84.46	83.56	88.29	71.55	92.33	0
		SPE	92.74	97.57	**100**	99.27	98.73	99.65	86.48
Dynamic SICA($l = 2$)	Misdetection rate(%)	I_d^2	0	10.90	18.95	0	0	0	–
		SPE	27.65	43.16	47.78	0	0	0	39.00
Dynamic SICA($l = 1$)	Misdetection rate(%)	I_d^2	8.63	20.62	31.11	0	0	0	–
		SPE	32.96	52.05	47.16	0	0	0	0
SICA	Misdetection rate(%)	I_d^2	0	20.57	14.58	0	0	0	–
		SPE	37.54	54.29	43.72	0	0	0	13.39

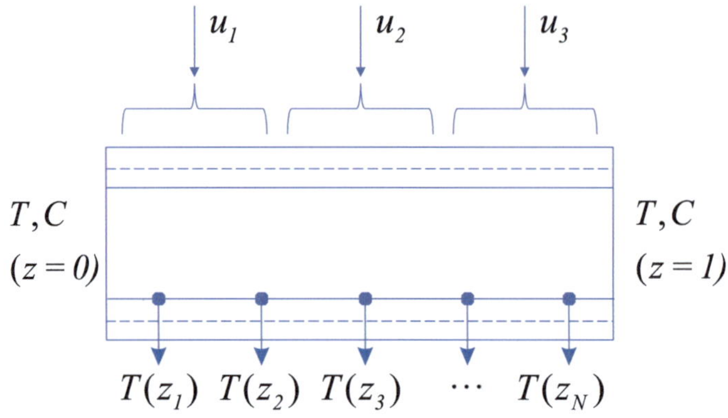

Fig. 12.8 Tubular reactor with a chemical reaction

$$z=0,\ \begin{cases} -\dfrac{\partial T}{\partial z} = Pe_h[T_i(t) - T(z,t)], \\ -\dfrac{\partial C}{\partial z} = Pe_m[C_i(t) - C(z,t)], \end{cases}$$

$$z=1,\ \begin{cases} \dfrac{\partial T}{\partial z} = 0, \\ \dfrac{\partial C}{\partial z} = 0. \end{cases}$$

In this system, T and C represent the dimensionless temperature and concentration, respectively. The parameters Pe_h, Pe_m, Le, Da, γ, η, μ, T_i, and C_i correspond to the energy Peclet number, mass Peclet number, Lewis number, Damköhler number, activation energy, heat of reaction, heat-transfer coefficient, inlet temperature, and inlet concentration. This system is classified as a nonself-adjoint parabolic DPS with Neumann boundary conditions. The two DPSs are coupled, and in practice, only temperature measurements can be obtained. The parameter values used in the simulation are $Pe_h = 5.0$, $Pe_m = 5.0$, $Le = 1.0$, $Da = 0.875$, $\gamma = 15.0$, $\eta = 0.8375$, $\mu = 13$, $T_i = 1$, and $C_i = 1$.

In this system, the controlled variable is the wall temperature T_W, expressed as $T_W(z,t) = \boldsymbol{b}^{\mathrm{T}}(z)\boldsymbol{u}(t)$. Three actuators are positioned along the reactor, with temporal signals $\boldsymbol{u}(t)=[u_1(t), \cdots, u_3(t)]$ and spatial distributions $\boldsymbol{b}(z) = [b_1(z), \cdots, b_3(z)]$, where $b_i(z) = H(z-(i-1)\pi/3) - H(z-i\pi/3)$ for $i = 1, 2, 3$ and $H(\cdot)$ denotes the standard Heaviside function. The temporal inputs are applied as $u_i(t) = 1.0 + 0.4\sin[-t/10 + i/10]$ to excite the system. A total of 61 temperature sensors are uniformly distributed along the reactor to acquire measurements.

The sampling interval Δt is set to 0.0005 s, with a SNR of 50 dB. Normal operating data consisting of 800 temporal samples are employed for analysis, and the calculated cross-correlation order is $l = 2$. Details on determining this order can be found in Sect. A.3 of the appendix. Abnormalities 1, 2, 4, 5, and 6 correspond to variations in

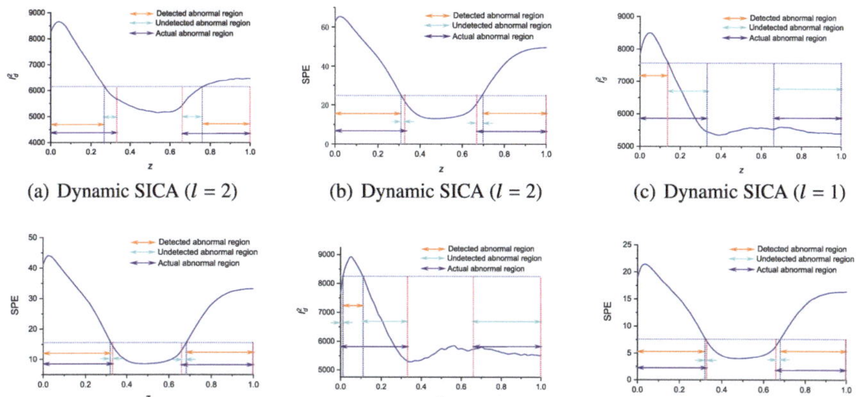

(a) Dynamic SICA (l = 2) (b) Dynamic SICA (l = 2) (c) Dynamic SICA (l = 1)

(d) Dynamic SICA (l = 1) (e) SICA (f) SICA

Fig. 12.9 Localization results in the case of Abnormality 2

Table 12.3 Settings of Abnormalities for tubular reactor

Abnormality	Type	Samples in the time domain
0	Normal condition	$1 \rightarrow 800$
1	$\boldsymbol{u}_1 \rightarrow 1.3 * \boldsymbol{u}_1$	$400 \rightarrow 600$
2	$\boldsymbol{u}_1 \rightarrow 1.3 * \boldsymbol{u}_1$, $\boldsymbol{u}_3 \rightarrow 1.4 * \boldsymbol{u}_3$	$400 \rightarrow 600$, $300 \rightarrow 500$
3	$\mu \rightarrow 2 + \mu$	$500 \rightarrow 800$
4	$\boldsymbol{u}_2 \rightarrow 8 * \boldsymbol{u}_2$	350 (Impulsive)
5	$\boldsymbol{u}_2 \rightarrow 10 * \boldsymbol{u}_2$, $\boldsymbol{u}_3 \rightarrow 1.4 * \boldsymbol{u}_3$	350 (Impulsive), $300 \rightarrow 500$
6	$\boldsymbol{u}_2 \rightarrow 10 * \boldsymbol{u}_2$, $\boldsymbol{u}_3 \rightarrow 15 * \boldsymbol{u}_3$	350 (Impulsive), 400 (Impulsive)

the actuators' temporal inputs, whereas abnormality 3 represents a change in the heat-transfer coefficient μ. Notably, abnormalities 4, 5, and 6 involve impulsive changes in the actuators' temporal signals.

The SICA approach and the proposed method with varying cross-correlation orders ($l = 2$ and $l = 1$) were examined. The localization outcomes for abnormality 2 are illustrated in Fig. 12.9a–f. Detection and misdetection rates for all abnormalities are summarized in Table 12.4. Considering the performance of both I_d^2 and SPE statistics, the proposed dynamic SICA method outperforms the traditional SICA approach for all abnormalities except abnormality 3, which is caused by a system parameter change. The dynamic SICA with $l = 2$ shows better results than $l = 1$ for abnormalities 1, 2, 3, and 5, while slightly underperforming for abnormalities 4 and 6. This indicates that a low cross-correlation order is sufficient for impulsive-type abnormalities. Further robustness analysis for the tubular reactor can be found in Sect. A.3 of the appendix. The settings for abnormalities are defined in Table 12.3.

Table 12.4 Abnormality localization results on tubular reactor

Method		Index/Abnormality	1	2	3	4	5	6
Dynamic SICA(l = 2)	Detection rate(%)	I_d^2	80.78	76.16	51.85	0	48.57	0
		SPE	92.79	92.65	**100**	93.09	99.10	94.15
Dynamic SICA(l = 1)	Detection rate(%)	I_d^2	41.74	20.54	51.15	0	0	0
		SPE	96.40	95.95	**100**	97.30	**100**	**100**
SICA	Detection rate(%)	I_d^2	28.23	14.39	57.14	0	0	0
		SPE	96.99	96.70	**100**	93.39	99.55	99.85
Dynamic SICA(l = 2)	Misdetection rate(%)	I_d^2	0	0	0	–	0	–
		SPE	0	0	0	0	0	0
Dynamic SICA(l = 1)	Misdetection rate(%)	I_d^2	0	0	0	–	–	–
		SPE	0	0	0	0.91	0	0.60
SICA	Misdetection rate(%)	I_d^2	0	0	0	–	–	–
		SPE	0	0	0	0	0	0

12.6 Conclusion

This chapter introduces a data-driven dynamic SICA approach for abnormality localization in DPSs. The methodology consists of two main steps: constructing spatial references off-line and performing abnormality localization on-line. During the off-line stage, S-T data collected under normal operation are used to determine confidence bounds for the spatial reference statistics. These bounds then serve as spatial references during the on-line localization stage. Two experiments are conducted to assess the effectiveness of the proposed approach in locating abnormalities. The method is capable of identifying abnormal spatial locations across a wide range of industrial distributed processes.

Appendix

A.1 Independent Component Analysis

ICA [16, 38] is defined based on a statistical "latent variable" model as follows:

$$x_i = a_{i1}s_1 + a_{i2}s_2 + \cdots + a_{im}s_m, \text{ for all } i = 1, \cdots, n,$$

where $x_1, \cdots, x_n$ are n observed random variables, modeled as linear combinations of m statistically independent random variables $s_1, \cdots, s_m$. The coefficients a_{ij}, for $i = 1, \cdots, n$ and $j = 1, \cdots, m$, are real-valued. In this study, we focus on the over-determined scenario with $n \geq m$.

The above describes the basic ICA model. The observed values $x_i(t)$ represent samples of the random variable x_i, with p samples available for each x_i. This leads to the following mixing model:

$$\boldsymbol{X} = \boldsymbol{A}\boldsymbol{S},$$

where $\boldsymbol{X} = [\boldsymbol{x}(1), \cdots, \boldsymbol{x}(p)] \in \mathbb{R}^{n\times p}$ denotes the data matrix, $\boldsymbol{A} \in \mathbb{R}^{n\times m}$ is the unknown mixing matrix, and $\boldsymbol{S} = [\boldsymbol{s}(1), \cdots, \boldsymbol{s}(p)] \in \mathbb{R}^{m\times p}$ contains the independent components. Since the independent components cannot be directly observed, they are treated as latent variables. The primary goal of ICA is to estimate both $\boldsymbol{A}$ and $\boldsymbol{S}$ using only the observed data $\boldsymbol{X}$. Specifically, a de-mixing matrix $\boldsymbol{W}$ is sought such that the estimated independent components matrix $\hat{\boldsymbol{S}} = \boldsymbol{W}\boldsymbol{X}$ has columns that are as statistically independent as possible, obtained via:

$$\hat{\boldsymbol{S}} = \boldsymbol{W}\boldsymbol{X}.$$

In ICA, two preprocessing steps are typically applied: centering and whitening. For simplicity, consider the n-dimensional random vector $\boldsymbol{x}(k)$ at sample k. Centering aims to transform $\boldsymbol{x}(k)$ into a zero-mean variable, which ensures that the independent components $\boldsymbol{s}(k)$ are also zero-mean. This is achieved by:

$$\boldsymbol{x}(k) \leftarrow \boldsymbol{x}(k) - E\{\boldsymbol{x}(k)\},$$

where E denotes the expectation operator. Once the estimation is performed on centered data, the original estimate of $\boldsymbol{s}(k)$ can be recovered by reintroducing the mean vector.

The other preprocessing step is whitening, which transforms $\boldsymbol{x}(k)$ into $\tilde{\boldsymbol{x}}(k)$ with uncorrelated components of unit variance. Thus,

$$E\{\tilde{\boldsymbol{x}}(k)\tilde{\boldsymbol{x}}^{\mathrm{T}}(k)\} = \boldsymbol{I} \in \mathbb{R}^{n \times n}.$$

This can be achieved using the eigenvalue decomposition (EVD) of the covariance matrix $\boldsymbol{C}_x$:

$$\boldsymbol{C}_x = E\{\boldsymbol{x}(k)\boldsymbol{x}^{\mathrm{T}}(k)\} = \boldsymbol{U}\boldsymbol{\Lambda}\boldsymbol{U}^{\mathrm{T}},$$

where $\boldsymbol{U}$ represents the orthogonal matrix of eigenvectors of the covariance matrix $\boldsymbol{C}_x$, and $\boldsymbol{\Lambda}$ is the diagonal matrix of its eigenvalues. The whitening transformation is then performed as:

$$\tilde{\boldsymbol{x}}(k) = \boldsymbol{U}\boldsymbol{\Lambda}^{-1/2}\boldsymbol{U}^{\mathrm{T}}\boldsymbol{x}(k).$$

Hence we have that

$$\tilde{\boldsymbol{x}}(k) = \boldsymbol{U}\boldsymbol{\Lambda}^{-1/2}\boldsymbol{U}^{\mathrm{T}}\boldsymbol{A}\boldsymbol{s}(k) = \tilde{\boldsymbol{A}}\boldsymbol{s}(k),$$

and

$$E\{\tilde{\boldsymbol{x}}(k)\tilde{\boldsymbol{x}}^{\mathrm{T}}(k)\} = \tilde{\boldsymbol{A}}E\{\boldsymbol{s}(k)\boldsymbol{s}^{\mathrm{T}}(k)\}\tilde{\boldsymbol{A}}^{\mathrm{T}} = \tilde{\boldsymbol{A}}\tilde{\boldsymbol{A}}^{\mathrm{T}} = \boldsymbol{I}.$$

Therefore the new mixing matrix $\tilde{\boldsymbol{A}}$ is orthogonal and $\boldsymbol{s}(k)$ can be estimated as follows:

$$\hat{\boldsymbol{s}}(k) = \tilde{\boldsymbol{A}}^{\mathrm{T}}\tilde{\boldsymbol{x}}(k) = \tilde{\boldsymbol{A}}^{\mathrm{T}}\boldsymbol{U}\boldsymbol{\Lambda}^{-1/2}\boldsymbol{U}^{\mathrm{T}}\boldsymbol{x}(k) = \tilde{\boldsymbol{A}}^{\mathrm{T}}\boldsymbol{Q}\boldsymbol{x}(k),$$

where $\boldsymbol{Q} = \boldsymbol{U}\boldsymbol{\Lambda}^{-1/2}\boldsymbol{U}^{\mathrm{T}}$ denotes the whitening matrix. Hence the de-mixing matrix $\boldsymbol{W}$ can be written as:

$$\boldsymbol{W} = \tilde{\boldsymbol{A}}^{\mathrm{T}}\boldsymbol{Q}.$$

The FastICA algorithm [39] is employed to estimate $\tilde{\boldsymbol{A}}$. It utilizes a fixed-point iteration method to maximize the non-Gaussianity of the independent components $\hat{\boldsymbol{s}}(k)$.

A.2 Kernel Density Estimation

Since the latent variables in ICA-based monitoring are generally non-Gaussian, KDE [35] is employed to determine the confidence limits for the I_d^2 and SPE statistics.

Here a univariate kernel density estimator with kernel K is utilized:

$$\hat{f}(x) = \frac{1}{nh} \sum_{i=1}^{n} K\left(\frac{x - X_i}{h}\right),$$

where h is the *window width*, also known as the *smoothing parameter* or *bandwidth*.

The kernel function K satisfies the condition:

$$\int_{-\infty}^{\infty} K(x)\mathrm{d}x = 1.$$

Here we adopt the most commonly used Gaussian Kernel:

$$K(t) = \frac{1}{\sqrt{2\pi}} e^{-(1/2)t^2}.$$

An essential step in KDE is selecting the smoothing parameter h, denoted here as h_{opt}. The degree of smoothing critically influences the density estimate: if h approaches zero, the estimate reduces to a series of Dirac delta spikes at the sample points, resulting in undersmoothing; conversely, a large h smooths out details, leading to oversmoothing. To address this, the LSCV method [35] is employed, which is fully automatic and computationally efficient.

Given any estimator $\hat{f}$ of a density f, the integrated square error can be written as:

$$\int \left(\hat{f} - f\right)^2 = \int \hat{f}^2 - 2\int \hat{f} f + \int f^2. \tag{12.14}$$

In Eq. 12.14, since the last term is independent of $\hat{f}$, the optimal window width—i.e., the one minimizing the integrated squared error—corresponds to that which minimizes the quantity R defined by

$$R(\hat{f}) = \int \hat{f}^2 - 2\int \hat{f} f.$$

The core idea of the LSCV method is to estimate $R(\hat{f})$ directly from the data and then select the window width h by minimizing this estimate. Since the term $\int \hat{f}^2$ can be computed from $\hat{f}$, let $\hat{f}_{-i}$ denote the density estimate obtained from all data points except X_i

$$\hat{f}_{-i}(x) = (n-1)^{-1}h^{-1}\sum_{j\neq i} K\left\{h^{-1}(x - X_j)\right\}.$$

Then define

$$M_0(h) = \int \hat{f}^2 - 2n^{-1}\sum_i \hat{f}_{-i}(X_i).$$

The score M_0 depends only on the data and the idea of LSCV is to minimize the score M_0 over h.

A.3 Cross-Correlation Order Calculation

The cross-correlation order l corresponds to the order of the DPSs in the spatial domain. For instance, in the commonly studied parabolic DPSs, the system equations involve first- or second-order spatial derivatives, so l is typically set to 1 or 2. By processing the spatial correlation data in batches, the spatial cross-correlation can be effectively incorporated.

As the proposed dynamic-SICA method is model-free, the cross-correlation order is determined directly from the collected data. Specifically, a cumulants-based identification approach is employed to compute the spatial cross-correlation order. Given the sampled system output $\boldsymbol{Y}^{\mathrm{T}} \in \mathbb{R}^{L\times N}$, for each row $\boldsymbol{y}_i \in \mathbb{R}^{N\times 1}$ $(i = 1, \cdots, L)$, the spatial cross-correlation order l_i is calculated using MATLAB's Higher-Order Spectral Analysis Toolbox [40] with default settings. The results for the catalytic rod and tubular reactor experiments are shown in Fig. 12.10a, b, respectively.

As illustrated in Fig. 12.10a, b, the cross-correlation order for the majority of temporal samples is 2; thus, l is set to 2 for both experiments.

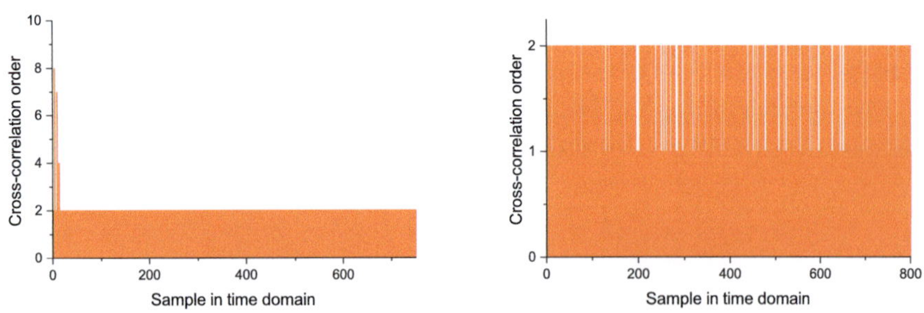

(a) Catalytic rod (SNR=20dB)   (b) Tubular reactor (SNR=50dB)

Fig. 12.10 Cross-correlation order analysis for catalytic rod and tubular reactor

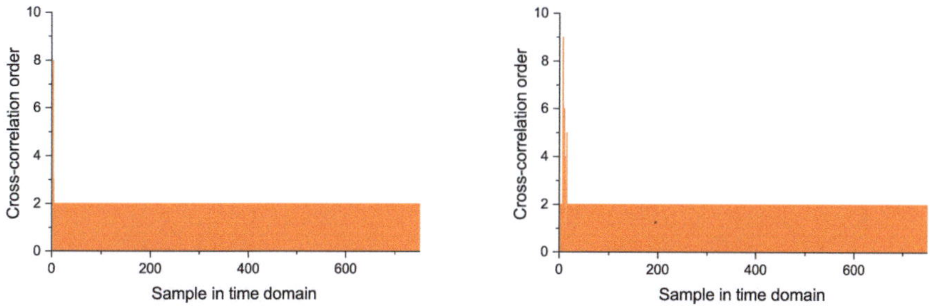

(a) Catalytic rod (SNR=30dB) (b) Catalytic rod (SNR=10dB)

Fig. 12.11 Cross-correlation order analysis for catalytic rod

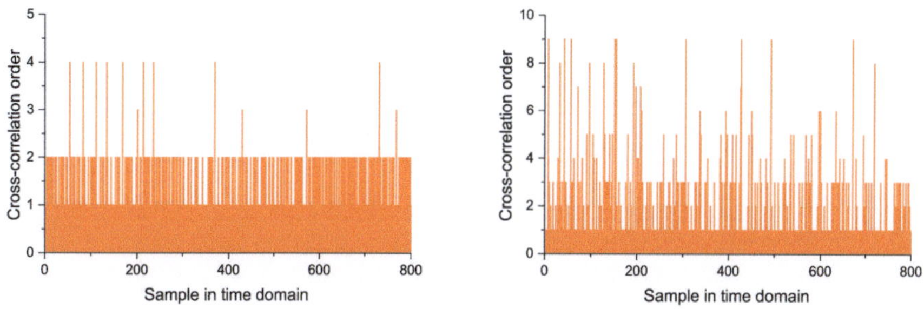

(a) Tubular reactor (SNR=40dB) (b) Tubular reactor (SNR=30dB)

Fig. 12.12 Cross-correlation order analysis for tubular reactor

A.4 Robustness Analysis

To further evaluate the robustness of the proposed dynamic-SICA method, additional comparison experiments were performed on both the catalytic rod and tubular reactor.

Specifically, for the catalytic rod, additional noise with SNRs of 30 dB and 10 dB was applied during both the off-line training and on-line localization stages. Similarly, for the tubular reactor, noise with SNRs of 40 dB and 30 dB was added to both stages.

A.4.1 On the Cross-Correlation Order Calculation

As the first step of the proposed dynamic-SICA method is to calculate the cross-correlation order l in the spatial domain, the robustness of this calculation should be investigated first.

For the catalytic rod, the cross-correlation order analysis is presented in Fig. 12.11a, b, showing that most temporal samples have a cross-correlation order of 2, consistent with the result in Fig. 12.10a at SNR = 20 dB. For the tubular reactor,

Table 12.5 Abnormality localization results by the proposed approach on catalytic rod with noises

Noises		Index/Abnormality	1	2	3	4	5	6	7
SNR=30 dB	Detection rate(%)	I_d^2	85.10	100	99.17	51.34	85.39	76.03	0
		SPE	100	100	100	100	100	100	100
SNR=20 dB	Detection rate(%)	I_d^2	83.93	100	100	78.29	81.76	66.00	0
		SPE	100	100	100	100	100	100	100
SNR=10 dB	Detection rate(%)	I_d^2	0	0	0	77.91	0	78.64	0
		SPE	90.70	100	100	100	100	100	0
SNR=30 dB	Misdetection rate(%)	I_d^2	0	3.09	18.86	0	0	0	–
		SPE	51.30	61.16	50.00	0	0	0	62.71
SNR=20 dB	Misdetection rate(%)	I_d^2	0	10.90	18.95	0	0	0	–
		SPE	27.65	43.16	47.78	0	0	0	39.00
SNR=10 dB	Misdetection rate(%)	I_d^2	-	-	–	0	–	0	–
		SPE	4.68	30.41	39.38	0	0	0	–

Table 12.6 Abnormality localization results by the proposed approach on tubular reactor with noises

Noise		Index/Abnormality	1	2	3	4	5	6
SNR=50 dB	Detection rate(%)	I_d^2	80.78	76.16	51.85	0	48.57	0
		SPE	92.79	92.65	100	93.09	99.10	94.15
SNR=40 dB	Detection rate(%)	I_d^2	100	97.90	0	0	100	86.49
		SPE	97.00	94.60	73.43	85.29	98.65	82.49
SNR=30 dB	Detection rate(%)	I_d^2	100	97.31	0	0	100	37.87
		SPE	100	96.86	1.90	60.96	100	66.32
SNR=50 dB	Misdetection rate(%)	I_d^2	0	0	0	–	0	–
		SPE	0	0	0	0	0	0
SNR=40 dB	Misdetection rate(%)	I_d^2	29.00	19.04	–	–	16.94	31.68
		SPE	0	0	0	0	0	0
SNR=30 dB	Misdetection rate(%)	I_d^2	25.00	16.58	–	–	14.05	28.61
		SPE	6.20	4.30	0	0	0.89	2.43

Fig. 12.12a, b indicate mean cross-correlation orders of 1.40 and 1.70 for SNRs of 40 dB and 30 dB, respectively. Therefore, the cross-correlation order l is set to 2 for both systems.

A.4.2 On the Abnormality Localization Performance

The abnormality localization performance under additional noise for the catalytic rod and tubular reactor is summarized in Tables 12.5 and 12.6, respectively, with the cross-correlation order set to $l = 2$.

As shown in Tables 12.5 and 12.6, the localization performance tends to deteriorate as the noise level increases, considering both the detection and misdetection rates for I_d^2 and SPE statistics. Specifically, higher noise levels lead to a decrease in the detection rate and a corresponding increase in the misdetection rate.

For the catalytic rod, as the SNR decreases from 30 dB to 20 dB, the localization performance remains largely unchanged or shows a slight decline across all abnormalities. However, at a lower SNR of 10 dB, the performance of the I_d^2 statistic deteriorates markedly, whereas the SPE statistic continues to perform satisfactorily for all abnormalities except abnormality 7, which is associated with impulsive variations.

For the tubular reactor, as the SNR decreases from 50 dB to 30 dB, the performance of both statistics shows a slight decline across all abnormalities, except for abnormality 3, which results from a variation in a system parameter.

Overall, the proposed dynamic-SICA method demonstrates satisfactory robustness in abnormality localization tasks.

References

1. Christofides PD (2012) Nonlinear and robust control of PDE systems: Methods and applications to transport-reaction processes. Springer Science+Business Media, Berlin
2. Wang B-C, Li H-X (2019) A sliding window based dynamic spatiotemporal modeling for distributed parameter systems with time-dependent boundary conditions. IEEE Trans Ind Inform 15(4):2044–2053
3. Wang Z, Li H-X (2019) Incremental spatiotemporal learning for online modeling of distributed parameter systems. IEEE Trans Syst Man Cybern Syst 49(12):2612–2622
4. Meng X-B, Li H-X, Yang H-D (2019) Evolutionary design of spatiotemporal learning model for thermal distribution in lithium-ion batteries. IEEE Trans Ind Inform 15(5):2838–2848
5. Xu K-K, Li H-X, Yang H-D (2019) Kernel-based random vector functional-link network for fast learning of spatiotemporal dynamic processes. IEEE Trans Syst Man Cybern Syst 49(5):1016–1026
6. Li H-X, Qi C (2010) Modeling of distributed parameter systems for applications-synthesized review from time-space separation. J Process Control 20(8):891–901
7. Wang J-W, Li H-X, Wu H-N (2017) A membership-function-dependent approach to design fuzzy pointwise state feedback controller for nonlinear parabolic distributed parameter systems with spatially discrete actuators. IEEE Trans Syst Man Cybern Syst 47(7):1486–1499

8. Ding SX (2008) Model-based fault diagnosis techniques: design schemes, algorithms, and tools. Springer Science+Business Media, Berlin
9. Ding SX (2014) Data-driven design of fault diagnosis and fault-tolerant control systems. Springer Science+Business Media, Berlin
10. Jackson JE (1985) Multivariate quality control. Commun Stat-Theory. Methods 14(11):2657–2688
11. Jackson JE (1959) Quality control methods for several related variables. Technometrics 1(4):359–377
12. Li G, Qin SJ, Zhou D (2010) Geometric properties of partial least squares for process monitoring. Automatica 46(1):204–210
13. Zhang Y, Zhou H, Qin SJ et al (2010) Decentralized fault diagnosis of large-scale processes using multiblock kernel partial least squares. IEEE Trans Ind Informatics 6(1):3–10
14. Liu Q, Qin SJ, Chai T (2017) Unevenly sampled dynamic data modeling and monitoring with an industrial application. IEEE Trans Ind Informatics 13(5):2203–2213
15. Yu W, Zhao C (2018) Sparse exponential discriminant analysis and its application to fault diagnosis. IEEE Trans Ind Electron 65(7):5931–5940
16. Hyvärinen A, Karhunen J, Oja E (2004) Independent component analysis, vol 46. John Wiley & Sons, Chichester
17. Lee J-M, Yoo CK, Lee I-B (2004) Statistical process monitoring with independent component analysis. J Process Control 14(5):467–485
18. Lee J-M, Yoo CK, Lee I-B (2004) Statistical monitoring of dynamic processes based on dynamic independent component analysis. Chem Eng Sci 59(14):2995–3006
19. Calhoun VD, Adali T, Pearlson GD et al (2001) Spatial and temporal independent component analysis of functional MRI data containing a pair of task-related waveforms. Hum Brain Mapp 13(1):43–53
20. Stone JV, Porrill J, Porter NR et al (2002) Spatiotemporal independent component analysis of event-related fMRI data using skewed probability density functions. Neuroimage 15(2):407–421
21. Demetriou MA (2002) A model-based fault detection and diagnosis scheme for distributed parameter systems: A learning systems approach. ESAIM: Control Optimisation Calc Variat 7:43–67
22. El-Farra NH (2006) Integrated fault detection and fault-tolerant control architectures for distributed processes. Ind Eng Chem Res 45(25):8338–8351
23. Deutscher J (2016) Fault detection for linear distributed-parameter systems using finite-dimensional functional observers. Int J Control 89(3):550–563
24. Fischer F, Deutscher J (2017) Fault detection for parabolic systems with distributed inputs and outputs using the modulation function approach. IFAC-PapersOnLine 50(1):6774–6779
25. Cai J, Ferdowsi H, Sarangapani J (2016) Model-based fault detection, estimation, and prediction for a class of linear distributed parameter systems. Automatica 66:122–131
26. Dong J, Wang Q, Wang M et al (2018) Data-driven quality monitoring techniques for distributed parameter systems with application to hot-rolled strip laminar cooling process. IEEE Access 6:16646–16654
27. Ferdowsi H, Jagannathan S (2018) Fault diagnosis of distributed parameter systems modeled by linear parabolic partial differential equations with state faults. J Dyn Syst Meas Control 140(1):011010
28. Feng Y, Li H-X (2018) Independent component analysis based fault detection and spatial localization of distributed parameter systems. Proceedings of the 2018 IEEE International Conference on Systems, Man, and Cybernetics (SMC). IEEE, Piscataway, p 2236–2241
29. Feng Y, Li H-X (2019) Detection and spatial identification of fault for parabolic distributed parameter systems. IEEE Trans Ind Electron 66(9):7300–7309
30. Giannakis GB, Mendel JM (1990) Cumulant-based order determination of non-Gaussian ARMA models. IEEE Trans Acoust Speech Signal Process 38(8):1411–1423
31. Ku W, Storer RH, Georgakis C (1995) Disturbance detection and isolation by dynamic principal component analysis. Chemom Intell Lab Syst 30(1):179–196

32. Ljung L (1998) System identification. In: Signal analysis and prediction. Springer, Berlin
33. Bell AJ, Sejnowski TJ (1995) An information-maximization approach to blind separation and blind deconvolution. Neural Comput 7(6):1129–1159
34. Martin EB, Morris AJ (1996) Non-parametric confidence bounds for process performance monitoring charts. J Process Control 6(6):349–358
35. Silverman BW (1986) Density estimation for statistics and data analysis. CRC Press, Boca Raton
36. Strikwerda JC (2004) Finite difference schemes and partial differential equations. SIAM, Philadelphia
37. Qi C, Li H-X, Zhang X et al (2010) Time/space-separation-based SVM modeling for nonlinear distributed parameter processes. Ind Eng Chem Res 50:332–341
38. Hyvarinen A, Oja E (2000) Independent component analysis: algorithms and applications. Neural Netw 13(4):411–430
39. Hyvarinen A (1999) Fast and robust fixed-point algorithms for independent component analysis. IEEE Trans Neural Networks 10(3):626–634
40. Fan J, Wang Y (2014) Fault detection and diagnosis of non-linear non-Gaussian dynamic processes using kernel dynamic independent component analysis. Inf Sci 259:369–379

MIX
Papier aus verantwortungsvollen Quellen
Paper from responsible sources
FSC® C105338

If you have any concerns about our products,
you can contact us on
ProductSafety@springernature.com

In case Publisher is established outside the EU,
the EU authorized representative is:
Springer Nature Customer Service Center GmbH
Europaplatz 3, 69115 Heidelberg, Germany

Printed by Libri Plureos GmbH
in Hamburg, Germany